Lecture Notes Editorial Policies

Lecture Notes in Statistics provides a format for the informal and quick publication of monographs, case studies, and workshops of theoretical or applied importance. Thus, in some instances, proofs may be merely outlined and results presented which will later be published in a different form.

Publication of the Lecture Notes is intended as a service to the international statistical community, in that a commercial publisher, Springer-Verlag, can provide efficient distribution of documents that would otherwise have a restricted readership. Once published and copyrighted, they can be documented and discussed in the scientific literature.

Lecture Notes are reprinted photographically from the copy delivered in camera-ready form by the author or editor. Springer-Verlag provides technical instructions for the preparation of manuscripts. Volumes should be no less than 100 pages and preferably no more than 400 pages. A subject index is expected for authored but not edited volumes. Proposals for volumes should be sent to one of the series editors or addressed to "Statistics Editor" at Springer-Verlag in New York.

Authors of monographs receive 50 free copies of their book. Editors receive 50 free copies and are responsible for distributing them to contributors. Authors, editors, and contributors may purchase additional copies at the publisher's discount. No reprints of individual contributions will be supplied and no royalties are paid on Lecture Notes volumes. Springer-Verlag secures the copyright for each volume.

Series Editors:

Professor P. Bickel
Department of Statistics
University of California
Berkeley, California 94720
USA

Professor P. Diggle
Department of Mathematics
Lancaster University
Lancaster LA1 4YL
England

Professor S. Fienberg
Department of Statistics
Carnegie Mellon University
Pittsburgh, Pennsylvania 15213
USA

Professor K. Krickeberg
3 Rue de L'Estrapade
75005 Paris
France

Professor I. Olkin
Department of Statistics
Stanford University
Stanford, California 94305
USA

Professor N. Wermuth
Department of Psychology
Johannes Gutenberg University
Postfach 3980
D-6500 Mainz
Germany

Professor S. Zeger
Department of Biostatistics
The Johns Hopkins University
615 N. Wolfe Street
Baltimore, Maryland 21205-2103
USA

Lecture Notes in Statistics 165

Edited by P. Bickel, P. Diggle, S. Fienberg, K. Krickeberg,
I. Olkin, N. Wermuth, and S. Zeger

Springer
New York
Berlin
Heidelberg
Barcelona
Hong Kong
London
Milan
Paris
Singapore
Tokyo

Karl Mosler

Multivariate Dispersion, Central Regions and Depth

The Lift Zonoid Approach

Springer

Karl Mosler
Seminar für Wirtschafts- und Sozialstatistik
Universität Köln
D-50923 Köln
Germany
mosler@wiso.uni-koeln.de

Library of Congress Cataloging-in-Publication Data
Mosler, Karl C., 1947-
Multivariate dispersion, central regions, and depth : the lift zonoid approach / Karl Mosler.
p. cm. — (Lecture notes in statistics ; 165)
Includes bibliographical references and index.
ISBN 0-387-95412-0 (acid-free paper)
1. Probability measures. I. Title. II. Lecture notes in statistics (Springer-Verlag) ; v. 165.
QA273.6 .M68 2002
519.2—dc21 2002067536

ISBN 0-387-95412-0 Printed on acid-free paper.

Printed in the United States of America.

9 8 7 6 5 4 3 2 1 SPIN 10861563

Typesetting: Pages created by the author using a Springer $\TeX$ macro package.

www.springer-ny.com

Springer-Verlag New York Berlin Heidelberg
A member of BertelsmannSpringer Science+Business Media GmbH

Preface

This book introduces a new representation of probability measures, the lift zonoid representation, and demonstrates its usefulness in statistical applications.

The material divides into nine chapters. Chapter 1 exhibits the main idea of the lift zonoid representation and surveys the principal results of later chapters without proofs. Chapter 2 provides a thorough investigation into the theory of the lift zonoid. All principal properties of the lift zonoid are collected here for later reference. The remaining chapters present applications of the lift zonoid approach to various fields of multivariate analysis. Chapter 3 introduces a family of central regions, the zonoid trimmed regions, by which a distribution is characterized. Its sample version proves to be useful in describing data. Chapter 4 is devoted to a new notion of data depth, zonoid depth, which has applications in data analysis as well as in inference. In Chapter 5 nonparametric multivariate tests for location and scale are investigated; their test statistics are based on notions of data depth, including the zonoid depth. Chapter 6 introduces the depth of a hyperplane and tests which are built on it. Chapter 7 is about volume statistics, the volume of the lift zonoid and the volumes of zonoid trimmed regions; they serve as multivariate measures of dispersion and dependency. Chapter 8 treats the lift zonoid order, which is a stochastic order to compare distributions for their dispersion, and also indices and related orderings. The final Chapter 9 presents further orderings of dispersion, which are particularly suited for the analysis of economic disparity and concentration.

The chapters are, to a large extent, self-contained. Cross-references between Chapters 3 to 9 have been kept to a minimum. A reader who wants to learn the theory of the lift zonoid approach may browse through the introductory survey in Chapter 1 and then study Chapter 2 carefully. A reader who is primarily interested in applications should read Chapter 1, proceed to any of the later chapters and go back to relevant parts of Chapter 2 for proofs and theoretical details when needed. Some standard notions from probability and convex analysis are found in Appendix A.

Acknowledgments
The research which is reported in this book started as a joint work with Gleb Koshevoy, Russian Academy of Sciences, in the mid-nineties. Large parts of the manuscript are based on papers I have coauthored with him. So, I am heavily indebted to his ideas and scholarship. I also thank Rainer Dyckerhoff for contributing the Chapter 5 on nonparametric statistical inference with data depths.

Several people have read parts of the manuscript. I am very grateful to Jean Averous, Alfred Müller, and Wolfgang Weil, as well as to Katharina Cramer, Rainer Dyckerhoff, Richard Hoberg, and Thomas Möller for many helpful comments and hints to the literature.

Thanks are also to the Deutsche Forschungsgemeinschaft for funding Gleb Koshevoy's stays in Hamburg and Köln. Last not least I thank the editors of this series and John Kimmel of Springer Verlag for his continual encouragement and patience.

Köln, Germany Karl Mosler
February 2002

Contents

Preface v

1 Introduction **1**
1.1 The brief 1
1.2 Representing a probability measure 2
1.3 Lift zonoids 4
1.4 Examples of lift zonoids 9
1.5 Representing distributions by convex compacts 14
1.6 Ordering distributions 16
1.7 Central regions and data depth 19
1.8 Statistical inference 22

2 Zonoids and lift zonoids **25**
2.1 Zonotopes and zonoids 27
2.1.1 Zonoid of a measure 27
2.1.2 Equivalent definitions of the zonoid of a measure 30
2.1.3 Support function of a zonoid 32
2.1.4 Zonoids as expected random segments 34
2.1.5 Volume of a zonoid 35
2.1.6 Measures with equal zonoids 38
2.2 Lift zonoid of a measure 40
2.2.1 Definition and first properties 40
2.2.2 Lift zonotope 43
2.2.3 Univariate case 43
2.3 Embedding into convex compacts 48
2.3.1 Inclusion of lift zonoids 49
2.3.2 Uniqueness of the representation 50
2.3.3 Lift zonoid metric 51
2.3.4 Linear transformations and projections 52
2.3.5 Lift zonoids of spherical and elliptical distributions 55

2.4 Continuity and approximation . . . 58
2.4.1 Convergence of lift zonoids . . . 59
2.4.2 Monotone approximation of measures . . . 65
2.4.3 Volume of a lift zonoid . . . 66
2.5 Limit theorems . . . 68
2.6 Representation of measures by a functional . . . 71
2.6.1 Statistical representations . . . 74
2.6.2 Lift zonoids and the empirical process . . . 77
2.7 Notes . . . 77

3 Central regions **79**
3.1 Zonoid trimmed regions . . . 81
3.2 Properties . . . 84
3.3 Univariate central regions . . . 85
3.4 Examples of zonoid trimmed regions . . . 88
3.5 Notions of central regions . . . 93
3.6 Continuity and law of large numbers . . . 96
3.7 Further properties . . . 97
3.8 Trimming of empirical measures . . . 100
3.9 Computation of zonoid trimmed regions . . . 102
3.10 Notes . . . 103

4 Data depth **105**
4.1 Zonoid depth . . . 108
4.2 Properties of the zonoid depth . . . 111
4.3 Different notions of data depth . . . 116
4.4 Combinatorial invariance . . . 123
4.5 Computation of the zonoid depth . . . 127
4.6 Notes . . . 130

5 Inference based on data depth (by Rainer Dyckerhoff) **133**
5.1 General notion of data depth . . . 134
5.2 Two-sample depth test for scale . . . 136
5.3 Two-sample rank test for location and scale . . . 139
5.4 Classical two-sample tests . . . 141
5.4.1 Box's M test . . . 141
5.4.2 Friedman-Rafsky test . . . 142
5.4.3 Hotelling's T^2 test . . . 144
5.4.4 Puri-Sen test . . . 145
5.5 A new Wilcoxon distance test . . . 147
5.6 Power comparison . . . 149

5.7 Notes 163

6 Depth of hyperplanes 165
6.1 Depth of a hyperplane and MHD of a sample 166
6.2 Properties of MHD and majority depth 168
6.3 Combinatorial invariance 171
6.4 Measuring combinatorial dispersion 173
6.5 MHD statistics 174
6.6 Significance tests and their power 174
6.7 Notes 179

7 Volume statistics 181
7.1 Univariate Gini index 182
7.2 Lift zonoid volume 186
7.3 Expected volume of a random convex hull 188
7.4 The multivariate volume-Gini index 191
7.5 Volume statistics in cluster analysis 197
7.6 Measuring dependency 198
7.7 Notes 205

8 Orderings and indices of dispersion 207
8.1 Lift zonoid order 208
8.2 Order of marginals and independence 213
8.3 Order of convolutions 214
8.4 Lift zonoid order vs. convex order 216
8.5 Volume inequalities and random determinants 219
8.6 Increasing, scaled, and centered orders 219
8.7 Properties of dispersion orders 222
8.8 Multivariate indices of dispersion 224
8.9 Notes 228

9 Economic disparity and concentration 229
9.1 Measuring economic inequality 230
9.2 Inverse Lorenz function (ILF) 232
9.3 Price Lorenz order 238
9.4 Majorizations of absolute endowments 242
9.5 Other inequality orderings 245
9.6 Measuring industrial concentration 248
9.7 Multivariate concentration function 252
9.8 Multivariate concentration indices 255
9.9 Notes 257

Appendix A: Basic notions 259

Appendix B: Lift zonoids of bivariate normals **265**

Bibliography **274**

Index **288**

1

Introduction

1.1 The brief

The book introduces a new representation of probability measures – the lift zonoid representation – and demonstrates its usefulness in multivariate analysis. A measure on the Euclidean d-space is represented by a convex set in $(d+1)$-space, its lift zonoid. This yields an embedding of the d-variate measures into the space of symmetric convex compacts in $\mathbb{R}^{d+1}$. The embedding map is positive homogeneous, additive, and continuous. It has many applications in data analysis as well as in inference and in the comparison of random vectors.

First, lift zonoids are useful in multivariate data analysis in order to describe an empirical distribution by central (so-called trimmed) regions. The lift zonoid trimmed regions range from the convex hull of the data points to their mean and characterize the distribution uniquely. They give rise to a concept of data depth related to the mean. Both, the trimmed regions and the depth, have nice analytic and computational properties. They lend themselves to multivariate statistical inference, including nonparametric tests for location and scale.

Secondly, for comparing random vectors, the Hausdorff distance between lift zonoids defines a measure metric and the set inclusion of lift zonoids defines a stochastic order. The lift zonoid order reflects the dispersion of random vectors and is slightly weaker than the multivariate convex order. The lift zonoid order and related orderings have a broad range of applications in stochastic comparison problems in economics and other fields.

The lift zonoid has been introduced by Koshevoy and Mosler (1998) and developed in a number of papers by these authors. This introductory chapter exhibits the definition and main properties of the lift zonoid, which are illustrated by many examples. It explains the lift zonoid order and the zonoid data depth and surveys their principal applications. In later chapters the material will be discussed in more detail, with proofs and notes on the literature.

1.2 Representing a probability measure

A d-variate probability measure μ is a mapping from the Borel sets $\mathcal{B}^d$ into the reals that is nonnegative, σ-additive, and normed. It can be described and represented in many ways. A simple and therefore the most widely used representation of μ is its cumulative *distribution function (c.d.f.)* F_μ,

$$F_\mu(\boldsymbol{x}) = \mu(\boldsymbol{x} + \mathbb{R}^d_-), \quad \boldsymbol{x} \in \mathbb{R}^d,$$

where $\boldsymbol{x} + \mathbb{R}^d_- = \{\boldsymbol{y} \in \mathbb{R}^d : \boldsymbol{y} \leq \boldsymbol{x}\}$ stands for the *lower orthant* at $\boldsymbol{x}$ and $\leq$ is the usual componentwise ordering of $\mathbb{R}^d$. Equivalently, μ can be represented by a set, its *distribution epigraph*,

$$\Gamma_\mu = \{(x_0, \boldsymbol{x}) \in \mathbb{R}^{d+1} : x_0 \geq F_\mu(\boldsymbol{x}), \boldsymbol{x} \in \mathbb{R}^d\}.$$

In this way, the class $\mathcal{M}$ of all d-variate probability measures is *embedded* into the space $\mathcal{F}^d$ of nondecreasing d-variate real functions,

$$\begin{aligned} \mathcal{M} &\to \mathcal{F}^d, \\ \mu &\mapsto F_\mu, \end{aligned}$$

or the class $\mathcal{U}^{d+1}$ of upper sets in $\mathbb{R}^{d+1}$,

$$\begin{aligned} \mathcal{M} &\to \mathcal{U}^{d+1}, \\ \mu &\mapsto \Gamma_\mu. \end{aligned}$$

The class $\mathcal{M}$ corresponds to the subclass of d-variate real functions that are Δ-monotone[1] and right continuous, or to the subclass of their epigraphs.

A probability distribution can also be represented by some generating function. Most of the representations satisfy limit theorems for the empirical

[1] For a function $f : \mathbb{R}^d \to \mathbb{R}$, define $\Delta_i^\varepsilon f(a_1, \ldots, a_i, \ldots, a_d) = f(a_1, \ldots, a_i + \varepsilon, \ldots, a_d) - f(a_1, \ldots, a_i, \ldots, a_d)$. f is called *Δ-monotone* if for any $\{i_1, \ldots, i_k\} \subset \{1, \ldots, d\}$, $1 \leq k \leq d$, holds $\Delta_{i_1}^\varepsilon \ldots \Delta_{i_k}^\varepsilon f(a_1, \ldots, a_d) \geq 0$ for every $\varepsilon > 0$. In particular, every Δ-monotone function is nondecreasing.

distribution of a random sample, which are basic for statistical inference. Each representation provides a specific view on the probability distribution and lends itself to the analysis of certain probabilistic structures, to which it is particularly adapted. For example, the *characteristic function* transforms convolutions into products of integrals. Therefore, the representation of distributions by their characteristic functions allows a simple treatment of sums of independent random variables. The representation by c.d.f.s, among others, is well suited for the analysis of mixtures of distributions.

The probably most important aspect of c.d.f.s is their visuality for univariate or bivariate distributions: A c.d.f. depicts, in a simple way, the underlying probability distribution. Several features of the distribution can be immediately seen from the graph or epigraph of its c.d.f.: its general location, its support, and its atoms, if they exist. Other features, concerning the regions of relatively dense or sparse probability mass and eventual symmetries, can be recognized through a more careful inspection of the distribution epigraph.

The pointwise *ordering* of c.d.f.s, that is the inclusion of distribution epigraphs, induces a *stochastic order* (reflexive, transitive, and antisymmetric) among probability distributions: Define

$$\mu \leq_{lo} \nu \quad \text{if} \quad F_\mu(\boldsymbol{x}) \geq F_\nu(\boldsymbol{x}) \; \forall\, \boldsymbol{x} \in \mathbb{R}^d$$

or, equivalently, $\mu \leq_{lo} \nu$ if $\Gamma_\mu \subset \Gamma_\nu$. In dimension one $\mu \leq_{lo} \nu$ means that ν is *stochastically larger* than μ, which is the *usual stochastic order*. In higher dimensions this order is called the *lower orthant order*. It compares two multivariate distributions not only with respect to their location; provided the two distributions have identical marginals, the lower orthant order compares them also in terms of their positive dependency.

A *measure metric* for probability distributions is provided by the supremum norm of c.d.f.s,

$$\delta_{KS}(\mu, \nu) = \sup_{\boldsymbol{x} \in \mathbb{R}^d} |F_\mu(\boldsymbol{x}) - F_\nu(\boldsymbol{x})|.$$

The metric δ_{KS} is named the *Kolmogorov-Smirnov metric* due to its applications in statistical inference.

The empirical distribution function of a random sample satisfies limit theorems which are fundamental to statistical inference: First, the Glivenko-Cantelli theorem, saying that, with probability one, the empirical c.d.f. of a random sample from μ converges uniformly to the c.d.f. of μ,

$$\lim_{n \to \infty} \sup_{\boldsymbol{x} \in \mathbb{R}^d} |F_n(\boldsymbol{x}) - F_\mu(\boldsymbol{x})| = 0 \quad \mu\text{-almost surely}.$$

Second, the Donsker central limit theorem, which identifies the asymptotic distribution of the empirical c.d.f., saying that

$$\sqrt{n} \sup_{x \in \mathbb{R}^d} |F_n(x) - F_\mu(x)|$$

converges to a Brownian bridge.

However, despite its ubiquitous use in probability theory and statistics, the representation by c.d.f.s has certain drawbacks. One is that the domain (and hence the graph) of the c.d.f. is not bounded, which impairs the visuality of this representation. Another is that a multivariate c.d.f. does not transform nicely under an affine transformation of its arguments.

In this book another pictorial representation of probability measures is investigated, the lift zonoid representation. Like the representation by c.d.f.s, the lift zonoid representation provides an order and a metric, it satisfies proper Glivenko-Cantelli and Donsker theorems, and proves to be useful in many fields of application.

1.3 Lift zonoids

In this section the central notion of the book is introduced, the lift zonoid of a probability measure. We present the principal definitions and properties and illustrate them by examples. For more properties, technical details and proofs, the reader is referred to the subsequent Chapter 2.

The lift zonoid of a d-variate probability measure represents the measure, like the distribution epigraph, by a set in $\mathbb{R}^{d+1}$. It takes a particularly simple form when μ is an *empirical distribution*, say, $\mu = 1/n \sum_{i=1}^n \delta_{a_i}$, where δ_{a_i} denotes the singleton (= Dirac) measure on a_i. An empirical distribution imposes equal mass to a finite number of, not necessarily different, points in $\mathbb{R}^d$. Then the *lift zonoid* of μ is defined by

$$\widehat{Z}(\mu) = \frac{1}{n} \sum_{i=1}^{n} [(0, \mathbf{0}), (1, a_i)] \,. \tag{1.1}$$

Here $[(0, \mathbf{0}), (1, a_i)]$ stands for the line segment that extends from the origin $(0, \mathbf{0})$ to the point $(1, a_i)$ in $\mathbb{R}^{d+1}$. The sum is the Minkowski sum[2] of sets.

To illustrate the definition, three simple examples of univariate and bivariate distributions are presented.

[2]That is, the sets are added point by point.

Example 1.1 Consider $d = 1$ and the measure μ with $\mu(\{1.2\}) = \mu(\{2.8\}) = 1/2$. The lift zonoid of this empirical distribution is

$$\widehat{Z}(\mu) = [(0,0),(0.5,0.6)] + [(0,0),(0.5,1.4)];$$

it is depicted in Figure 1.1(a).

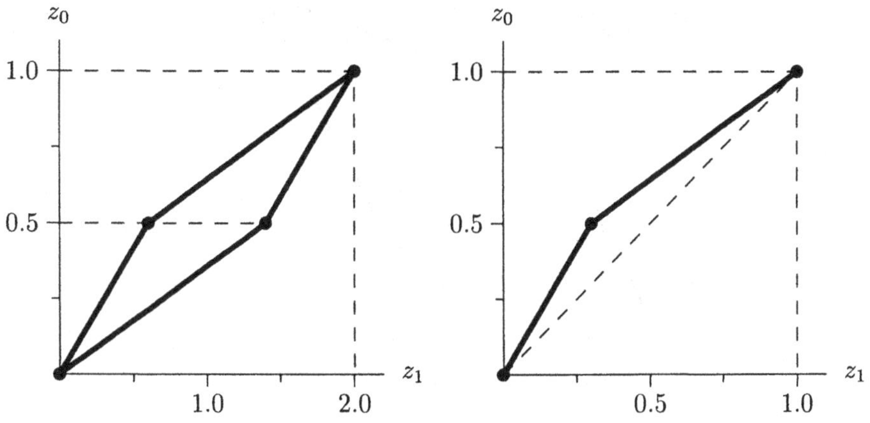

(a) Lift Zonoid (b) Lorenz curve

FIGURE 1.1: Lift zonoid (a) and Lorenz curve (b) of a univariate empirical distribution μ, given by $\mu(\{1.2\}) = \mu(\{2.8\}) = 1/2$.

Example 1.2 Let $d = 2$ and the measure ν be given by $\nu(\{(9,0)\}) = 1/3$, $\nu(\{(0,3)\}) = 2/3$. This is an empirical distribution that has the lift zonoid

$$\widehat{Z}(\nu) = \frac{1}{3}[(0,0,0),(1,9,0)] + \frac{2}{3}[(0,0,0),(1,0,3)],$$

as shown in Figure 1.2(a). Note that in both figures the first coordinate of the lift zonoid is depicted in vertical direction.

Example 1.3 For general dimension d let μ be a singleton measure having its total mass at some $a \in \mathbb{R}^d$, $\mu = \delta_a$. The lift zonoid of μ is the line segment that connects the origin of $\mathbb{R}^{d+1}$ with the point $(1,a)$, $\widehat{Z}(\delta_a) = [0,(1,a)]$.

The lift zonoid (1.1) of a univariate empirical distribution shows a close relation to its *Lorenz curve*, that is to the piecewise linear connection of the following points in $\mathbb{R}^2$,

$$\frac{1}{n}\left(j, \frac{1}{\bar{a}}\sum_{i=1}^{j} a_{(i)}\right), \quad j = 0,1,\ldots,n,$$

where $a_{(i)}$ stands for the i-th observation ordered from below and $\overline{a} = n^{-1}\sum_{i=1}^{n} a_i$. Observe that the Lorenz curve corresponds to the north-west border of the lift zonoid if the abscissa is rescaled by the factor $\overline{a}^{-1}$; see Figure 1.1.

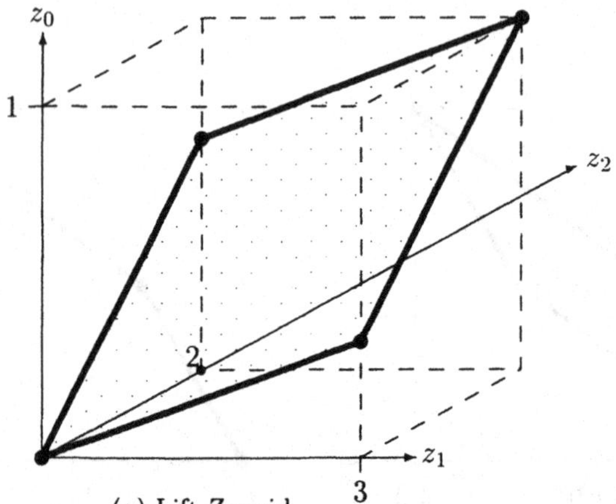

(a) Lift Zonoid

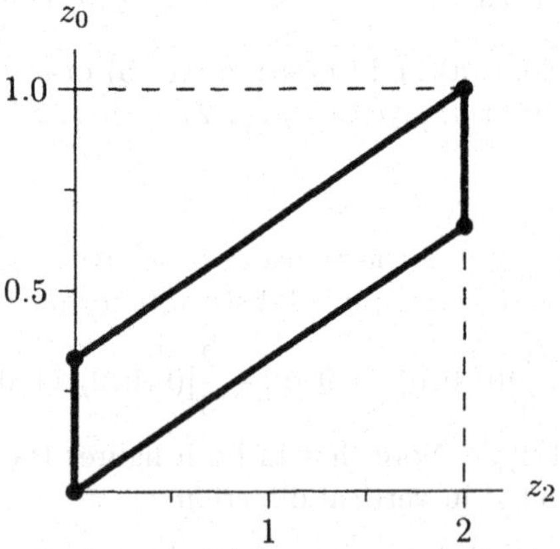

(b) Projection to second coordinate
1/3[(0,0),(1,0)] + 2/3[(0,0),(1,3)]

FIGURE 1.2: Lift zonoid (a) and its projection (b) for a bivariate empirical distribution ν, given by $\nu(\{(9,0)\}) = 1/3$, $\nu(\{(0,3)\}) = 2/3$.

In higher dimensions there exists a simple relation between the lift zonoid $\widehat{Z}(\mu)$ of a multivariate distribution and the lift zonoids of its marginal distributions: The lift zonoid of a marginal is the respective projection of $\widehat{Z}(\mu)$.

Figure 1.2(b) exhibits the projection to the (z_0, z_2)-plane, which is the lift zonoid of the second marginal ν_2 of ν, $\nu_2(\{0\}) = 1/3, \nu_2(\{3\}) = 2/3$.

Now we proceed to a general definition of the lift zonoid of a probability distribution. It is defined as the expectation of a certain random compact set. A *random compact set* is a random variable in the space of nonempty, compact subsets of the Euclidean space, endowed with a proper σ-algebra. It has a set valued expectation[3]. In particular, if $\boldsymbol{X}$ is a d-variate random vector then the random line segment $[\mathbf{0}, \boldsymbol{X}]$, that extends from the origin to the random point $\boldsymbol{X}$, is a random compact set in $\mathbb{R}^d$.

Consider the class of general d-variate probability distributions which have finite first moments,

$$\mathcal{P}_1 = \left\{ \mu : \mathcal{B}^d \to [0,1] \ : \ \mu \text{ probability measure and } \int_{\mathbb{R}^d} ||x|| \, \mu(dx) < \infty \right\}.$$

For any $\mu \in \mathcal{P}_1$ the lift zonoid is defined as follows: Let $\boldsymbol{X}$ be a random variable in $\mathbb{R}^d$ distributed as μ and consider the random line segment $[(0, \mathbf{0}), (1, \boldsymbol{X})]$ in $\mathbb{R}^{d+1}$. The lift zonoid is the set valued expectation of this random line segment,

$$\widehat{Z}(\mu) = \mathrm{E}\left([(0,\mathbf{0}), (1, \boldsymbol{X})]\right).$$

Several equivalent definitions can be stated:

Theorem (Equivalent definitions of the lift zonoid)

For a probability measure $\mu \in \mathcal{P}_1$ holds

$$\begin{aligned}
\widehat{Z}(\mu) &= \left\{ \left(\int_{\mathbb{R}^d} g(x)\, \mu(dx), \int_{\mathbb{R}^d} x\, g(x)\, \mu(dx) \right) \ : \ g : \mathbb{R}^d \to [0,1] \right\} \\
&= \left\{ \left(\int_{\mathbb{R}^d} \nu(dx), \int_{\mathbb{R}^d} x\, \nu(dx) \right) \ : \right. \\
&\qquad\qquad \left. \nu(B) \le \mu(B) \text{ for all } B \in \mathcal{B}^d, \nu \in \mathcal{P}_1 \right\} \\
&= \mathit{conv} \left\{ \left(\int_B \mu(dx), \int_B x\, \mu(dx) \right) \ : \ B \in \mathcal{B}^d \right\}.
\end{aligned}$$

[3]For a definition of set valued expectations, see Appendix A 3 and, e.g., Weil and Wieacker (1993).

Here the convex hull of a set is denoted by *conv*. As is seen from the theorem, the lift zonoid $\widehat{Z}(\mu)$ consists of points in $\mathbb{R}^{d+1}$ that can be written as expectations in several ways. For a discussion of these expectations and a proof of the theorem see Chapter 2.

If $d = 1$, the lift zonoid of a probability measure with positive mean is essentially given by its Lorenz curve: It has vertices at the origin of $\mathbb{R}^2$ and at the point $(1, \epsilon(\mu))$, where $\epsilon(\mu)$ stands for the expectation of μ. Its border consists of two curves, the generalized Lorenz curve and its symmetric counterpart, the dual generalized Lorenz curve[4].

The lift zonoid of a distribution that has finite support is a zonotope, called the *lift zonotope*. A *zonotope* is a sum of line segments, and a *zonoid* is the limit, in the Hausdorff[5] sense, of zonotopes. The lift zonoid of a given probability distribution with finite expectation is a zonoid.

If the support of μ is $\{\boldsymbol{a}_1, \boldsymbol{a}_2, \ldots, \boldsymbol{a}_n\}$ and $\mu(\{\boldsymbol{a}_j\}) = p_j$, the definition specializes to

$$\begin{aligned} \widehat{Z}(\mu) &= \sum_{j=1}^{n} p_j \left[(0, \boldsymbol{0}), (1, \boldsymbol{a}_j)\right] \\ &= \left\{ \left(\sum_{j=1}^{n} q_j, \sum_{j=1}^{n} q_j \boldsymbol{a}_j \right) \ : \ 0 \le q_j \le p_j \quad \text{for all } j \right\} \qquad (1.2) \\ &= conv \left\{ \left(\sum_{j \in J} p_j, \sum_{j \in J} p_j \boldsymbol{a}_j \right) \ : \ J \subset \{1, 2, \ldots, n\} \right\}. \end{aligned}$$

To illustrate (1.2) consider an empirical probability measure μ on the points $\boldsymbol{a}_1, \ldots, \boldsymbol{a}_m$. If ν is not larger than μ, $\nu(B) \le \mu(B)$ for all B, the support of ν is contained in the set $\{\boldsymbol{a}_1, \ldots, \boldsymbol{a}_n\}$ and there holds $q_j = \nu(\boldsymbol{a}_j) \le 1/n$ for each j, $\boldsymbol{\epsilon}(\nu) = \sum_{j=1}^{n} q_j \boldsymbol{a}_j$. Hence the point $(\sum_{j=1}^{m} q_j, \boldsymbol{\epsilon}(\nu))$ is in $1/n \sum_{j=1}^{n} [(0, \boldsymbol{0}), (1, \boldsymbol{a}_j)]$. On the other hand, for any point $\boldsymbol{z} \in 1/n \sum_{j=1}^{n} [(0, \boldsymbol{0}), (1, \boldsymbol{a}_j)]$ there exists such a ν with $\nu \le \mu$ and $\boldsymbol{z} = (\sum_{j=1}^{m} q_j, \boldsymbol{\epsilon}(\nu))$.

[4]Define $GL(t) = \int_0^t Q_\mu(s)ds$, $t \in [0, 1]$, where Q_μ is the quantile function of μ. Then $(t, GL(t)), 0 \le t \le 1$, is the *generalized Lorenz curve* and $(t, 1 - GL(1-t)), 0 \le t \le 1$, is the *dual generalized Lorenz curve* of μ. If $\epsilon(\mu)$ is positive, $(t, 1/\epsilon(\mu) GL(t)), 0 \le t \le 1$, is the usual Lorenz curve of μ.

[5]Throughout this book, convergence of sets in the Euclidean space is meant as convergence with respect to the Hausdorff distance. The Hausdorff distance $\delta_H(C, D)$ of two compact sets C and D is the smallest ε such that C plus the ε-ball includes D and vice versa.

The lift zonoid satisfies two limit theorems which form the basis of its use in statistical inference: a law of large numbers and a central limit theorem.

Consider a sequence $\boldsymbol{X}_1$, $\boldsymbol{X}_2, \ldots$ of i.i.d. random vectors in $\mathbb{R}^d$ which are all distributed as μ and let $\widetilde{\mu}^n$ be the random empirical distribution of $\boldsymbol{X}_1$, $\boldsymbol{X}_2, \ldots, \boldsymbol{X}_n$. Then the lift zonoid of $\widetilde{\mu}^n$ is the arithmetic mean of n random line segments. The law of large numbers (which is proved in Chapter 2 as Theorem 2.38) says that the lift zonoid of the empirical distribution converges to the lift zonoid of the true distribution:

Theorem (Law of large numbers). *For a probability measure $\mu \in \mathcal{P}_1$ holds*

$$\widehat{Z}(\widetilde{\mu}^n) = \frac{1}{n}\sum_{i=1}^{n}[(0,\boldsymbol{0}),(1,\boldsymbol{X}_i)] \xrightarrow{H} \widehat{Z}(\mu) \quad \mu\text{-almost surely},$$

with convergence in the Hausdorff distance.

The central limit theorem (Theorem 2.39 below) gives the asymptotic distribution of the Hausdorff distance between the lift zonoids of the empirical and the true distribution:

Theorem (Central limit theorem). *Assume that $\mu \in \mathcal{P}_1$ and that the second moments are finite, i.e., $\int_{\mathbb{R}^d} \|\boldsymbol{x}\|^2 \mu(d\boldsymbol{x}) < \infty$. Then*

$$\sqrt{n}\,\delta_H(\widehat{Z}(\widetilde{\mu}^n), \widehat{Z}(\mu)) \quad \xrightarrow{d} \quad \max_{(p_0,\boldsymbol{p})\in S^d} |\mathbb{G}\,(p_0,\boldsymbol{p})|\,,$$

where $\mathbb{G} = \{\mathbb{G}\,(p_0,\boldsymbol{p})\}_{(p_0,\boldsymbol{p})\in S^d}$ is a zero-mean continuous Gaussian process on the sphere and $\xrightarrow{d}$ means convergence in distribution.

For details on the covariance function of $\mathbb{G}$ and for proof, see Theorem 2.39.

1.4 Examples of lift zonoids

Examples 1.1 to 1.3 concerned the lift zonoids of simple empirical distributions. This section lists further examples, including the lift zonoids of various parametric probability distributions in dimension one and more. The calculation of the lift zonoid of a univariate distribution and several examples will be presented in Section 2.2.3. More detailed examples with multivariate distributions will be given in Section 2.3.5.

Example 1.4 (Binomial distribution) The binomial distribution $Bin(n, \pi)$ is a discrete probability measure having finite support in $\mathbb{R}$ with

$\mu(\{j\}) = \binom{n}{j}\pi^j(1-\pi)^{n-j}, j = 0, 1, \ldots, n$. Its lift zonoid is a polytope[6] in $\mathbb{R}^2$,

$$\widehat{Z}(\mu) = \sum_{j=0}^{n} \binom{n}{j} \pi^j (1-\pi)^{n-j} [(0,0),(1,j)] . \tag{1.3}$$

The polytope has $2(n+1)$ vertices. For example the lift zonoid of $\mu = Bin(3, 0.5)$ has vertices

$$(0,0), (0.125, 0.375), (0.5, 1.125), (0.875, 1.5),$$

$$(1.0, 1.5), (0.875, 1.125), (0.5, 0.375), (0.125, 0) .$$

Figure 1.3 exhibits the lift zonoids of $Bin(3,\pi)$ for $\pi = 0.1, 0.5$, and 0.9, respectively.

Example 1.5 (Exponential distribution) The univariate exponential distribution $Exp(\lambda)$, with some $\lambda > 0$, has density function $f(x) = \lambda \exp(-\lambda x)$ if $x \geq 0$, and $f(x) = 0$ otherwise. The expectation is $1/\lambda$. Figure 1.4 shows the lift zonoids of three exponential distributions, with $\lambda = 1, 2$, and 3. Observe that, if λ increases, the upper vertices of the lift zonoids move from right to left (due to decreasing expectations) and the volumes become smaller. The latter reflects the decrease in the variance $1/\lambda^2$.

Example 1.6 (Univariate normal distribution) Figures 1.5 and 1.6 exhibit the lift zonoids of univariate normal distributions with various expectations a and variances σ^2. Note, again, that the volumes increase with the variance (Figure 1.5).

[6] A *polytope* is the convex hull of a finite number of points.

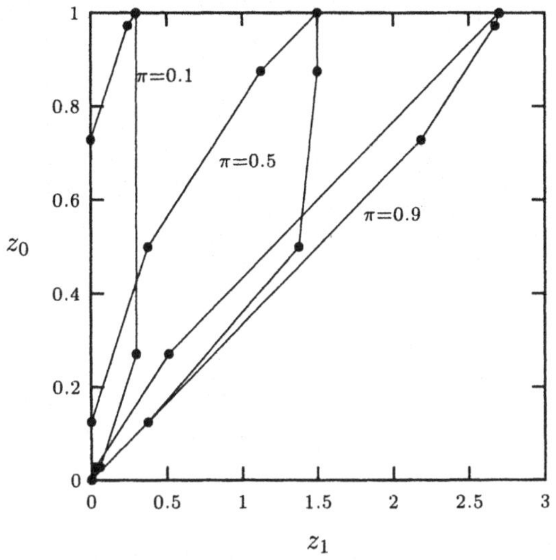

FIGURE 1.3: Lift zonoids of binomial distributions $Bin(3, \pi)$ with $\pi = 0.1, 0.5, 0.9$.

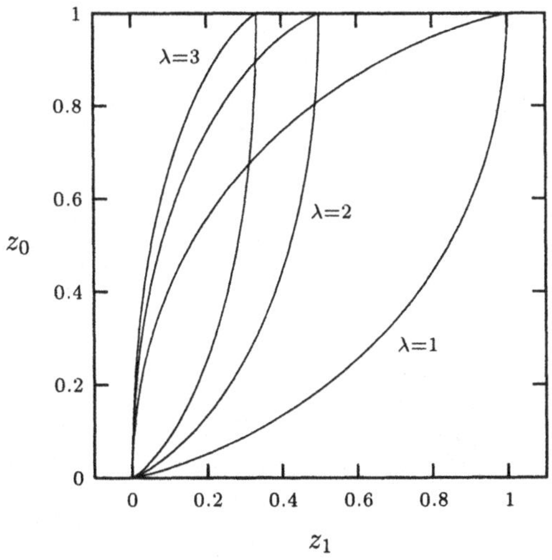

FIGURE 1.4: Lift zonoids of exponential distributions $Exp(\lambda)$ with $\lambda = 1, 2, 3$.

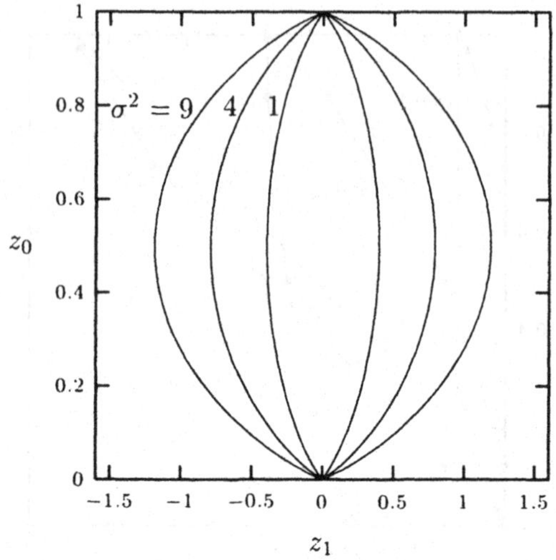

FIGURE 1.5: Lift zonoids of univariate normal distributions $N(0, \sigma^2)$ with $\sigma = 1, 4, 9$.

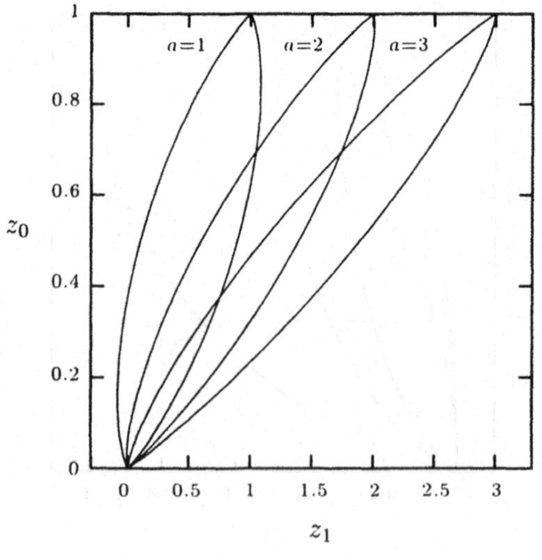

FIGURE 1.6: Lift zonoids of univariate normal distributions $N(a, 1)$ with $a = 1, 2, 3$.

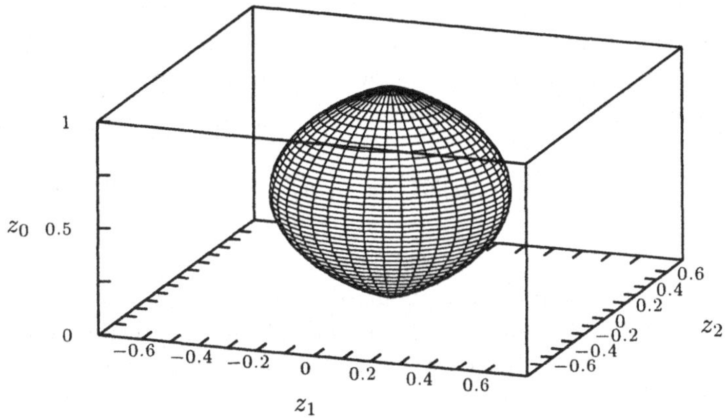

FIGURE 1.7: Lift zonoid of a bivariate standard normal distribution, with $a_1 = a_2 = 0$, $\sigma_1 = \sigma_2 = 1$, $\rho = 0$.

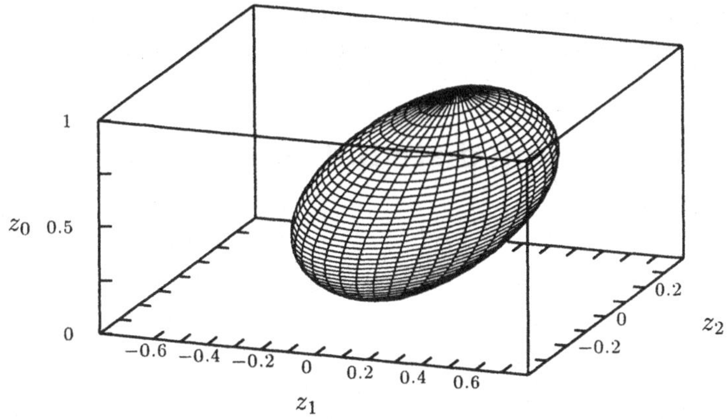

FIGURE 1.8: Lift zonoid of a bivariate normal distribution, with $a_1 = 0.25, a_2 = 0$, $\sigma_1 = \sigma_2 = 1$, $\rho = 0$.

Example 1.7 (Multivariate normal distribution) The lift zonoid of a d-variate normal distribution $N(\mathbf{a}, \Sigma)$, with expectation vector $\mathbf{a}$ and covariance matrix Σ, can be described as follows: For any $\alpha \in]0,1[$ the intersection of $\widehat{Z}(\mu)$ with the hyperplane $\{(z_0, \boldsymbol{z}) \in \mathbb{R}^{d+1} : z_0 = \alpha\}$ is an ellipsoid. The lift zonoid is built of these ellipsoids and two additional points, the origin and the point $(1, \mathbf{a})$ in $\mathbb{R}^{d+1}$. It is given by

$$\widehat{Z}(N(\boldsymbol{a}, \Sigma)) = \left\{(\alpha, \boldsymbol{x}) : (\boldsymbol{x} - \alpha\boldsymbol{a})\Sigma^{-1}(\boldsymbol{x} - \alpha\boldsymbol{a})' \leq \frac{1}{2\pi}e^{-u_\alpha^2}, \quad 0 \leq \alpha \leq 1\right\}, \tag{1.4}$$

where u_α is the α-quantile of the standard normal distribution, $u_0 = -\infty$, $u_1 = \infty$. Figures 1.7 and 1.8 show the lift zonoids of two bivariate normals[7] $N(a_1, a_2, \sigma_1^2, \sigma_2^2, \rho)$, the bivariate standard normal with $a_1 = a_2 = 0$ and a bivariate normal with $a_1 = 0.25, a_2 = 0$, both with $\sigma_1^2 = \sigma_2^2 = 1$ and $\rho = 0$. The lift zonoids of several other bivariate normal distributions, also with non-zero covariances, are found in the Appendix B.

Example 1.8 (Uniform distribution on the unit sphere) Let us consider a probability measure μ that is uniform on the unit sphere in $\mathbb{R}^d$. Again we describe the lift zonoid by its intersections with hyperplanes. For any $\alpha \in]0,1[$ the intersection of $\widehat{Z}(\mu)$ with the hyperplane $\{(z_0, \boldsymbol{z}) \in \mathbb{R}^{d+1} : z_0 = \alpha\}$ is a d-dimensional ball around the origin with radius $(\sin \pi\alpha)/\pi$. The lift zonoid is the union of these balls and the points $(0, \mathbf{0})$ and $(1, \mathbf{0})$.

1.5 Representing distributions by convex compacts

Several properties of the lift zonoid of a probability distribution are immediately seen from the definition and its equivalent formulations. The lift zonoid

- is convex,
- is compact,
- contains the origin and the point $\left(1, \int \boldsymbol{x}\mu(d\boldsymbol{x})\right)$,
- is symmetric around the point $\frac{1}{2}\left(1, \int \boldsymbol{x}\mu(d\boldsymbol{x})\right)$,

[7] We notate a bivariate normal by $N(a_1, a_2, \sigma_1^2, \sigma_2^2, \rho)$, where $\boldsymbol{a} = (a_1, a_2)$ is the expectation, σ_1^2, σ_2^2 are the marginal variances, and ρ is the correlation coefficient.

- is contained in the nonnegative orthant of $\mathbb{R}^{d+1}$ if μ has nonnegative support in $\mathbb{R}^d$.

Besides these properties, it can be shown that different distributions have different lift zonoids:

Theorem (Uniqueness). *If $\mu, \nu \in \mathcal{P}_1$, $\mu \neq \nu$, then $\widehat{Z}(\mu) \neq \widehat{Z}(\nu)$.*

The theorem states a key result: Every $\mu \in \mathcal{P}_1$ is uniquely determined by its lift zonoid, which allows the representation of distributions by their lift zonoids. It will be proven in Chapter 2 as Theorem 2.21. We conclude that the lift zonoid yields a representation of probability distributions on $\mathbb{R}^d$ by certain convex subsets of $\mathbb{R}^{d+1}$. It is the class of convex compacts which contain the origin and, for some $\boldsymbol{b} \in \mathbb{R}^d$, the point $(1, \boldsymbol{b})$ and are symmetric around $1/2(1, \boldsymbol{b})$. Every d-variate probability measure having finite first moment corresponds, in a unique way, to such a convex set.

The lift zonoid embedding is continuous with respect to the weak convergence of uniformly integrable measures[8] and the Hausdorff distance of sets:

Theorem (Continuity). *Let $(\mu^n)_{n\in\mathbb{N}}$ and μ be in $\mathcal{P}_1$.*

(i) *If $(\mu^n)_{n\in\mathbb{N}}$ is uniformly integrable and converges weakly to μ, the sequence $\widehat{Z}(\mu^n)$ of lift zonoids converges, in the Hausdorff distance, to the lift zonoid $\widehat{Z}(\mu)$ of μ.*

(ii) *If all μ^n and μ have densities with respect to Lebesgue measure, the reverse implication holds, too.*

The continuity theorem will be proven as Theorem 2.30 in Chapter 2. There are several, easy to check, sufficient conditions for uniform integrability. In particular, (μ^n) is uniformly integrable if there exists a compact set which includes all the supports. Another sufficient condition says: There exists a distribution $\nu \in \mathcal{P}_1$ such that $\widehat{Z}(\nu)$ contains all $\widehat{Z}(\mu^n)$.

The definitions of the lift zonoid and the lift zonoid embedding extend in a straightforward way from probability measures to finite measures on d-space. This extended embedding is positive homogeneous and additive: The lift zonoid of the scalar multiple of a measure is the multiple of its lift zonoid, and the lift zonoid of a sum of two measures is the Minkowski sum of their lift zonoids.

---

[8]A sequence (μ^n) in $\mathcal{P}_1$ is *uniformly integrable* if $\sup_{n\in\mathbb{N}} \int_{\|\boldsymbol{x}\|\geq\beta} \|\boldsymbol{x}\| \, \mu^n(d\boldsymbol{x}) \to 0$ for $\beta \to \infty$.

1.6 Ordering distributions

As a first application of the lift zonoid embedding, a stochastic order between multivariate probability distributions, which measures their dispersion, is introduced. The idea is to represent each probability distribution by its lift zonoid and to consider the order between distributions that is defined by the inclusion of their lift zonoids. This order has nice geometric properties.

Define, for μ and $\nu \in \mathcal{P}_1$,

$$\mu \leq_{lz} \nu \quad \text{if} \quad \widehat{Z}(\mu) \subset \widehat{Z}(\nu).$$

The relation is reflexive and, by the uniqueness theorem, antisymmetric. Moreover, due to its restriction to $\mathcal{P}_1$, the relation is transitive and hence a stochastic order on $\mathcal{P}_1$. We call it the ***lift zonoid order***. Like any stochastic order among probability distributions, the lift zonoid order induces a preorder (transitive and reflexive) among random variables. Define

$$\boldsymbol{X} \leq_{lz} \boldsymbol{Y} \quad \text{if} \quad \mu \leq_{lz} \nu,$$

if $\boldsymbol{X}$ and $\boldsymbol{Y}$ are random variables in $\mathbb{R}^d$ and distributed as μ and ν in $\mathcal{P}_1$, respectively.

We know that $(1, \mathrm{E}(\boldsymbol{X})) \in \widehat{Z}(\mu)$. If $\widehat{Z}(\mu) \subset \widehat{Z}(\nu)$, because of the theorem in Section 1.3 there exists a function $g : \mathbb{R}^d \to [0,1]$ such that

$$(1, \mathrm{E}(\boldsymbol{X})) = \left(\int_{\mathbb{R}^d} g(\boldsymbol{x})\, \nu(d\boldsymbol{x}), \int_{\mathbb{R}^d} \boldsymbol{x}\, g(\boldsymbol{x})\, \nu(d\boldsymbol{x}) \right).$$

From the equality of the first components conclude that $g(\boldsymbol{x}) = 1$ for all $\boldsymbol{x}$ in the support of ν and, therefore, from the equality of the remaining components, that $\mathrm{E}(\boldsymbol{X}) = \int_{\mathbb{R}^d} \boldsymbol{x}\, \nu(d\boldsymbol{x}) = \mathrm{E}(\boldsymbol{Y})$. Thus,

$$\boldsymbol{X} \leq_{lz} \boldsymbol{Y} \quad \text{implies that} \quad \mathrm{E}(\boldsymbol{X}) = \mathrm{E}(\boldsymbol{Y}).$$

The lift zonoid order is an ***integral order***. It can be characterized by expectations over a proper class of functions:

Theorem (Convex-linear order). *Define*

$$\mathcal{F}_{cl} = \{ f : \mathbb{R}^d \to \mathbb{R} \,:\, f(\boldsymbol{x}) = g(\langle \boldsymbol{p}, \boldsymbol{x} \rangle), g : \mathbb{R} \to \mathbb{R} \text{ convex}, \boldsymbol{p} \in \mathbb{R}^d \}$$

and let $\boldsymbol{X}$ *and* $\boldsymbol{Y}$ *be random vectors with distributions in* $\mathcal{P}_1$. *Then*

$$\boldsymbol{X} \leq_{lz} \boldsymbol{Y} \quad \text{if and only if} \quad \mathrm{E}(f(\boldsymbol{X})) \leq \mathrm{E}(f(\boldsymbol{Y})) \quad \text{holds for all } f \in \mathcal{F}_{cl}, \tag{1.5}$$

as far as both expectations exist.

The class $\mathcal{F}_{cl}$ is the class of *convex-linear functions*. Due to this characterization, the lift zonoid order is also named the *convex-linear order*. When g is increasing, $f(\boldsymbol{x}) = g(\langle \boldsymbol{p}, \boldsymbol{x} \rangle)$ can be interpreted as a *von Neumann-Morgenstern utility function* on commodity bundles $\boldsymbol{x}$, and $\boldsymbol{p}$ as a vector of prices, at least if the components of $\boldsymbol{p}$ are positive. The integral order generated by the class of such functions is an unanimous preference ordering in expected utility. A detailed treatment of the lift zonoid order is found in Chapter 8, where the above theorem is proven as Theorem 8.5.

Since any convex-linear function is convex, the restriction

$$\mathrm{E}(f(\boldsymbol{X})) \leq \mathrm{E}(f(\boldsymbol{Y})) \quad \text{for all convex } f : \mathbb{R}^d \to \mathbb{R}, \tag{1.6}$$

as far as both expectations exist, is sufficient for $\boldsymbol{X} \leq_{lz} \boldsymbol{Y}$.

The restriction (1.6) defines a wellknown dispersion order among random vectors (and among their distributions), the *convex order* which is also called *dilation order*. In many fields of applied probability and statistics, the convex order has been used to derive inequalities and to compare distributions with respect to their dispersion, mostly for $d = 1$. Depending on the area of application, the convex order reflects increases of dispersion, risk, or inequality and therefore has got many names: dilation order in probability and statistics, convex order in applied probability and operations research, *increasing risk* in economics. A related univariate order, where (1.6) is assumed for all increasing convex f, is the *increasing convex order*, which is also called *second order stochastic dominance* in decision making and *stop-loss order* in insurance.

Due to (1.6), the lift zonoid order is implied by the convex order. If $d = 1$, the reverse implication is obvious. Thus, both orders describe *multivariate dilations*, that is extensions of the univariate dilation order. In higher dimensions the lift zonoid order is slightly weaker than the convex order and can be used in similar applications. Within proper parametric classes, for example multivariate normals, the two stochastic orders coincide. In Section 7.1 the multivariate *Lorenz order* is introduced, by which distributions of shares are compared with respect to dispersion. In Chapters 8 and 9 the lift zonoid order and related orderings are investigated in detail.

A random vector $\boldsymbol{Y}$ is larger than another random vector $\boldsymbol{X}$ in the convex order if and only if $\boldsymbol{Y}$ is distributed like $\boldsymbol{X}$ plus noise, that is, plus a random vector $\boldsymbol{Z}$ that has conditional expectation $\mathrm{E}(\boldsymbol{Z}|\boldsymbol{X}) = 0$. As the lift zonoid is implied by the convex order, there holds

$$\boldsymbol{X} \leq_{lz} \boldsymbol{X} + \boldsymbol{Z} \quad \text{if} \quad \mathrm{E}(\boldsymbol{Z}|\boldsymbol{X}) = 0\,. \tag{1.7}$$

That is, an additive disturbance yields an increase in lift zonoid order.

It will be demonstrated in Chapter 8 that the lift zonoid order is continuous and persists under mixtures and convolutions. The order persists also under arbitrary affine transformations: Let $\boldsymbol{A}$ be a $d \times k$ matrix and $\boldsymbol{b} \in \mathbb{R}^k$; then

$$\boldsymbol{X}\boldsymbol{A} + \boldsymbol{b} \leq_{lz} \boldsymbol{Y}\boldsymbol{A} + \boldsymbol{b} \quad \text{if} \quad \boldsymbol{X} \leq_{lz} \boldsymbol{Y}.$$

In particular, $\boldsymbol{X} \leq_{lz} \boldsymbol{Y}$ implies that all marginals are ordered in the same way; but the reverse is in general not true.

The lift zonoid order, like other stochastic orders, is a partial order. It allows the comparison of some pairs of distributions but not of all. A complete ordering of distributions is obtained from a real-valued index. Many commonly used indexes of variability are consistent with the lift zonoid order, that is, they increase with respect to this order.

The *volume of the lift zonoid* is a statistic which reflects the dispersion of the distribution and is, by definition, consistent with the lift zonoid order. In dimension $d = 1$ the volume equals the *Gini mean difference*, which measures the dispersion of a probability distribution. In higher dimensions multivariate versions of the Gini mean difference and the *Gini index*, which is a rescaling of the mean difference in order to measure disparity, have been developed on the basis of the lift zonoid volume.

Chapter 7 is devoted to statistics that are based on the lift zonoid volume and on the volumes of central regions. These statistics measure dispersion and have many applications, e.g. in cluster analysis. They also turn out to be useful in measuring dependency. In Section 8.8 another multivariate extension of the Gini index is investigated. It comes as half the expected distance between two independent, identically distributed random vectors.

By slightly modifying the lift zonoid order, it is possible to measure multivariate *economic inequality* and *industrial concentration*.

If in condition (1.5) the class $\mathcal{F}_{cl}$ is replaced by a proper subclass of functions one obtains a stochastic order which is weaker than the lift zonoid order. Several of such orders have interesting economic applications. Especially, if the parameter vector $\boldsymbol{p}$ is restricted to the nonnegative cone $\mathbb{R}^d_+$, the *price majorization*[9] $\preceq_P$ arises:

$$\mu \preceq_P \nu \quad \text{if} \quad \int_{\mathbb{R}^d} f(x)\mu(dx) \leq \int_{\mathbb{R}^d} f(x)\nu(dx) \quad \text{holds for all } f \in \mathcal{F}_P,$$

[9] also called *convex-posilinear order*

as far as both integrals exist, where

$$\mathcal{F}_P = \{f : \mathbb{R}^d \to \mathbb{R} : f(\boldsymbol{x}) = \psi(\langle \boldsymbol{p}, \boldsymbol{x} \rangle),\ \psi : \mathbb{R} \to \mathbb{R} \text{convex},\ \boldsymbol{p} \in \mathbb{R}^d_+\}.$$

For this and similar orderings, which are economically meaningful, see Chapter 9 below.

The last part of Chapter 9 is about the measurement of multivariate economic concentration. Horizontal concentration of firms within an industry is commonly measured by the concentration curve of a size characteristic and by indices derived from it. Among such indices the concentration rates and the indices by Rosenbluth and Herfindahl are in most popular use. The market power of each firm is measured by some characteristic of size. This can be an output quantity, like sales or value added, an input quantity, like investments or number of workers, or a financial quantity, like assets or stock market value. Also the physical presence in the market as given by the number and site of plants or sale points can serve as a relevant attribute. A stochastic order and several indexes for measuring industrial concentration are given. The order is related to the convex-posilinear order and the indexes generalize the usual univariate Rosenbluth and Herfindahl indexes.

1.7 Central regions and data depth

Our second stream of applications concerns a new notion of data depth and central regions in multivariate data analysis.

An important task of data analysis is to identify sets of points that are central in a given data cloud. Such a *central region* includes a properly defined center and, in this way, describes the typical location of the data. Moreover, a central region is designed to contain a relatively large part of the data such that its size and shape reflect their dispersion in space. The complement of a central region is an *outlying region*; its elements can be seen as *outliers*. (Here by an outlier we understand a "good" outlier, that is a true but less typical result of the assumed data generating process.) Also families of nested central regions are of interest.

If the family is rich enough it can contain the full information on the data. E.g., in dimension one the interval between two quantiles $Q(\alpha)$ and $Q(1-\alpha)$ for $\alpha \in [0, 1/2]$ is a central region, and the family of these intervals fully determines the distribution.

In multivariate data analysis a concept of central regions should be affine equivariant. That is, if the data points are subject to some affine transformation the central region transforms in the same way.

The lift zonoid relates to a notion of central regions, the ***zonoid trimmed regions*** , which are affine equivariant and have many other useful properties.

To introduce the idea, first the zonoid trimmed regions of an empirical distribution are defined. For $\boldsymbol{x}_1, \boldsymbol{x}_2, \ldots, \boldsymbol{x}_n \in \mathbb{R}^d$,

$$D_\alpha(\boldsymbol{x}_1, \ldots, \boldsymbol{x}_n) = \left\{ \sum_{i=1}^{n} \lambda_i \boldsymbol{x}_i : \sum_{i=1}^{n} \lambda_i = 1, 0 \le \lambda_i, \alpha\lambda_i \le \frac{1}{n} \text{ for all } i \right\} \quad (1.8)$$

is the ***α-trimmed region*** of the empirical distribution on $\boldsymbol{x}_1, \boldsymbol{x}_2, \ldots, \boldsymbol{x}_n$. Clearly, D_α is convex for every α. If $0 \le \alpha \le 1/n$, D_α is the convex hull of the data points. D_1 is a singleton containing their mean $\overline{\boldsymbol{x}} = 1/n \sum_{i=1}^{n} \boldsymbol{x}_i$. Moreover, D_α is monotone decreasing on α, that is, $D_\alpha \subset D_\beta$ if $\alpha > \beta$.

Figure 1.9 exhibits zonoid trimmed regions for a sample of ten data points in two-space and several values of α.

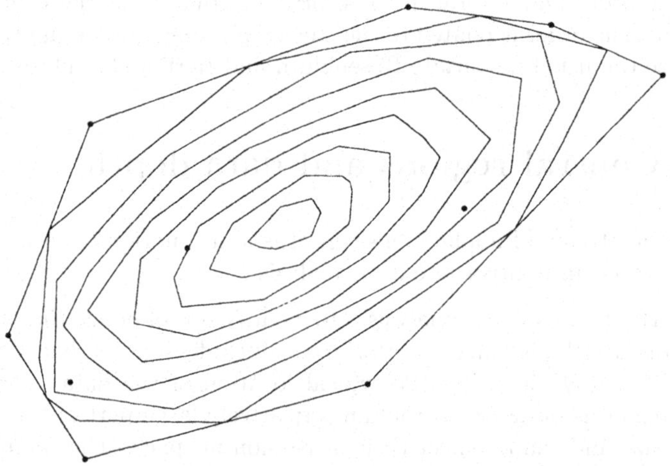

FIGURE 1.9: Zonoid trimming regions with $n = 10$ and $d = 2$. The trimming regions are drawn for $\alpha = 0.1, 0.2, \ldots, 0.9$.

For a general probability distribution similar zonoid trimmed regions are defined as well. This is called the ***population version*** of zonoid trimmed regions and will be presented in Chapter 3.

In order to derive many important properties of the trimming, a close relation between the above defined α-trimmed regions and the lift-zonoid of μ is established. This is the reason for the name *zonoid trimming*. For some $\alpha \in]0,1]$, consider the α-cut of the lift zonoid of μ, i.e., its intersection with the hyperplane $G_\alpha = \{z \in \mathbb{R}^{d+1} : z_0 = \alpha\}$. $D_\alpha(\mu)$ comes out to be the projection of the α-cut on the last d coordinates, enlarged by the factor $1/\alpha$ (Proposition 3.2).

Closely related to the concept of central regions is that of data depth. A data depth is a real-valued function that indicates how deep – in some sense – a point is located in a given data cloud. The depth defines a center of the cloud, that is the set of deepest points, and measures how far apart a point is located from the center. Thus, a depth is a kind of mid rank which allows to order multivariate data points by their centrality. It can be employed in multivariate rank tests for homogeneity, as well as in cluster analysis and in outlier detection.

Each data depth defines a family of central regions for a given distribution: A central region is the set of all points whose depth has at last some given value. A general notion of data depth and central regions is provided in Section 5.1.

The trimmed region with maximum depth is the *set of deepest points*. It is of particular interest in describing the location of the distribution. If this region is a singleton, there exists just one deepest point, which is called the *median* of the distribution with respect to this depth.

We introduce an affine equivariant data depth, named *zonoid depth*, as follows. Let $\boldsymbol{y}, \boldsymbol{x}_1, \boldsymbol{x}_2, \dots, \boldsymbol{x}_n \in \mathbb{R}^d$. The number

$$d_Z(\boldsymbol{y}|\boldsymbol{x}_1, \dots, \boldsymbol{x}_n) = \sup\{\alpha : \boldsymbol{y} \in D_\alpha(\boldsymbol{x}_1, \boldsymbol{x}_2, \dots, \boldsymbol{x}_n)\} \tag{1.9}$$

is called the zonoid depth of $\boldsymbol{y}$ with respect to $\boldsymbol{x}_1, \boldsymbol{x}_2, \dots, \boldsymbol{x}_n$. Here the convention $\sup \emptyset = 0$ is used.

The zonoid depth is, as a function of $\boldsymbol{y}$, unity at the mean, $\overline{\boldsymbol{x}}$, of the data points and vanishes outside their convex hull. The deepest point is unique and equal to the mean. There holds

$$d_Z(\boldsymbol{y}|\boldsymbol{x}_1, \dots, \boldsymbol{x}_n) \begin{cases} = 0 & \text{if } \boldsymbol{y} \notin conv(\boldsymbol{x}_1, \boldsymbol{x}_2, \dots, \boldsymbol{x}_n), \\ \geq \frac{1}{n} & \text{if } \boldsymbol{y} \in conv(\boldsymbol{x}_1, \boldsymbol{x}_2, \dots, \boldsymbol{x}_n). \end{cases}$$

Theorem (Properties of the zonoid depth). *The zonoid depth is*

- **affine equivariant:** *For any $d \times k$ matrix $\boldsymbol{A}$ having full rank d and any $\boldsymbol{c} \in \mathbb{R}^k$ holds*
  $d_Z(\boldsymbol{y}\boldsymbol{A} + \boldsymbol{c}|\boldsymbol{x}_1\boldsymbol{A} + \boldsymbol{c}, \dots, \boldsymbol{x}_n\boldsymbol{A} + \boldsymbol{c}) = d_Z(\boldsymbol{y}|\boldsymbol{x}_1, \dots, \boldsymbol{x}_n)$.

- **continuous on y:** *The function $y \mapsto d_Z(y|x_1, \ldots, x_n)$ is continuous on $y \in conv\{x_1, x_2, \ldots, x_n\}$.*
- **decreasing on rays:** *The function $\gamma \mapsto d_Z(\gamma y + \overline{x}|x_1, \ldots, x_n)$ is decreasing for $\gamma \geq 0$, $y \in \mathbb{R}^d$.*
- **continuous on the data:** *The function $(x_1, \ldots, x_n) \mapsto d_Z(y|x_1, \ldots, x_n)$ is continuous at any $(x_1, \ldots, x_n)$ for which y is in the relative interior of $conv\{x_1, \ldots, x_n\}$.*
- **increasing on dilation:** *If $(z_1, \ldots, z_n)$ is a dilation of $(x_1, \ldots, x_n)$ then $d_Z(y|x_1, \ldots, x_n) \leq d_Z(y|z_1, \ldots, z_n)$.*

The theorem follows from results proven in Section 4.2.

The zonoid data depth as well as the zonoid trimmed regions can be efficiently calculated. For details see Chapter 4, where the zonoid depth is also contrasted with other popular data depths.

1.8 Statistical inference

A data depth measures the degree of centrality of y in a data cloud $x_1, x_2, \ldots, x_n$. This can be used not only to describe and analyse statistical data but also to construct nonparametric tests for multivariate statistical inference. In particular, two-sample rank tests for homogeneity against scale and shift alternatives have been developed (Dyckerhoff, 1998).

In Chapter 5 various concepts of data depth are used to construct tests for *homogeneity* of two multivariate distributions *against scale alternatives*. Given two samples, the idea is to rank each data point according to its depth with respect to the pooled sample. The point with the lowest depth gets rank one and the point with the highest depth gets rank $m + n$, where m and n are the sizes of the two samples. The sum of ranks of the first sample serves as a test statistic.

Some depth functions, when being applied to an empirical distribution, are noncontinuous and constant on regions of positive Lebesgue measure; an example is the halfspace depth. If this is the case, several data points will often have the same depth. To resolve ties a *random tie-breaking scheme* can be used. Then, under the null hypothesis of homogeneity, the ranks of the first sample are distributed as a sample without replacement from the integers $1, \ldots, m + n$. Therefore the null-distribution of the test statistic is simply the distribution of the test statistic of the usual (univariate) *Wilcoxon test.* Small values of the test statistic indicate that most points of the first sample are far from the center of the pooled data cloud, whereas large values of the test statistic indicate that they are near the center. Thus, the null hypothesis is rejected if the test statistic assumes either very small or very large values. In Section 5.2 such tests are investigated with the *Mahalanobis depth*, the *halfspace depth* and the *zonoid depth*.

While the null distribution of the test statistic is the same for all depths, the *power of the test* depends on which depth is used in assigning the ranks. Simulation studies have been performed with multivariate normal and Cauchy samples. They show that for normal samples the three depth tests have fairly the same power and that this power comes close to that of Box's M-test, which is optimal under normality. For Cauchy samples the M-test breaks down, but the tests based on the halfspace depth and the zonoid depth have satisfactory power.

For homogeneity against location alternatives, Liu and Singh (1993) have proposed another test that rests on data depths; see Section 5.3. Here another scheme of assigning ranks has to be used. To each point of the *second sample* assign a rank according to its depth in the *first sample.* If the sum of these ranks is low, this indicates that most of the points in the second sample are far from the center of the first sample, which hints to a shift in location. However, this test is not distribution free. Its null distribution is not easy to find. It depends on the sample distribution and the depth used.

Liu and Singh (1993) prove that the test statistic is asymptotically normal if either $d = 1$ or the Mahalanobis depth is used. Extensive simulations (see Section 5.3) support the conjecture that asymptotical normality holds also for $d > 1$ and for the other depths. However, the simulations show also that another nonparametric test, the Puri-Sen test, outperforms the depth test, at least if the zonoid depth is employed.

In Chapter 6 the *depth of a hyperplane* is introduced and studied. Given a hyperplane H in $\mathbb{R}^d$ and a finite set of points $\widetilde{V}$ in $\mathbb{R}^d$, consider the minimal portion of points of $\widetilde{V}$ which are in one of the two closed halfspaces bordered by H. Call this number $d_{Hyp}(\widetilde{V} \mid H)$ the depth of H with respect to $\widetilde{V}$. This notion of depth is easily extended to the depth of a hyperplane with respect to a probability measure.

The depth of a hyperplane is easily calculated. It is invariant under affine transformations of $\mathbb{R}^d$ and gives rise to sample functions that are useful in multivariate data analysis, descriptive as well as inferential.

Two samples can be compared by the *mean hyperplane depth* (MHD) of the first sample with respect to second. Let $\boldsymbol{X}_1, \ldots, \boldsymbol{X}_n, \boldsymbol{Y}_1, \ldots, \boldsymbol{Y}_m$ be independent random vectors in $\mathbb{R}$, the $\boldsymbol{X}_i$ distributed as F, the $\boldsymbol{Y}_j$ distributed as G. To test for homogeneity, $H_0 : F = G$, use the sample function g,

$$g(\boldsymbol{x}_1, \dots, \boldsymbol{x}_n; \boldsymbol{y}_1, \dots, \boldsymbol{y}_m) = \tag{1.10}$$

$$\frac{1}{\binom{n}{d}} \sum d_{Hyp}(\{\boldsymbol{x}_1, \dots, \boldsymbol{x}_n; \boldsymbol{y}_1, \dots, \boldsymbol{y}_m\} \,|\, \mathrm{aff}(\boldsymbol{x}_{i_1} \dots, \boldsymbol{x}_{i_d})),$$

where the summation extends over all $\{i_1, \dots, i_d\}$ for which $(\boldsymbol{x}_{i_1} \dots, \boldsymbol{x}_{i_d})$ has full rank and $\mathrm{aff}(\boldsymbol{x}_{i_1} \dots, \boldsymbol{x}_{i_d})$ denotes the hyperplane which contains $\boldsymbol{x}_{i_1} \dots, \boldsymbol{x}_{i_d}$. g is the mean hyperplane depth (MHD) of the first sample with respect to the pooled data. We use the MHD as a basic statistic to measure how close the data of the first sample are located to the center of the second sample.

The MHD of a sample with respect to itself has a particular meaning; see 6.3. It takes its maximum if all points in the sample are extreme points, and it becomes smaller the more points are located in inner 'chambers', that is, polytopes spanned by other sample points as vertices. Thus, the MHD measures an aspect of the data which is called their *combinatorial dispersion.*

As an inferential tool, the MHD is used in two-sample tests for homogeneity against location and scale, respectively. Some properties of these tests, including power results, are established in Section 6.6.

2

Zonoids and lift zonoids

This chapter contains the general theory of lift zonoids. The lift zonoid of a given measure on $\mathbb{R}^d$ is the zonoid of a related measure on $\mathbb{R}^{d+1}$. Therefore, first the zonoid of a measure is investigated in Sections 2.1.1 and 2.1.2. Three definitions are provided and their equivalence is proved. The zonoid of a measure whose first moment exists is a convex compact in $\mathbb{R}^d$; it is centrally symmetric and contains the origin. If the measure has finite support the zonoid comes out to be a polytope and is named the zonotope of the measure.

Convex compacts can be conveniently described and analyzed by their support functions. Thus, as a fundamental tool of our further investigations, the support function of the zonoid of a measure is derived and the extreme points of the zonoid are characterized (Section 2.1.3). Next, in Section 2.1.4 the zonoid of a probability measure is shown to be the set valued expectation of a random segment, which provides still another definition of the zonoid.

The volume of the zonoid is an interesting parameter of the underlying measure. In Section 2.1.5 formulae are derived for the volume of the zonotope of an empirical measure as well as of the zonoid of a general measure.

But different measures can have the same zonoid. A necessary and sufficient condition for the equality of zonoids is proven and examples of such measures are given in Section 2.1.6.

In contrast to the zonoid, the lift zonoid characterizes the underlying measure uniquely. The lift zonoid of a given measure on $\mathbb{R}^d$ is the zonoid of a measure on $\mathbb{R}^{d+1}$, which is the product of the univariate Dirac measure at $z_0 = 1$ and the given measure on $\mathbb{R}^d$. Geometrically, this means that the measure is 'lifted' to a measure on $\mathbb{R}^{d+1}$, with support in the hyperplane at $z_0 = 1$.

As every lift zonoid is a zonoid, many properties and results – including the support function – are easily derived from those previously obtained (Section 2.2.1). Section 2.2.2 shortly comments on the lift zonotope, which is the lift zonoid of a measure with finite support. In Section 2.2.3 the univariate case is discussed: We show that the border of a lift zonoid consists of the generalized Lorenz curve and its dual and calculate the lift zonoids of several known parametric distributions.

By the lift zonoid, the space of d-variate measures having finite first moments is mapped into the space of centrally symmetric convex compacts in $\mathbb{R}^{d+1}$. First, the mapping is shown to be positively homogeneous and additive, that is positively linear. The set inclusion of lift zonoids yields a relation between d-variate measures which is transitive and reflexive. It is called the *lift zonoid order*. In Section 2.3.1 the lift zonoid order is characterized by univariate orderings. We prove that the lift zonoid order among two d-variate measures is equivalent to the same order of their images under all linear mappings into $\mathbb{R}$. Through this characterization, all aspects of multivariate measures, that are described by the inclusion of their lift zonoids, can be investigated by looking at the lift zonoids of univariate measures. Section 2.3.2 contains the proof that every measure is uniquely determined by its lift zonoid.

To summarize, the lift zonoid order is antisymmetric, further, the lift zonoid mapping is an embedding, positively homogeneous and linear, of measures into convex compacts. Before we demonstrate (in Section 2.4.1) that the mapping is also continuous, the lift zonoid metric is discussed, which is equal to the Hausdorff distance between lift zonoids (Section 2.3.3), and the behavior of the lift zonoid under linear transformations of the measure is investigated (Section 2.3.4). In particular, it is shown that the lift zonoid of a marginal is the corresponding projection of the measure's lift zonoid.

Next, a number of convergence results for lift zonoids are established. The main theorem of Section 2.4.1 says that the weak convergence of uniformly integrable measures implies the Hausdorff convergence of the corresponding lift zonoids. In other words, for uniformly integrable measures, weak convergence is sufficient for convergence in the lift zonoid metric. If the measures are L-continuous, the reverse holds, too. We show by an example that uniform integrability cannot be dispensed with. Other simple conditions are given that are sufffficient for uniform integrability and, thus, convergence of lift zonoids. Section 2.4.2 treats the monotone approximation of measures with regard to the lift zonoid order. It is demonstrated that a given measure can be approximated by empirical measures which are smaller and by L-continuous measures which are larger in lift zonoid order. In Section 2.4.3 the formula for the volume of the lift zonoid of a general measure is proven.

Further, the asymptotic behavior of the lift zonoid is investigated in a sampling context. We draw a sample of independent, identically distributed random vectors and consider the lift zonoid of its empirical distribution. This lift zonoid is a random set, namely, a finite sum of random segments. It is named the sample lift zonotope. In Section 2.5 two limit theorems are established, a law of large numbers and a central limit theorem. The first states that the sample lift zonotope approaches the distribution's lift zonoid almost surely, and the second says that this convergence is governed by a Gaussian process on the sphere. The lift zonoid representation can be regarded as a representation of measures by means of a set of d-variate functions, which form the integrands of the support functions of the lift zonoid. In Section 2.6 this view is extended from the Euclidean to more general spaces and the representation of measures by functionals defined on a certain set of functions is discussed.

2.1 Zonotopes and zonoids

In this section several equivalent definitions of the zonoid of a measure are introduced and discussed. By definition, a *zonotope*[1] is a finite sum of line segments in $\mathbb{R}^d$, and a *zonoid* is a limit, in the Hausdorff sense, of zonotopes.

The zonoid of a measure can be defined either as the range of a vector measure or as the set of gravity centers of a set of measures or as the set-valued expectation of a random segment. The zonotope of a measure is a zonoid that arises from a discrete measure. In the sequel the basic properties of zonoids and zonotopes and a number of new ones are reported. Our main analytical tool will be the support function of a zonoid.

2.1.1 Zonoid of a measure

As a prerequisite, we need the notions of Minkowski sum and Hausdorff distance between sets.

Definition 2.1 (Minkowski sum, Hausdorff distance) *Let C and D be sets in $\mathbb{R}^d$. The set*

$$C + D = \{z : z = x + y, x \in C, y \in D\}$$

is the Minkowski sum *of C and D.*

$$\delta_H(C, D) = \inf\{\delta : C \subset D + B(\delta), D \subset C + B(\delta), \delta > 0\}$$

[1] Some facts about zonotopes are collected in Appendix A 3.

is the Hausdorff distance *between C and D. Here $B(\delta)$ denotes the ball around the origin with radius $\delta > 0$.*

Let $\mathcal{M}_1$ be the class of nonnegative measures μ on $\mathcal{B}^d$ whose first moment is finite, $\int_{\mathbb{R}^d} ||x||\mu(dx) < \infty$. Denote the total mass of μ by $\alpha(\mu) = \int_{\mathbb{R}^d} \mu(dx)$ and the gravity center by $\epsilon(\mu) = \int_{\mathbb{R}^d} x\mu(dx)$. Let $\mathcal{P}_1$ be the subclass of probability measures $\mu \in \mathcal{M}_1$. We define the *zonoid of a measure* as follows.

Definition 2.2 (Zonoid of a measure) *Let $\mu \in \mathcal{M}_1$ be given. The set*

$$Z(\mu) = \left\{ \int_{\mathbb{R}^d} x\, g(x)\mu(dx) \;:\; g : \mathbb{R}^d \to [0,1] \text{ measurable} \right\}$$

is called the zonoid of the measure μ.

Before this definition is illustrated by simple examples, we mention several properties of zonoids of measures which can be immediately derived from the definition.

Proposition 2.1 (Properties of zonoids of measures)
Let $\mu \in \mathcal{M}_1$.

(i) *The zonoid of μ is convex and bounded.*

(ii) *The zonoid of μ contains $\mathbf{0} \in \mathbb{R}^d$ and is centrally symmetric around $\frac{1}{2}\epsilon(\mu)$, i.e.,*

$$(\epsilon(\mu) - z) \in Z(\mu) \quad \text{holds whenever} \quad z \in Z(\mu).$$

(iii) *$Z(\mu)$ is contained in the d-dimensional rectangle between $\mathbf{0}$ and $\epsilon(\mu)$, $\mathbf{0} \le z \le \epsilon(\mu)$ for any $z \in Z(\mu)$ if and only if the support of μ is in $\mathbb{R}^d_+$.*

Proof. (i): The class of measurable functions $g : \mathbb{R}^d \to [0,1]$ is convex and bounded. It follows that $Z(\mu)$ is convex. μ has finite first moment, hence $||z|| \le \int ||x||\mu(x) < \infty$ for any $z \in Z(\mu)$. That is, $Z(\mu)$ is bounded.
(ii): If a point $z \in Z(\mu)$ is generated by some function g, the point $\epsilon(\mu) - z$ is generated by the function $1 - g$.
(iii): Obvious. Q.E.D.

As will be seen in the next section, the zonoid of a measure in $\mathcal{M}_1$ is also closed, hence a centrally symmetric convex compact set.

Let us first consider the case of univariate measures. In dimension $d = 1$ the zonoid of a measure μ is an interval. Choosing in the definition the indicator functions $g = 1_{[0,\infty[}$ and $g = 1_{]-\infty,0]}$ one obtains $\int_0^\infty x\mu(dx)$ and $\int_{-\infty}^0 x\mu(dx)$, respectively, which obviously are extreme points of $Z(\mu)$. As the zonoid is convex, it follows that

$$Z(\mu) = \left[\int_{-\infty}^{0} x\mu(dx), \int_{0}^{\infty} x\mu(dx)\right].$$

If μ has support in $\mathbb{R}_+$ we obtain $Z(\mu) = [0, \epsilon(\mu)]$. Therefore, two measures on $\mathbb{R}_+$ have the same zonoid if they have the same gravity center.

For example, if μ is a probability and *uniformly distributed* on the interval $[0, 2\beta]$ there holds $Z(\mu) = [0, \beta]$. If μ is uniformly distributed on $[-\beta, \beta]$ then $Z(\mu) = [-\beta/4, \beta/4]$.

Now we turn to simple special cases of d-variate measures. The zonoid of the *Dirac measure* $\delta_{\boldsymbol{a}}$ at a point $\boldsymbol{a} \in \mathbb{R}^d$ is a segment,

$$Z(\delta_{\boldsymbol{a}}) = [\boldsymbol{0}, \boldsymbol{a}].$$

An *empirical measure* is a measure that gives equal mass to a finite number of, not necessarily different, points. Let $\mathcal{M}^n$ denote those empirical measures in $\mathcal{M}$ which give equal mass to n points in $\mathbb{R}^d$, and let $\mathcal{P}^n = \mathcal{M}^n \cap \mathcal{P}$ be the set of empirical probability measures on n points in $\mathbb{R}^d$. Obviously, $\mathcal{M}^n \subset \mathcal{M}_1$ holds for any n. A measure $\mu \in \mathcal{M}^n$ can be written as

$$\mu = \frac{\alpha(\mu)}{n} \sum_{i=1}^{n} \delta_{\boldsymbol{a}_i},$$

with some $\boldsymbol{a}_1, \boldsymbol{a}_2, \ldots, \boldsymbol{a}_n \in \mathbb{R}^d$. The zonoid of an empirical measure is a zonotope. It is the sum of line segments,

$$Z(\mu) = \sum_{i=1}^{n} \left[\boldsymbol{0}, \frac{\alpha(\mu)}{n} \boldsymbol{a}_i\right] = \frac{\alpha(\mu)}{n} \sum_{i=1}^{n} \{\boldsymbol{z} \in \mathbb{R}^d \ : \ \boldsymbol{z} = \lambda \boldsymbol{a}_i,\ 0 \le \lambda \le 1\}. \quad (2.1)$$

Let us consider the zonoid of an empirical measure more closely. For this we need some notions from the theory of polytopes, see Appendix A 3. $Z(\mu)$ is a zonotope, and its faces are zonotopes as well. Every face of $Z(\mu)$ corresponds to a proper subset J of $\{1, 2, \ldots, n\}$: the face is a shift of $\sum_{i \in J} [\boldsymbol{0}, \alpha(\mu)/n\, \boldsymbol{a}_i]$. The face is a facet if and only if the linear span of $\{\boldsymbol{a}_i\}_{i \in J}$ has dimension $d - 1$.

Example 2.1 (Unit square) The unit square in $\mathbb{R}^2$ (see Figure 2.1) is the zonoid of the empirical probability distribution on the points $(2, 0)$ and $(0, 2)$,

$$Z\left(\frac{1}{2}\delta_{(2,0)} + \frac{1}{2}\delta_{(0,2)}\right) = [(0, 0), (1, 0)] + [(0, 0), (0, 1)].$$

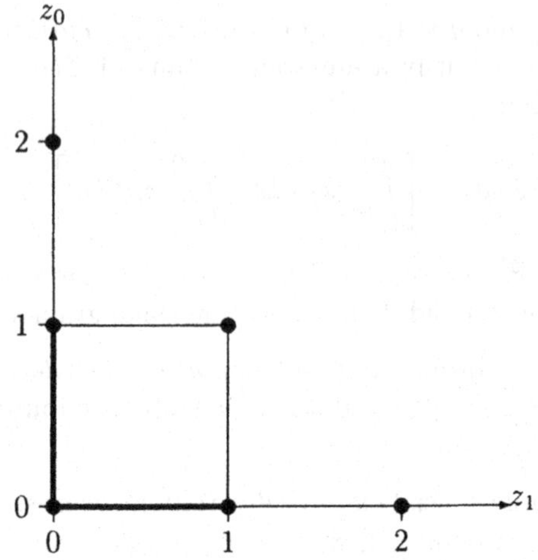

FIGURE 2.1: The unit square is the zonoid of an empirical probability measure on points $(2, 0)$ and $(0, 2)$ in $\mathbb{R}^2$.

2.1.2 Equivalent definitions of the zonoid of a measure

The zonoid of a measure can be introduced in different ways. Besides the above definition it can be defined either as a set of gravity centers of certain measures or as the range of a vector measure related to μ.

A *vector measure* is a σ-additive function defined on some σ-algebra and having values in $\mathbb{R}^d$. In particular, given μ, the integral $\tau_\mu(B) = \int_B \boldsymbol{x}\mu(d\boldsymbol{x})$ can be seen as an incomplete first moment, for any $B \in \mathcal{B}^d$. It defines a vector measure $\tau_\mu : \mathcal{B}^d \to \mathbb{R}^d$, which is nonnegative and finite. The set

$$Z_L(\mu) = \left\{ \int_B \boldsymbol{x}\ \mu(d\boldsymbol{x}) \ : \ B \in \mathcal{B}^d \right\}$$

is the range of this vector measure. In general, $Z_L(\mu)$ is not convex. E.g., for the Dirac measure on a point $\boldsymbol{a}$ which is not the origin, the set $Z_L(\mu)$ consists of just two points, $Z_L(\delta_{\boldsymbol{a}}) = \{\boldsymbol{0}, \boldsymbol{a}\}$. A famous theorem of Lyapunov (for proofs see Halmos (1948) and Lindenstrauss (1966)) provides a sufficient condition for convexity:

Proposition 2.2 (Lyapunov's theorem) *The range of a vector measure is closed; the range of a nonatomic vector measure is convex and closed.*

A vector measure τ is *nonatomic* if $\tau(B) > 0$ implies the existence of a set A, $A \subset B$, with $0 < \tau(A) < \tau(B)$. In particular, τ_μ is nonatomic if μ is L-continuous. Now we come to two equivalent characterizations of the zonoid of a measure.

Theorem 2.3 (Convex body) *For any $\mu \in \mathcal{M}_1$, $Z(\mu)$ is a compact set. There holds*

$$Z(\mu) = conv\left\{\int_B x\ \mu(dx)\ :\ B \in \mathcal{B}^d\right\} \tag{2.2}$$

$$= \left\{\int_{\mathbb{R}^d} x\ \nu(dx)\ :\ \nu \le \mu\right\}. \tag{2.3}$$

If μ is L-continuous,

$$Z(\mu) = \left\{\int_B x\ \mu(dx)\ :\ B \in \mathcal{B}^d\right\}.$$

Recall that $\nu \le \mu$ means $\nu(B) \le \mu(B)$ for every $B \in \mathcal{B}^d$. The second equation of Theorem 2.3 says that $Z(\mu)$ is the set of gravity centers of all measures which are smaller than μ in that they give less or equal mass to each Borel set.

Proof. Denote $Z_L(\mu) = \{\int_B x\ \mu(dx)\ :\ B \in \mathcal{B}^d\}$ and $Z_C(\mu) = \{\int_{\mathbb{R}^d} x\ \nu(dx)\ :\ \nu \le \mu\}$. By the first part of Lyapunov's theorem $Z_L(\mu)$ is closed and, since $\mu \in \mathcal{M}_1$, $Z_L(\mu)$ is bounded. It follows that $conv\ Z_L(\mu)$ is closed and bounded as well. We prove $Z(\mu) = conv\ Z_L(\mu) = Z_C(\mu)$ in three steps. Then the assertion concerning an L-continuous measure follows from the second part of Lyapunov's theorem.

(i) $Z(\mu) \subset conv\ Z_L(\mu)$: Let $z \in Z(\mu)$. Then $z = \int_{\mathbb{R}^d} x\ g(x)\mu(dx)$ with some $g : \mathbb{R}^d \to [0,1]$. There exists a sequence $g_n \nearrow g$,

$$g_n(x) = \sum_{i=1}^{n} \alpha_{i,n}\, 1_{A_{i,n}}(x)\,, \quad x \in \mathbb{R}^d\,,$$

with $\alpha_{i,n} \ge 0$, $A_{i,n} \in \mathcal{B}^d$, $A_{i-1,n} \subset A_{i,n}$ for $i = 1, 2, \ldots, n$. Recall that 1_S denotes the indicator function of a set S. Then $\sum_{i=1}^n \alpha_{i,n} \le 1$ holds since $0 \le g \le 1$. We define $\alpha_{0,n} = 1 - \sum_{i=1}^n \alpha_{i,n}$, $A_{0,n} = \emptyset$, and obtain that

$$g_n(x) = \sum_{i=0}^{n} \alpha_{i,n} 1_{A_{i,n}}(x) \quad \text{and} \quad \sum_{i=0}^{n} \alpha_{i,n} = 1\,.$$

Then

$$\begin{aligned} z &= \int_{\mathbb{R}^d} x\, g(x)\, \mu(dx) = \int_{\mathbb{R}^d} x \lim_{n\to\infty} \sum_{i=0}^{n} \alpha_{i,n} 1_{A_{i,n}}(x)\, \mu(dx) \\ &= \lim_{n\to\infty} \sum_{i=0}^{n} \alpha_{i,n} \int_{A_{i,n}} x\, \mu(dx) \end{aligned}$$

by the majorized convergence theorem and, therefore, $z \in conv\, Z_L(\mu)$.

(ii) $conv\, Z_L(\mu) \subset Z_C(\mu)$: Since $Z_C(\mu)$ is convex it suffices to show that $conv\, Z_L(\mu) \subset Z_C(\mu)$. For this, assume that $z = \int_A x\, \mu(dx)$ with some A. Let $\nu(B) = \mu(A \cap B)$. ν is a Borel measure; there holds $z = \int x\, \nu(dx)$ and $\nu(B) \leq \mu(B)$ for all $B \in \mathcal{B}^d$. Hence $z \in Z_C(\mu)$.

(iii) $Z_C(\mu) \subset Z(\mu)$: Let $z \in Z_C(\mu)$. $z = \int_{\mathbb{R}^d} x\, \nu(dx)$ with some ν, $\nu \leq \mu$. It follows that ν is μ-continuous. Let $g = d\nu/d\mu$ be the Radon-Nikodym derivative. We have $z = \int x\, g(x)\mu(dx)$ and, as $\nu \leq \mu$, $0 \leq g \leq 1$. Hence $z \in Z(\mu)$. Q.E.D.

2.1.3 Support function of a zonoid

Zonoids can be characterized via their support functions. Many of the results on zonoids in subsequent sections will be established by support function arguments. Some definitions and facts about support functions, extreme points and supporting points are collected in Part 3 of Appendix A.

The following theorem and corollary provide the support function of the zonoid of a measure and its extremal points.

Theorem 2.4 (Support function of zonoid) *For $\mu \in \mathcal{M}_1$, the support function of $Z(\mu)$ is given by*

$$\begin{aligned} h(Z(\mu), p) &= \int_{\langle x, p\rangle \geq 0} \langle x, p\rangle \mu(dx) \\ &= \int_{\mathbb{R}^d} \max\{0, \langle x, p\rangle\}\mu(dx), \quad p \in \mathbb{R}^d. \end{aligned} \tag{2.4}$$

Proof. Define for x and $p \in \mathbb{R}^d$

$$1_p(x) = \begin{cases} 1 & \text{if } \langle x, p\rangle \geq 0, \\ 0 & \text{otherwise,} \end{cases} \tag{2.5}$$

and for any measurable function $g : \mathbb{R}^d \to [0,1]$

$$\zeta(\mu, g) = \int_{\mathbb{R}^d} g(\boldsymbol{x})\, \boldsymbol{x}\, \mu(d\boldsymbol{x})\,. \tag{2.6}$$

We obtain

$$g(\boldsymbol{x})\langle \boldsymbol{x}, \boldsymbol{p}\rangle \le 1_{\boldsymbol{p}}(\boldsymbol{x})\langle \boldsymbol{x}, \boldsymbol{p}\rangle,$$

and therefore

$$\begin{aligned} \langle \zeta(\mu, g), \boldsymbol{p}\rangle &= \int_{\mathbb{R}^d} g(\boldsymbol{x})\langle \boldsymbol{x}, \boldsymbol{p}\rangle \mu(d\boldsymbol{x}) \\ &\le \int_{\mathbb{R}^d} 1_{\boldsymbol{p}}(\boldsymbol{x})\langle \boldsymbol{x}, \boldsymbol{p}\rangle \mu(d\boldsymbol{x}) = \langle \zeta(\mu, 1_{\boldsymbol{p}}), \boldsymbol{p}\rangle \\ &= \int_{\langle \boldsymbol{x}, \boldsymbol{p}\rangle \ge 0} \langle \boldsymbol{x}, \boldsymbol{p}\rangle \mu(d\boldsymbol{x}) = \int_{\mathbb{R}^d} \max\{0, \langle \boldsymbol{x}, \boldsymbol{p}\rangle\} \mu(d\boldsymbol{x})\,. \end{aligned} \tag{2.7}$$

Because of $\zeta(\mu, 1_{\boldsymbol{p}}) \in Z(\mu)$, conclude (2.4) . Q.E.D.

The following corollary characterizes the extreme points of the zoonoid of a measure and the points which support it.

Corollary 2.5 (Extreme points of a zonoid) *Let $\boldsymbol{y} \in \mathbb{R}^d$ and $\boldsymbol{p} \in S^{d-1}$.*

(i) *$\boldsymbol{y}$ supports $Z(\mu)$ in direction $\mathbf{p}$ if and only if there exists a function g such that $\boldsymbol{y} = \int_{\mathbb{R}^d} \boldsymbol{x}\, g(\boldsymbol{x})\, \mu(d\boldsymbol{x})$,*

$$g(\boldsymbol{x}) = 1 \quad \textit{if} \quad \langle \boldsymbol{x}, \boldsymbol{p}\rangle > 0 \quad \textit{and} \quad g(\boldsymbol{x}) = 0 \quad \textit{if} \quad \langle \boldsymbol{x}, \boldsymbol{p}\rangle < 0\,.$$

(ii) *$\boldsymbol{y}$ is an extreme point of $Z(\mu)$ in direction $\mathbf{p}$ if and only if*

$$\boldsymbol{y} = \int_{\langle \boldsymbol{x}, \boldsymbol{p}\rangle > 0} \boldsymbol{x}\, \mu(d\boldsymbol{x}) + \int_{\langle \boldsymbol{x}, \boldsymbol{p}\rangle = 0} 1_B(\boldsymbol{x})\, \boldsymbol{x}\, \mu(d\boldsymbol{x}), \tag{2.8}$$

with some $B \in \mathcal{B}^d$.

Note that, if $\boldsymbol{y}$ is a unique extreme point in direction $\boldsymbol{p}$, the second integral in (2.8) vanishes.

Proof. (i): The 'if' part is obvious. Let $\boldsymbol{y}$ support $Z(\mu)$ in direction $\boldsymbol{p}$. Then, since $Z(\mu)$ is closed, $\boldsymbol{y} \in Z(\mu)$ and there exists a function $g : \mathbb{R}^d \to [0,1]$ such that $\boldsymbol{y} = \int_{\mathbb{R}^d} \boldsymbol{x} g(\boldsymbol{x}) \mu(d\boldsymbol{x})$ and $\langle \boldsymbol{y}, \boldsymbol{p}\rangle = h(Z(\mu), \boldsymbol{p})$. From Theorem 2.4 follows that

$$\langle \boldsymbol{y}, \boldsymbol{p}\rangle = \int_{\mathbb{R}^d} \langle \boldsymbol{x}, \boldsymbol{p}\rangle g(\boldsymbol{x}) \mu(d\boldsymbol{x}) = \int_{\langle \boldsymbol{x}, \boldsymbol{p}\rangle \ge 0} \langle \boldsymbol{x}, \boldsymbol{p}\rangle \mu(d\boldsymbol{x})$$

and therefore

$$g(\boldsymbol{x}) = \begin{cases} 1 & \text{if } \langle \boldsymbol{x}, \boldsymbol{p} \rangle > 0, \\ g_0(\boldsymbol{x}) & \text{if } \langle \boldsymbol{x}, \boldsymbol{p} \rangle = 0, \\ 0 & \text{if } \langle \boldsymbol{x}, \boldsymbol{p} \rangle < 0, \end{cases}$$

where g_0 is the restriction of g to $\{\boldsymbol{x} : \langle \boldsymbol{x}, \boldsymbol{p} \rangle = 0\}$. This proves part (i). Such a $\boldsymbol{y}$ is, in addition, extreme only iff g_0 is binary, having values in $\{0, 1\}$, that is, if g_0 is the indicator function of some set $B \in \mathcal{B}^d$. Then $\boldsymbol{y}$ has the form (2.8). Q.E.D.

Corollary 2.6 (Support function of the centered zonoid) *The support function of $Z(\mu) - \frac{1}{2}\epsilon(\mu)$ is given by*

$$h\left(Z(\mu) - \frac{1}{2}\epsilon(\mu), \boldsymbol{p}\right) = \frac{1}{2}\int_{\mathbb{R}^d} |\langle \boldsymbol{x}, \boldsymbol{p} \rangle|\, \mu(d\boldsymbol{x}), \quad \boldsymbol{p} \in \mathbb{R}^d.$$

Proof. For any $\boldsymbol{p} \in \mathbb{R}^d$, the support function of the singleton $\{-\frac{1}{2}\epsilon(\mu)\}$ equals $-\frac{1}{2}\int \langle \boldsymbol{x}, \boldsymbol{p} \rangle \mu(d\boldsymbol{x})$. Hence, by the linearity of the support function,

$$\begin{aligned} h(Z(\mu) - \frac{1}{2}\epsilon(\mu), \boldsymbol{p}) &= \int_{\mathbb{R}^d} \max\{0, \langle \boldsymbol{x}, \boldsymbol{p} \rangle\} \mu(d\boldsymbol{x}) - \frac{1}{2}\int_{\mathbb{R}^d} \langle \boldsymbol{x}, \boldsymbol{p} \rangle \mu(d\boldsymbol{x}) \\ &= \frac{1}{2}\int_{\mathbb{R}^d} |\langle \boldsymbol{x}, \boldsymbol{p} \rangle|\, \mu(d\boldsymbol{x}). \end{aligned}$$

The last equation holds because $|\beta| = 2\max\{0, \beta\} - \beta$ for any $\beta \in \mathbb{R}$. Q.E.D.

Corollary 2.7 (Bounded support) *For any $\mu \in \mathcal{M}_1$ the support function of $Z(\mu)$ is uniformly bounded on the sphere S^{d-1},*

$$0 \le h(Z(\mu), \boldsymbol{p}) = \int_{\mathbb{R}^d} \max\{0, \langle \boldsymbol{x}, \boldsymbol{p} \rangle\} \mu(d\boldsymbol{x}) \le \int_{\mathbb{R}^d} ||\boldsymbol{x}|| \mu(d\boldsymbol{x}), \quad \boldsymbol{p} \in S^{d-1}. \tag{2.9}$$

Proof. For $\boldsymbol{p} \in S^{d-1}$ we have $\langle \boldsymbol{x}, \boldsymbol{p} \rangle \le ||\boldsymbol{x}||$, and hence (2.9). The right hand side is finite since $\mu \in \mathcal{M}_1$. Q.E.D.

2.1.4 Zonoids as expected random segments

The zonoid of a probability measure is the expectation of a random set. If $\boldsymbol{X}$ is a random vector distributed as μ, the zonoid of μ equals the set valued expectation of the random segment that extends from the origin to $\boldsymbol{X}$. This result can now be demonstrated with the help of Theorem 2.4. For the definition of a random compact set and its expectation, see Part 3 of Appendix A.

Theorem 2.8 (Zonoid as expected random segment) *Let* $\mu \in \mathcal{P}_1$ *and* $\boldsymbol{X}$ *be a random vector distributed as* μ. *Then* $Z(\mu) = \mathrm{E}([\mathbf{0}, \boldsymbol{X}])$.

Proof. As $Z(\mu)$ is convex and closed, $Z(\mu)$ and $\mathrm{E}([\mathbf{0}, \boldsymbol{X}])$ coincide if and only if their support functions coincide. According to (2.4), we have $h(Z(\mu), \boldsymbol{p}) = \int_{\mathbb{R}^d} \max\{0, \langle \boldsymbol{x}, \boldsymbol{p}\rangle\}\mu(d\boldsymbol{x})$. On the other hand, $h([\mathbf{0}, \boldsymbol{X}], \boldsymbol{p}) = \max\{0, \langle \boldsymbol{X}, \boldsymbol{p}\rangle\}$. Hence

$$\begin{aligned} h(\mathrm{E}([\mathbf{0}, \boldsymbol{X}]), \boldsymbol{p}) &= \mathrm{E}(h([\mathbf{0}, \boldsymbol{X}], \boldsymbol{p})) \\ &= \int_{\mathbb{R}^d} \max\{0, \langle \boldsymbol{x}, \boldsymbol{p}\rangle\}\mu(d\boldsymbol{x}) \;=\; h(Z(\mu), \boldsymbol{p}) \,. \quad (2.10) \end{aligned}$$

Q.E.D.

2.1.5 Volume of a zonoid

The volume of a zonoid can be considered as a parameter of the underlying measure. In this section a formula for the volume of the zonoid of an empirical measure is proved and a theorem about the zonoid volume of a general measure is stated.

If μ is an empirical measure giving equal mass to each of the points $\boldsymbol{x}_1, \boldsymbol{x}_2, \ldots, \boldsymbol{x}_n$, its zonoid is a zonotope, $Z(\mu) = n^{-1}\alpha(\mu)\sum_{i=1}^n [\mathbf{0}, \boldsymbol{x}_i]$. The volume of the zonotope $Z(\mu)$ can be easily calculated from the points $\boldsymbol{x}_1, \boldsymbol{x}_2, \ldots, \boldsymbol{x}_n$.

Example 2.2 (Convergent zonotopes) Consider the sequence of zonotopes

$$Z_n = \sum_{i=1}^n \left[(0,0)\,, \left(\frac{1}{n}, \frac{2(i-1)}{n^2}\right)\right], \qquad n \in \mathbb{N}.$$

For each n, Z_n is the zonoid of the empirical probability measure at points $\boldsymbol{a}_i^n = (1, 2(i-1)/n)$, $i = 1, 2, \ldots, n$. It can be shown (see Section 2.4 and Example 2.3 below) that Z_n converges in the Hausdorff distance to the set

$$Z_\infty = \{(x_1, x_2) \in \mathbb{R}^2 : x_2^2 \le x_1 \le 2x_2 - x_2^2, 0 \le x_2 \le 1\}.$$

Consider the zonotope Z_3. It has vertices at

$$(0,0), \left(\frac{1}{3}, 0\right), \left(\frac{1}{3}, \frac{4}{9}\right), \left(\frac{2}{3}, \frac{2}{9}\right), \left(\frac{2}{3}, \frac{6}{9}\right), \left(1, \frac{6}{9}\right),$$

see Figure 2.2. An elementary calculation yields $\mathrm{vol}_2 Z_3 = 8/27$. In general, the volume of the zonotope of an empirical measure is calculated as follows.

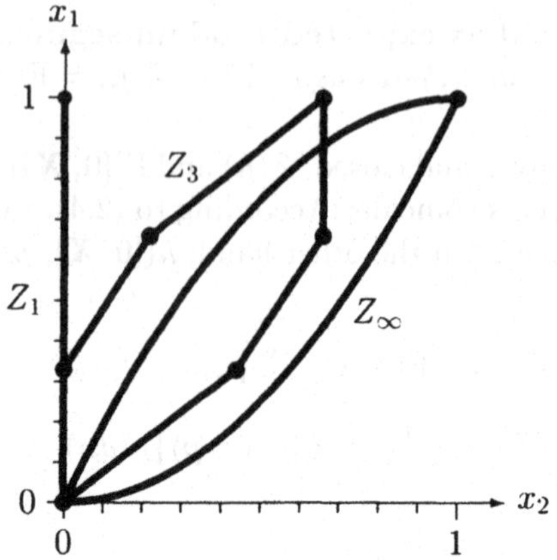

FIGURE 2.2: Zonotopes $Z_1 = [(0,0),(1,0)]$ and $Z_3 = [(0,0),(1/3,0)] + [(0,0),(1/3,2/9)] + [(0,0),(1/3,4/9)]$ and limiting zonoid Z_∞ from Example 2.2.

Proposition 2.9 (Volume of the zonoid of an empirical measure) *Let μ be an empirical measure on the points $\boldsymbol{a}_1, \boldsymbol{a}_2, \ldots, \boldsymbol{a}_n$ in $\mathbb{R}^d$. The d-dimensional volume of $Z(\mu)$ is given by*

$$\operatorname{vol}_d Z(\mu) = \frac{1}{d!}\left(\frac{\alpha(\mu)}{n}\right)^d \sum_{i_1=1}^{n} \cdots \sum_{i_d=1}^{n} |\det[\boldsymbol{a}_{i_1}, \ldots, \boldsymbol{a}_{i_d}]| \qquad (2.11)$$

$$= \left(\frac{\alpha(\mu)}{n}\right)^d \sum_{i_1<\cdots<i_d} |\det[\boldsymbol{a}_{i_1}, \ldots, \boldsymbol{a}_{i_d}]| \,. \qquad (2.12)$$

Proof. The expressions (2.11) and (2.12) are obviously the same. Formula (2.11) is a consequence of (2.1) and a well known equation,

$$\operatorname{vol}_d \sum_{i=1}^{n} [\boldsymbol{0}, \boldsymbol{x}_i] = \frac{1}{d!} \sum_{i_1=1}^{n} \cdots \sum_{i_d=1}^{n} |\det[\boldsymbol{x}_{i_1}, \ldots, \boldsymbol{x}_{i_d}]| \,, \qquad (2.13)$$

which follows from a result of Weil (1971); see also Shephard (1974), who attributes the equation to McMullen. Q.E.D.

Example 2.3 (Zonotope volume) The zonotope Z_n in Example 2.2 belongs to the empirical probability measure at points $\boldsymbol{a}_i^n = (1, 2(i-1)/n)$,

$i = 1, 2, \ldots, n$. By the volume formula (2.12) we obtain

$$\begin{aligned}
\mathrm{vol}_2 Z_n &= \frac{1}{n^2} \sum_{i<j} \left[\frac{2(j-1)}{n} - \frac{2(i-1)}{n} \right] \\
&= \frac{2}{n^3} \sum_{i=1}^{n} \sum_{j=i+1}^{n} (j-i) \\
&= \frac{2}{n^3} \sum_{k=1}^{n} \frac{(k-1)k}{2} = \frac{n(n+1)(2n-2)}{6n^3} .
\end{aligned}$$

Note that $\lim_{n\to\infty} \mathrm{vol}_2 Z_n = \frac{1}{3}$, which is the volume of the limiting zonoid Z_∞.

Proposition 2.9 provides the volume of the zonoid of an empirical measure. It can be extended to general measures as follows:

Theorem 2.10 (Volume of the zonoid of a general measure)
Consider independent random vectors $\boldsymbol{X}_1, \ldots, \boldsymbol{X}_d$, each distributed as $\frac{1}{\alpha(\mu)}\mu$ with some $\mu \in \mathcal{M}_1$. $\boldsymbol{Q}(\mu) = [\boldsymbol{X}_1, \ldots, \boldsymbol{X}_d]$ denotes the $d \times d$ matrix having rows $\boldsymbol{X}_1, \ldots, \boldsymbol{X}_d$. The d-dimensional volume of the zonoid $Z(\mu)$ is given by

$$\mathrm{vol}_d Z(\mu) = \frac{1}{d!} (\alpha(\mu))^d \, \mathrm{E}(|\det \boldsymbol{Q}(\mu)|) . \tag{2.14}$$

The theorem says that the volume of the zonoid of a measure is the expectation of the modulus of a random determinant multiplied by a factor which corresponds to the measure's total mass.

We do not prove Theorem 2.10 here because for its proof we need an approximation theorem (Theorem 2.36), which has still to be derived in the sequel. Theorem 2.10 parallels a similar result on the volume of the lift zonoid of a measure, which will be stated as Theorem 2.37 and proved at the end of this chapter. Theorem 2.10 can be demonstrated in just the same way.

Let us mention a few special cases of the volume formula. If μ is a discrete probability distribution with $\mu(\{\boldsymbol{a}_i\}) = \pi_i$, Equation (2.14) specializes to

$$\mathrm{vol}_d Z(\mu) = \sum_{i_1 < i_2 < \cdots < i_d} |\det[\boldsymbol{a}_{i_1}, \ldots, \boldsymbol{a}_{i_d}]| \prod_{j=1}^{d} \pi_{i_j} . \tag{2.15}$$

If $\mu \in \mathcal{M}_1$ and $d = 2$, denote $\boldsymbol{X}^{(1)} = (U^{(1)}, V^{(1)})$, $\boldsymbol{X}^{(2)} = (U^{(2)}, V^{(2)})$. Then

$$\mathrm{vol}_d Z(\mu) = \frac{1}{2} (\alpha(\mu))^2 \, \mathrm{E}(|U^{(1)} V^{(2)} - U^{(2)} V^{(1)}|) . \tag{2.16}$$

In particular, let $d = 2$ and $\mu = \delta_1 \otimes \tilde{\mu}$. Then $U^{(1)} = U^{(2)} = 1$ almost surely and

$$\text{vol}_d Z(\mu) = \frac{1}{2}(\alpha(\mu))^2 \, \mathrm{E}(|V^{(2)} - V^{(1)}|), \tag{2.17}$$

which is the well known Gini mean difference of the measure $\tilde{\mu}$ on $\mathbb{R}$.

The Gini mean difference of a univariate probability distribution and related parameters like the Gini coefficient and the Rosenbluth concentration index are useful devices to describe the variability of a distribution. These parameters have numerous applications in the measurement of disparity and concentration. The Gini mean difference (2.17) is the volume of the lift zonoid of $\tilde{\mu}$. In Chapter 7, by considering lift zonoids of multivariate distributions and their volumes, these parameters of variability will be extended in a natural way. Also the volume of the lift zonoid is used as a measure of homogeneity in clustering procedures.

2.1.6 Measures with equal zonoids

As is seen above in Section 2.1.1, measures with support in $\mathbb{R}_+$ have the same zonoid if they have the same gravity center. Thus, different measures can have equal zonoids, which means that the zonoid does not determine the underlying measure in a unique way.

In the present section this problem is investigated in detail and the classes of measures that have the same zonoids are characterized. Like any closed convex sets, the zonoids of two measures coincide if and only if they have equal support functions. By Theorem 2.4, $Z(\mu) = Z(\nu)$ if and only if

$$\int_{\langle \boldsymbol{x},\boldsymbol{p}\rangle \geq 0} \langle \boldsymbol{x},\boldsymbol{p}\rangle \mu(d\boldsymbol{x}) = \int_{\langle \boldsymbol{x},\boldsymbol{p}\rangle \geq 0} \langle \boldsymbol{x},\boldsymbol{p}\rangle \nu(d\boldsymbol{x}), \quad \boldsymbol{p} \in \mathbb{R}^d. \tag{2.18}$$

First a necessary condition is established: If the zonoids of two measures coincide, their gravity centers are equal.

Proposition 2.11 (Equal expectations) *For $\mu, \nu \in \mathcal{M}_1$, $Z(\mu) = Z(\nu)$ implies $\boldsymbol{\epsilon}(\mu) = \boldsymbol{\epsilon}(\nu)$.*

Proof. Let $\boldsymbol{e}_j, j = 1, \ldots d$, denote the canonical unit vectors of $\mathbb{R}^d$, hence $x_j = \langle \boldsymbol{x}, \boldsymbol{e}_j\rangle$ for any $\boldsymbol{x}$, $j = 1, \ldots, d$. Inserting $\boldsymbol{p} = \boldsymbol{e}_j$ and $\boldsymbol{p} = -\boldsymbol{e}_j$ in Equation (2.18) one obtains

$$\int_{x_j \geq 0} x_j \, \mu(d\boldsymbol{x}) = \int_{x_j \geq 0} x_j \, \nu(d\boldsymbol{x}), \qquad -\int_{x_j \leq 0} x_j \, \mu(d\boldsymbol{x}) = -\int_{x_j \leq 0} x_j \, \nu(d\boldsymbol{x}),$$

and by substraction

$$\int_{\mathbb{R}^d} x_j\, \mu(dx) = \int_{\mathbb{R}^d} x_j\, \nu(dx),$$

for $j = 1, 2, \ldots, d$. Hence $\boldsymbol{\epsilon}(\mu) = \boldsymbol{\epsilon}(\nu)$. Q.E.D.

Next, a necessary and sufficient condition for the equality of two zonoids is presented, which obviously implies the equality of support functions, (2.18).

Theorem 2.12 (Equality of zonoids) *For $\mu, \nu \in \mathcal{M}_1$, $Z(\mu) = Z(\nu)$ if and only if*

$$\int_{\langle \boldsymbol{x}, \boldsymbol{p} \rangle \geq 0} \boldsymbol{x}\, \mu(d\boldsymbol{x}) = \int_{\langle \boldsymbol{x}, \boldsymbol{p} \rangle \geq 0} \boldsymbol{x}\, \nu(d\boldsymbol{x}) \tag{2.19}$$

for every $\boldsymbol{p} \in \mathbb{R}^d$.

Proof. The theorem is another consequence of Theorem 2.4. Multiplying Equation (2.19) with $\boldsymbol{p}$ gives

$$\int_{\langle \boldsymbol{x}, \boldsymbol{p} \rangle \geq 0} \langle \boldsymbol{x}, \boldsymbol{p} \rangle\, \mu(d\boldsymbol{x}) = \int_{\langle \boldsymbol{x}, \boldsymbol{p} \rangle \geq 0} \langle \boldsymbol{x}, \boldsymbol{p} \rangle\, \nu(d\boldsymbol{x}),$$

for every $\boldsymbol{p} \in \mathbb{R}^d$. That is, $h(Z(\mu), \boldsymbol{p}) = h(Z(\nu), \boldsymbol{p})$ for all $\boldsymbol{p}$ and, by Theorem 2.4, $Z(\mu) = Z(\nu)$. On the reverse, assume that $Z(\mu) = Z(\nu)$. Then in every direction $\boldsymbol{p}$ the extreme points of $Z(\mu)$ and $Z(\nu)$ must be the same. They have the form (2.8),

$$\int_{\langle \boldsymbol{x}, \boldsymbol{p} \rangle > 0} \boldsymbol{x}\, \mu(d\boldsymbol{x}) + \int_{\langle \boldsymbol{x}, \boldsymbol{p} \rangle = 0} 1_B(\boldsymbol{x})\, \boldsymbol{x}\, \mu(d\boldsymbol{x}).$$

with some $B \in \mathcal{B}^d$. But for all $\boldsymbol{p}$, besides a set which has Lebesgue measure[2] zero on S^{d-1}, the extreme point in direction $\boldsymbol{p}$ is unique, hence the second integral vanishes, and (2.19) is satisfied. For the remaining $\boldsymbol{p}$, Equation (2.19) is established by a limit argument. Q.E.D.

Example 2.4 (Univariate measures) In case $d = 1$ the measures μ and ν have the same zonoid if $\int_0^\infty x\mu(dx) = \int_0^\infty x\nu(dx)$ and $\int_{-\infty}^0 x\mu(dx) = \int_{-\infty}^0 x\nu(dx)$.

Example 2.5 (Discrete measures) Let $d \geq 1$, $\mu, \nu \in \mathcal{P}_1$. Assume that $\{\boldsymbol{x}_1, \boldsymbol{x}_2, \ldots\}$ and $\{\boldsymbol{y}_1, \boldsymbol{y}_2, \ldots\}$ are the supports of μ and ν, respectively, and

[2]The Lebesgue measure on S^{d-1} is defined in a natural way; see Part A 2 of the Appendix.

that $\mu(\{\boldsymbol{x}_i\}) = q_i$ and $\nu(\{\boldsymbol{y}_j\}) = r_j$ for all i and j. Assume further that, for every $\boldsymbol{p} \in S^{d-1}$,

$$\sum_{\boldsymbol{x}_i \in R_{\boldsymbol{p}}} q_i \boldsymbol{x}_i = \sum_{\boldsymbol{y}_j \in R_{\boldsymbol{p}}} r_j \boldsymbol{y}_j, \tag{2.20}$$

where $R_{\boldsymbol{p}} = \{\lambda \boldsymbol{p} : \lambda > 0\}$ is the ray passing through $\boldsymbol{p}$. Equation (2.20) means that for every $\boldsymbol{p}$ the ray through $\boldsymbol{p}$ contributes the same amount to the gravity center of μ as to the gravity center of ν. Then, for any $\tilde{\boldsymbol{p}} \in S^{d-1}$,

$$\int_{\langle \boldsymbol{x}, \tilde{\boldsymbol{p}} \rangle \geq 0} \boldsymbol{x}\, \mu(d\boldsymbol{x}) = \sum_{\langle \boldsymbol{x}_i, \tilde{\boldsymbol{p}} \rangle \geq 0} q_i \boldsymbol{x}_i = \sum_{\boldsymbol{p} \in S^{d-1} : \langle \boldsymbol{p}, \tilde{\boldsymbol{p}} \rangle \geq 0} \; \sum_{\boldsymbol{x}_i \in R_{\boldsymbol{p}}} q_i \boldsymbol{x}_i .$$

Therefore, by Theorem 2.12, restriction (2.20) implies that $Z(\mu) = Z(\nu)$.

2.2 Lift zonoid of a measure

Now we are prepared to introduce and discuss our central notion, the *lift zonoid* of a measure in $\mathcal{M}_1$. The lift zonoid of a measure on $\mathbb{R}^d$ is a zonoid which belongs to a properly lifted measure on $\mathbb{R}^{d+1}$. The fundamental difference between the zonoid and the lift zonoid of a measure is that the latter characterizes the measure in a unique way while the former does not. Given a measure $\mu \in \mathcal{M}_1$, take its product with the univariate Dirac measure at 1. This is a measure on $\mathbb{R}^{d+1}$ denoted by $\widehat{\mu}$. Its support is contained in the hyperplane of $\mathbb{R}^{d+1}$ with first coordinate equal to one and, on this hyperplane, $\widehat{\mu}$ is the same as μ. It has an intuitive geometric meaning: The measure μ is first embedded as a measure on $\mathbb{R}^{d+1}$ with support in the hyperplane at 0 and then lifted to the hyperplane at 1. By that, $\widehat{\mu}$ is called a *lifted measure*. This gives the name to the lift zonoid, which is the zonoid of the lifted measure.

2.2.1 Definition and first properties

Definition 2.3 (Lift zonoid) *For $\mu \in \mathcal{M}_1$, define $\widehat{\mu} = \delta_1 \otimes \mu$.*

$$\widehat{Z}(\mu) = Z(\delta_1 \otimes \mu) = Z(\widehat{\mu})$$

is called the lift zonoid of μ.

Recall that a point $\boldsymbol{z}$ in $\mathbb{R}^{d+1}$ is denoted by $\boldsymbol{z} = (z_0, z_1, \ldots, z_d)$ and that, for any $J = \{i_1, i_2, \ldots, i_k\} \subset \{0, 1, \ldots, d\}$, we denote $pr_J(\boldsymbol{z}) = (z_{i_1}, z_{i_2}, \ldots, z_{i_k})$.

In particular, $pr_{\{1,\dots,d\}}$ is the projection of points in $\mathbb{R}^{d+1}$ to their last d coordinates. There holds

$$\widehat{\mu}(B) = \mu(pr_{\{1,\dots,d\}}(B \cap G_1)), \quad B \in \mathcal{B}^{d+1}, \tag{2.21}$$

where $G_1 = \{z \in \mathbb{R}^{d+1} : z_0 = 1\}$ denotes the hyperplane at $z_0 = 1$.

As is seen from the definition, the zonoid of μ equals the projection of the lift zonoid on its last d coordinates,

$$Z(\mu) = pr_{\{1,\dots,d\}}(\widehat{Z}(\mu)). \tag{2.22}$$

If μ is a probability measure, the lift zonoid equals the expectation of a random segment in $\mathbb{R}^{d+1}$:

Theorem 2.13 (Expected random segment) *Let $\mu \in \mathcal{P}_1$ and $\boldsymbol{X}$ be a random vector distributed as μ. Then*

$$\widehat{Z}(\mu) = \mathrm{E}([(0,\mathbf{0}),(1,\boldsymbol{X})]). \tag{2.23}$$

Proof. This follows from Theorem 2.8. Q.E.D.

Theorem 2.14 (Equivalent definitions of the lift zonoid) *For $\mu \in \mathcal{M}_1$ holds*

$$\widehat{Z}(\mu) = \left\{\int_{\mathbb{R}^d} (1,\boldsymbol{x})\, g(\boldsymbol{x})\, \mu(d\boldsymbol{x}) : g : \mathbb{R}^d \to [0,1] \text{ measurable}\right\} \tag{2.24}$$

$$= \left\{\int_{\mathbb{R}^d} (1,\boldsymbol{x})\, \nu(d\boldsymbol{x}) \;:\; \nu \leq \mu\right\} \tag{2.25}$$

$$= \mathit{conv}\left\{\int_B (1,\boldsymbol{x})\, \mu(d\boldsymbol{x}) \;:\; B \in \mathcal{B}^d\right\}. \tag{2.26}$$

If μ is L-continuous,

$$\widehat{Z}(\mu) = \left\{\int_B (1,\boldsymbol{x})\, \mu(d\boldsymbol{x}) \;:\; B \in \mathcal{B}^d\right\}. \tag{2.27}$$

Proof. Let $z \in \widehat{Z}(\mu)$. Then, by the definition of the zonoid of a measure (Definition 2.2), with some measurable $\widehat{g} : \mathbb{R}^{d+1} \to [0,1]$,

$$\begin{aligned} z &= \int_{\mathbb{R}^{d+1}} (x_0,\boldsymbol{x})\, \widehat{g}(x_0,\boldsymbol{x})\, \widehat{\mu}(d(x_0,\boldsymbol{x})) \\ &= \int_{\mathbb{R}^d}\int_{\mathbb{R}} (x_0,\boldsymbol{x})\, \widehat{g}(x_0,\boldsymbol{x})\, \delta_1(dx_0)\, \mu(d\boldsymbol{x}) \\ &= \int_{\mathbb{R}^d} (1,\boldsymbol{x})\, \widehat{g}(1,\boldsymbol{x})\, \mu(d\boldsymbol{x}). \end{aligned}$$

Choosing $g(\boldsymbol{x}) = \widehat{g}(1, \boldsymbol{x})$ we obtain (2.24). The rest is similarly shown; see Theorem 2.3. Q.E.D.

Every lift zonoid is the zonoid of a measure, but not vice versa. For example, the unit square in $\mathbb{R}^2$ is the zonoid of a measure but no lift zonoid; see Example 2.1. Since a lift zonoid is a zonoid, all properties derived for zonoids of measures hold for lift zonoids as well. Here only a few properties are stated for further reference. They follow immediately from the definition and Proposition 2.1.

Proposition 2.15 (Properties of the lift zonoid) *Let* $\mu \in \mathcal{M}_1$.

(i) $\widehat{Z}(\mu)$ *is convex and compact.*

(ii) $\widehat{Z}(\mu)$ *contains* $(0, \mathbf{0}) \in \mathbb{R}^{d+1}$ *and is centrally symmetric around* $(\frac{1}{2}\alpha(\mu), \frac{1}{2}\epsilon(\mu))$.

(iii) *In general,*

$$\widehat{Z}(\mu) \subset \{(z_0, z_1, \ldots, z_d) \ : 0 \le z_0 \le \alpha(\mu)\}.$$

(iv) μ *has support in* $\mathbb{R}^d_+$ *if and only if*

$$\widehat{Z}(\mu) \subset \{(z_0, z_1, \ldots, z_d) \ : 0 \le z_0 \le \alpha(\mu), \mathbf{0} \le (z_1, \ldots, z_d) \le \epsilon(\mu)\}. \tag{2.28}$$

In (iv), (2.28) says that $\widehat{Z}(\mu)$ is contained in the $(d+1)$-dimensional rectangle between $(0, \mathbf{0})$ and $(\alpha(\mu), \epsilon(\mu))$.

From Theorem 2.4 the support function of a lift zonoid is obtained. We write shortly $h(\widehat{Z}(\mu), p_0, \boldsymbol{p})$ in place of $h(\widehat{Z}(\mu), (p_0, \boldsymbol{p}))$.

Proposition 2.16 (Support function of the lift zonoid) *Let* $\mu \in \mathcal{M}_1$. *The support function of* $\widehat{Z}(\mu)$ *is given by*

$$\begin{aligned} h(\widehat{Z}(\mu), p_0, \boldsymbol{p}) &= \int_{p_0 + \langle \boldsymbol{x}, \boldsymbol{p} \rangle \ge 0} p_0 + \langle \boldsymbol{x}, \boldsymbol{p} \rangle \mu(d\boldsymbol{x}) \\ &= \int_{\mathbb{R}^d} \max\{0, p_0 + \langle \boldsymbol{x}, \boldsymbol{p} \rangle\} \mu(d\boldsymbol{x}) \end{aligned} \tag{2.29}$$

for $(p_0, \boldsymbol{p}) \in \mathbb{R}^{d+1}$.

Remark. The support function of $\widehat{Z}(\mu)$ is completely determined by its restriction to the sphere S^d in $\mathbb{R}^{d+1}$. Observe that, by Corollary 2.7, the support function of a lift zonoid is uniformly bounded on S^d,

$$\begin{aligned} 0 \le h(\widehat{Z}(\mu), p_0, \boldsymbol{p}) &\le \int_{\mathbb{R}^d} \|(1,\boldsymbol{x})\| \mu(d\boldsymbol{x}) \\ &\le \alpha(\mu) + \int_{\mathbb{R}^d} \|\boldsymbol{x}\| \mu(d\boldsymbol{x}), \quad (p_0, \boldsymbol{p}) \in S^d . \end{aligned} \tag{2.30}$$

2.2.2 Lift zonotope

An important special case arises if μ has finite support in $\mathbb{R}^d$. Then the lift zonoid becomes a polytope, that is the convex hull of a finite number of points. It is named the *lift zonotope* of μ.

Let $\boldsymbol{A} = [a_{ik}]$ be a matrix in $\mathbb{R}^{n\times(d+1)}$ with nonnegative first column. The first column is indexed with zero. Let $\mu_{\boldsymbol{A}} \in \mathcal{M}_1$ be the measure that gives mass a_{i0} to the point $\boldsymbol{a}_i = (a_{i1}, \dots, a_{id})$, $i = 1, \dots, n$. Note that $\mu_{\boldsymbol{A}}$ is an empirical measure if the a_{i0} are all equal, $a_{i0} = \alpha(\mu_{\boldsymbol{A}})/n$. The lift zonotope of $\mu_{\boldsymbol{A}}$ is given by

$$\widehat{Z}(\mu_{\boldsymbol{A}}) = \left\{ (x_0, \boldsymbol{x}) \in \mathbb{R}^{d+1} : x_0 = \sum_{i=1}^n \lambda_i a_{i0},\ \boldsymbol{x} = \sum_{i=1}^n \lambda_i a_{i0}\, \boldsymbol{a}_i,\ 0 \le \lambda_i \le 1 \text{ for all } i \right\}.$$

This can also be written as a convex combination of line segments,

$$\widehat{Z}(\mu_{\boldsymbol{A}}) = a_{10} \cdot \left[(0,\boldsymbol{0}), (1, \boldsymbol{a}_1)\right] + \dots + a_{n0} \cdot \left[(0,\boldsymbol{0}), (1, \boldsymbol{a}_n)\right]. \tag{2.31}$$

2.2.3 Univariate case

In case $d = 1$ the lift zonoid is easily described. The following theorem shows that the lift zonoid is the convex hull of certain, conveniently parameterized points in the plane.

Theorem 2.17 (Lift zonoid of a univariate distribution) *Assume that* $d = 1$, $\mu \in \mathcal{M}_1$.

(i) *Then* $\widehat{Z}(\mu)$ *is the convex hull of the two points* $(0,0)$, $(\alpha(\mu), \epsilon(\mu))$ *and the following points in* $\mathbb{R}^2$ *:*

$$\left(\int_{]-\infty,y]} \mu(dx), \int_{]-\infty,y]} x\, \mu(dx) \right), \quad y \in \mathbb{R}, \tag{2.32}$$

and

$$\left(\int_{[y,\infty[} \mu(dx), \int_{[y,\infty[} x\,\mu(dx)\right), \quad y \in \mathbb{R}. \tag{2.33}$$

(ii) *Equivalently,* $\widehat{Z}(\mu)$ *is the convex hull of*

$$\left(t\,\alpha(\mu), \int_0^{t\,\alpha(\mu)} Q_\mu(s)ds\right), \quad 0 \le t \le 1, \tag{2.34}$$

and

$$\left(t\,\alpha(\mu), \int_{\alpha(\mu)(1-t)}^{\alpha(\mu)} Q_\mu(s)ds\right), \quad 0 \le t \le 1, \tag{2.35}$$

where $Q_\mu(s) = \min\{x \in \mathbb{R} : \mu(]-\infty, x]) \ge s\}$ *is the quantile function of* μ.

In economics, for a probability measure μ on $\mathbb{R}_+$ with positive mean, the function

$$\widehat{L} : t \mapsto \int_0^t Q_\mu(s)\,ds, \quad 0 \le t \le 1,$$

is known as the generalized Lorenz function, and its graph as the *generalized Lorenz curve* (see e.g. Nygård and Sandström (1981), Section 7.2.3), while the graph of its symmetric counterpart,

$$\widehat{DL} : t \mapsto \int_{1-t}^1 Q_\mu(s)ds, \quad 0 \le t \le 1,$$

is the *dual generalized Lorenz curve.* The *usual Lorenz function* L is the generalized one, rescaled by a factor $1/\epsilon(\mu)$, provided the expectation $\epsilon(\mu)$ is positive. That is

$$L(t) = \frac{1}{\epsilon(\mu)} \int_0^t Q_\mu(s)\,ds, \quad 0 \le t \le 1. \tag{2.36}$$

Thus, the theorem says that the lift zonoid of μ is bordered by two curves, the generalized Lorenz curve and the dual generalized Lorenz curve.

Proof. (i): By Proposition 2.16 the support function of $\widehat{Z}(\mu)$ reads as follows. For $(p_0, p_1) \in \mathbb{R}^2$,

$$\begin{aligned} h(\widehat{Z}(\mu), p_0, p_1) &= \int_{p_0+p_1x\ge 0} (p_0 + p_1x)\,d\mu(x) \\ &= \left\langle \int_{p_1x\ge -p_0} (1, x)\,d\mu(x), (p_0, p_1)\right\rangle . \end{aligned}$$

We obtain

$$h(\widehat{Z}(\mu), p_0, p_1) = \begin{cases} \langle \int_{]-\infty,y]}(1,x)\,d\mu(x), (p_0,p_1)\rangle & \text{if } p_1 < 0\,, \\ \langle \int_{[y,\infty[}(1,x)\,d\mu(x), (p_0,p_1)\rangle & \text{if } p_1 > 0\,, \\ \langle \int_{\mathbb{R}}(1,x)\,d\mu(x), (p_0,0)\rangle & \text{if } p_1 = 0, p_0 \geq 0\,, \\ \langle (0,0), (p_0,0)\rangle & \text{if } p_1 = 0, p_0 < 0\,, \end{cases}$$

where $y = -p_0/p_1$ if $p_1 \neq 0$. As $\widehat{Z}(\mu)$ is closed and convex, its boundary equals

$$\begin{aligned} \partial(\widehat{Z}(\mu)) = \{ & (z_0, z_1) \in \widehat{Z}(\mu) : \\ & \langle (z_0,z_1),(p_0,p_1)\rangle = h(\widehat{Z}(\mu),p_0,p_1),\ (p_0,p_1) \neq (0,0)\}\,. \end{aligned} \tag{2.37}$$

This is precisely the set of points given in (2.32) and (2.33), which proves part (i).

(ii): With

$$t = \frac{1}{\alpha(\mu)} \int_{]-\infty,y]} \mu(dx)\,, \quad y \in \mathbb{R},$$

there holds

$$\int_{]-\infty,y]} x\mu(dx) = \int_0^{t\,\alpha(\mu)} Q_\mu(s)ds\,.$$

It follows that (2.34) for $0 \leq t < 1$ and (2.32) for $y \in \mathbb{R}$ describe the same points. (2.34) with $t = 1$ yields the point $(\alpha(\mu), \epsilon(\mu))$. Similarly, we see that (2.35) corresponds to (2.33) and the point $(0,0)$. Q.E.D.

Example 2.6 (Uniform distribution) If μ is the uniform probability distribution on the interval $[0,\beta]$, the quantile function amounts to $Q_\mu(s) = \beta s$, $0 < s \leq 1$. From (2.34) and (2.35) one obtains the points

$$\left(t, \frac{\beta t^2}{2}\right) \quad \text{and} \quad \left(t, \frac{\beta(2t-t^2)}{2}\right), \quad 0 \leq t \leq 1\,.$$

Their convex hull is the lift zonoid of μ. For $\beta = 2$ it is equal to Z_∞ in Example 2.2 (see Figure 2.2).

Example 2.7 (Pareto distribution) The Pareto distribution is a univariate probability distribution that has historically originated from the measurement of economic inequality. It has the distribution function $F(x) = 1 - (c/x)^\alpha$ if $x \geq c$ and $F(x) = 0$ otherwise, where c and α are positive parameters. The quantile function is $Q(t) = c(1-t)^{-1/\alpha}, 0 \leq t \leq 1$. If $\alpha > 1$

the Pareto distribution possesses a finite expectation $\alpha c/(\alpha-1)$. Then the Lorenz curve and the dual Lorenz curve are given by

$$\begin{aligned} L(t) &= 1-(1-t)^{\frac{\alpha-1}{\alpha}}, & 0 \le t \le 1, \\ 1-L(1-t) &= t^{\frac{\alpha-1}{\alpha}}, & 0 \le t \le 1, \end{aligned}$$

respectively, and the lift zonoid is the set

$$\left\{(t,x) \in \mathbb{R}^2 \ : 1-(1-t)^{\frac{\alpha-1}{\alpha}} \le \frac{\alpha-1}{\alpha c} x \le t^{\frac{\alpha-1}{\alpha}}, 0 \le t \le 1\right\}.$$

Example 2.8 (Exponential distribution) The univariate exponential distribution $Exp(\lambda)$, with some $\lambda > 0$, has the distribution function $F(x) = 1-\exp(-\lambda x)$ if $x \ge 0$ and $F(x) = 0$ otherwise. The expectation equals $1/\lambda$. We obtain, for $0 \le t < 1$, $Q(t) = -\lambda^{-1}\ln(1-t)$, $L(t) = t + (1-t)\ln(1-t)$, $1-L(1-t) = t - t\ln(t)$, and the lift zonoid equals

$$\{(t,x) \in \mathbb{R}^2 : t+(1-t)\ln(1-t) \le \lambda x \le t - t\ln(t),\ 0 \le t \le 1\}.$$

The lift zonoids of several exponential distributions are depicted in Chapter 1, Figure 1.4.

Lift zonoid of a distribution type

We use Theorem 2.17 in order to calculate the lift zonoid of a distribution type. A univariate *distribution type* is a family of probability distributions, $\{\mu_{a,b}\}_{a\in\mathbb{R},b>0}$, such that

$$F_{a,b}(y) = F_{0,1}\left(\frac{y-a}{b}\right), \quad y \in \mathbb{R}, \tag{2.38}$$

holds, where $F_{a,b}$ denotes the distribution function of $\mu_{a,b}$. Then, if a random variable X has distribution $\mu_{a,b}$, the *standardized variable*, $(X-a)/b$, has distribution $\mu_{0,1}$ and

$$\begin{aligned} \int_{]-\infty,y]} x\ \mu_{a,b}(dx) &= \int_{]-\infty,(y-a)/b]} a+bu\ \mu_{0,1}(du) \\ &= a\ F_{0,1}\left(\frac{y-a}{b}\right) + b\int_{]-\infty,(y-a)/b]} u\ \mu_{0,1}(du)\,. \end{aligned}$$

Let $Q_{0,1}$ be the quantile function of $\mu_{0,1}$. By setting $t \equiv F_{0,1}((y-a)/b)$ and

$$E_{0,1}(Q_{0,1}(t)) \equiv \int_{]-\infty,Q_{0,1}(t)]} u\ \mu_{0,1}(du) = \int_{]-\infty,\frac{y-a}{b}]} u\ \mu_{0,1}(du)\,, \tag{2.39}$$

we obtain from (2.32) the points

$$\left(t, at + b\ E_{0,1}(Q_{0,1}(t))\right), \quad 0 < t < 1. \tag{2.40}$$

Similarly, (2.33) yields the points

$$\left(t, at + b[\epsilon(\mu_{0,1}) - E_{0,1}(Q_{0,1}(1-t))]\right), \quad 0 < t < 1. \tag{2.41}$$

Further, in the limits at $t = 0$ and $t = 1$, the points $(0,0)$ and $(1, a + b\epsilon(\mu_{0,1}))$ arise. We conclude:

Corollary 2.18 (Lift zonoid of a distribution type) *If $\mu_{a,b}$ belongs to a univariate distribution type, its lift zonoid is given by*

$$\begin{aligned}
\widehat{Z}(\mu_{a,b}) \quad = \quad & \left\{(0,0), (1, a + b\epsilon(\mu_{0,1}))\right\} \cup \\
& \Big\{(t,x) \in \mathbb{R}^2 : 0 < t < 1, \\
& E_{0,1}(Q_{0,1}(t)) \le \frac{x - at}{b} \le \epsilon(\mu_{0,1}) - E_{0,1}(Q_{0,1}(1-t))\Big\}.
\end{aligned}$$

Example 2.9 (Univariate normal distribution) In particular, consider the class of normal distributions with expectation a and variance b^2, $\mu_{a,b} = N(a, b^2)$. $\mu_{0,1}$ is the standard normal distribution, and $u_s = Q_{0,1}(s) = -Q_{0,1}(1-s)$ the standard normal s-quantile. Then

$$\begin{aligned}
E_{0,1}(y) \quad &= \quad \frac{1}{\sqrt{2\pi}} \int_{-\infty}^{y} x e^{-x^2/2} dx \quad = \quad -\frac{1}{\sqrt{2\pi}} e^{-y^2/2}, \\
E_{0,1}(Q_{0,1}(t)) \quad &= \quad E_{0,1}(Q_{0,1}(1-t)) \quad = \quad -\frac{1}{\sqrt{2\pi}} e^{-u_t^2/2},
\end{aligned}$$

$\epsilon(\mu_{0,1}) = 0$, hence

$$\widehat{Z}(N(a,b^2)) = \left\{(t,x) \in \mathbb{R}^2 : \left(\frac{x - at}{b}\right)^2 \le \frac{1}{2\pi} e^{-u_t^2}, \quad 0 \le t \le 1\right\}. \tag{2.42}$$

If the expectation a is zero, we see from (2.42) that the lift zonoid of $N(0, b^2)$ is not only symmetric around the point $(1/2, 0)$ but also symmetric with respect to the t-axis as well as to the line at $t = 1/2$. For any given expectation a, the lift zonoid increases with the standard deviation b. This is illustrated in Figures 1.5 and 1.6 in Chapter 1.

Further examples are found in the Sections 1.4 and 2.3.5; see also Part B of the Appendix.

2.3 Embedding measures into the set of convex compacts

The notion of the lift zonoid provides an embedding of the space of d-variate measures having finite first moments into the space of centrally symmetric convex compacts in $\mathbb{R}^{d+1}$. The embedding is injective, additive, positively homogenous, and continuous. First additivity and positive homogeneity, that is positive linearity, are substantiated. Then the inclusion ordering of lift zonoids is investigated and the injectivity of the embedding is proven. Later, in Section 2.4 it will be demonstrated that the embedding is also continuous with respect to weak majorized convergence of measures and Hausdorff convergence of sets.

Theorem 2.19 (Positive linearity) *Let μ and $\nu \in \mathcal{M}_1$, $\beta > 0$. Then*

$$\widehat{Z}(\beta\mu) = \beta\widehat{Z}(\mu), \qquad \widehat{Z}(\mu + \nu) = \widehat{Z}(\mu) + \widehat{Z}(\nu). \tag{2.43}$$

Proof. Positive homogeneity, $\widehat{Z}(\beta\mu) = \beta\widehat{Z}(\mu)$, is immediately seen from Definition 2.2. In order to prove additivity, recall that the support function of the Minkowski sum, $\widehat{Z}(\mu) + \widehat{Z}(\nu)$, is equal to the sum of the support functions of $\widehat{Z}(\mu)$ and $\widehat{Z}(\nu)$. So, we get for any $\boldsymbol{p} \in S^d$:

$$\begin{aligned} h(\widehat{Z}(\mu) + \widehat{Z}(\nu), \boldsymbol{p}) &= h(Z(\widehat{\mu}) + Z(\widehat{\nu}), \boldsymbol{p}) \\ &= h(Z(\widehat{\mu}), \boldsymbol{p}) + h(Z(\widehat{\nu}), \boldsymbol{p}) \\ &= \int_{\langle \boldsymbol{z}, \boldsymbol{p}\rangle \geq 0} \langle \boldsymbol{z}, \boldsymbol{p}\rangle\, \widehat{\mu}(d\boldsymbol{z}) + \int_{\langle \boldsymbol{z}, \boldsymbol{p}\rangle \geq 0} \langle \boldsymbol{z}, \boldsymbol{p}\rangle\, \widehat{\nu}(d\boldsymbol{z}) \\ &= \int_{\langle \boldsymbol{z}, \boldsymbol{p}\rangle \geq 0} \langle \boldsymbol{z}, \boldsymbol{p}\rangle (\widehat{\mu}(d\boldsymbol{z}) + \widehat{\nu}(d\boldsymbol{z})) \\ &= h(Z(\widehat{\mu} + \widehat{\nu}), \boldsymbol{p}) = h(\widehat{Z}(\mu + \nu), \boldsymbol{p}). \end{aligned}$$

We have used Theorem 2.4 and the fact that the adding and lifting of measures can be interchanged,

$$\delta_1 \otimes (\mu + \nu) = \delta_1 \otimes \mu + \delta_1 \otimes \nu\,,$$

hence $Z(\widehat{\mu} + \widehat{\nu}) = \widehat{Z}(\mu + \nu)$. From the equality of the support functions the equality of the sets is concluded. Q.E.D.

2.3.1 Inclusion of lift zonoids

The inclusion of lift zonoids yields a useful ordering of d-variate measures, the *lift zonoid order.*

Definition 2.4 (Lift zonoid order) *For* $\mu, \nu \in \mathcal{M}_1$, *the* lift zonoid order $\preceq_{lz}$ *is defined by*

$$\mu \preceq_{lz} \nu \quad \text{if} \quad \widehat{Z}(\mu) \subset \widehat{Z}(\nu).$$

It follows immediately from the definition and Theorem 2.15(ii) that:

- $\mu \preceq_{lz} \nu$ implies $\alpha(\mu) \leq \alpha(\nu)$ and $\epsilon(\mu) \leq \epsilon(\nu)$.

A detailed analysis of the lift zonoid order is postponed to Chapter 8. There will be also shown that another order of measures, the convex order[3], implies the lift zonoid order.

Here only a characterization of the lift zonoid order by univariate orderings is proven, which will be extremely useful in the sequel. It says that the lift zonoid order of two d-variate measures is equivalent to the same order of all their real-valued linear images.

By this characterization, all aspects of multivariate measures that are described by the inclusion of their lift zonoids can be investigated by looking at the lift zonoids of univariate measures.

For any $\boldsymbol{p} \in \mathbb{R}^d$ and $\mu \in \mathcal{M}_1$ let $l_{\boldsymbol{p}} : \boldsymbol{x} \mapsto \langle \boldsymbol{x}, \boldsymbol{p} \rangle$ and $\mu_{\boldsymbol{p}} = \mu \circ l_{\boldsymbol{p}}^{-1}$, hence

$$\mu_{\boldsymbol{p}}(]-\infty, t]) = \int_{\langle \boldsymbol{x}, \boldsymbol{p} \rangle \leq t} \mu(d\boldsymbol{x}), \quad t \in \mathbb{R}.$$

If μ is the distribution of a random vector $\boldsymbol{X}$, $\mu_{\boldsymbol{p}}$ is the distribution of the random variable $\langle \boldsymbol{X}, \boldsymbol{p} \rangle$.

Theorem 2.20 (Characterization of the lift zonoid inclusion) *Let* $\mu, \nu \in \mathcal{M}_1$. *Then*

$$\widehat{Z}(\mu) \subset \widehat{Z}(\nu) \quad \text{if and only if} \quad \widehat{Z}(\mu_{\boldsymbol{p}}) \subset \widehat{Z}(\nu_{\boldsymbol{p}}) \quad \text{for all } \boldsymbol{p} \in \mathbb{R}^d.$$

[3] For definition see (1.6).

Proof. By (2.29) the support function of a lift zonoid is

$$\begin{aligned} h(\widehat{Z}(\mu), p_0, \boldsymbol{p}) &= \int_{\mathbb{R}^d} \max\{0,\ p_0 + \langle \boldsymbol{x}, \boldsymbol{p} \rangle\} \mu(d\boldsymbol{x}) \\ &= \int_{\mathbb{R}} \max\{0,\ p_0 + t\} \mu_{\boldsymbol{p}}(dt)\,. \end{aligned} \tag{2.44}$$

Therefore, $\widehat{Z}(\mu) \subset \widehat{Z}(\nu)$ if and only if $h(\widehat{Z}(\mu), \cdot) \le h(\widehat{Z}(\nu), \cdot)$ if and only if

$$\int_{\mathbb{R}} \max\{0,\ p_0 + t\} \mu_{\boldsymbol{p}}(dt) \le \int_{\mathbb{R}} \max\{0,\ p_0 + t\} \nu_{\boldsymbol{p}}(dt) \tag{2.45}$$

for all $p_0 \in \mathbb{R}$, $\boldsymbol{p} \in \mathbb{R}^d$. As in (2.44), the support function of $\widehat{Z}(\mu_{\boldsymbol{p}})$ amounts to $h(\widehat{Z}(\mu_{\boldsymbol{p}}), q_0, q_1) = \int_{\mathbb{R}} \max\{0,\ q_0 + tq_1\} \mu_{\boldsymbol{p}}(dt)$, and similarly that of $\widehat{Z}(\nu_{\boldsymbol{p}})$. By that, the inequality (2.45) is equivalent to

$$h(\widehat{Z}(\mu_{\boldsymbol{p}}), p_0, 1) \le h(\widehat{Z}(\nu_{\boldsymbol{p}}), p_0, 1)\,,$$

for all $p_0 \in \mathbb{R}$. Since support functions are positively homogeneous, this is equivalent to $h(\widehat{Z}(\mu_{\boldsymbol{p}}), \cdot) \le h(\widehat{Z}(\nu_{\boldsymbol{p}}), \cdot)$, hence to $\widehat{Z}(\mu_{\boldsymbol{p}}) \subset \widehat{Z}(\nu_{\boldsymbol{p}})$. This proves the theorem. Q.E.D.

2.3.2 Uniqueness of the representation

We show that, as a consequence of Theorem 2.20, the lift zonoid determines the underlying measure in a unique way. This is the fundamental property by which d-variate measures can be embedded into the space of convex sets in $(d+1)$-space.

Theorem 2.21 (Uniqueness) *Every measure $\mu \in \mathcal{M}_1$ is uniquely determined by its lift zonoid.*

Proof. (i) First consider μ and $\nu \in \mathcal{P}_1$ and assume $\widehat{Z}(\mu) = \widehat{Z}(\nu)$. Let $\boldsymbol{X}$, $\boldsymbol{Y}$ be random vectors that are distributed according to μ and ν, respectively. From Theorem 2.20 and Theorem 2.17 follows that for every $\boldsymbol{p} \in \mathbb{R}^d$

$$\int_0^t Q_{\mu_{\boldsymbol{p}}}(s)ds = \int_0^t Q_{\nu_{\boldsymbol{p}}}(s)ds, \quad t \in [0,1],$$

and $Q_{\mu_{\boldsymbol{p}}}(t) = Q_{\nu_{\boldsymbol{p}}}(t)$ for all t. We conclude that $\mu_{\boldsymbol{p}} = \nu_{\boldsymbol{p}}$, i.e., $\langle \boldsymbol{X}, \boldsymbol{p} \rangle$ has the same distribution as $\langle \boldsymbol{Y}, \boldsymbol{p} \rangle$, for every $\boldsymbol{p} \in \mathbb{R}^d$. The Cramér-Wold Theorem

(e.g., Mardia et al. (1979)) then yields that $\boldsymbol{X}$ and $\boldsymbol{Y}$ have the same distribution. Hence the proposition holds for probability distributions.
(ii) Now, let μ and $\nu \in \mathcal{M}_1$. Then $\widehat{Z}(\mu) = \widehat{Z}(\nu)$ implies that

$$\alpha(\mu) = \alpha(\nu) \quad \text{and} \quad \widehat{Z}\left(\frac{1}{\alpha(\mu)}\mu\right) = \widehat{Z}\left(\frac{1}{\alpha(\nu)}\nu\right).$$

Owing to $\frac{1}{\alpha(\mu)}\mu$ and $\frac{1}{\alpha(\nu)}\nu \in \mathcal{P}_1$, we obtain $\frac{1}{\alpha(\mu)}\mu = \frac{1}{\alpha(\nu)}\nu$, and therefore $\mu = \nu$. Q.E.D.

Next, it is demonstrated that a lift zonoid is uniquely determined by certain two-dimensional projections.

Corollary 2.22 *For $\boldsymbol{p} \in \mathbb{R}^d$, let $\pi_{\boldsymbol{p}}$ be the projection of $\mathbb{R}^{d+1}$ to the two-dimensional plane $S_{\boldsymbol{p}}$ that is spanned by the points $(1, 0, \ldots, 0)$, $(0, \boldsymbol{p})$ and the origin in $\mathbb{R}^{d+1}$. Any lift zonoid $\widehat{Z}(\mu)$ is uniquely determined by the projections $\pi_{\boldsymbol{p}}(\widehat{Z}(\mu))$, $\boldsymbol{p} \in S^{d-1}$.*

Proof. This is a consequence of Theorems 2.21, 2.20 and the following remark. Q.E.D.

Remark. The lift zonoid $\widehat{Z}(\mu_{\boldsymbol{p}})$ is, in fact, a two-dimensional projection of $\widehat{Z}(\mu)$. For some $\boldsymbol{p} \in \mathbb{R}^d$ and $p_0 \in \mathbb{R}$, consider the extreme point $\boldsymbol{z} \in \widehat{Z}(\mu)$ in the direction $(p_0, \boldsymbol{p})$. Its projection on $S_{\boldsymbol{p}}$ equals, in coordinates related to the basis $\{(1, 0, \ldots, 0), (0, \boldsymbol{p})\}$ of $S_{\boldsymbol{p}}$,

$$\left(\int_{p_0 + \langle \boldsymbol{p}, \boldsymbol{x}\rangle \geq 0} \mu(d\boldsymbol{x}), \int_{p_0 + \langle \boldsymbol{p}, \boldsymbol{x}\rangle \geq 0} \langle \boldsymbol{p}, \boldsymbol{x}\rangle \mu(d\boldsymbol{x})\right)$$

$$= \left(\int_{p_0 + t \geq 0} \mu_{\boldsymbol{p}}(dt), \int_{p_0 + t \geq 0} t\mu_{\boldsymbol{p}}(dt)\right).$$

The latter is an extreme point of $\widehat{Z}(\mu_{\boldsymbol{p}})$ and has the form (2.32) or (2.33). For every p_0, the point $(p_0, \boldsymbol{p})$ belongs to the projection plane. Therefore, projections of other points of the lift zonoid belong to the convex hull of projections of such extreme points. According to Theorem 2.17 we get the result.

2.3.3 Lift zonoid metric

Our approach also yields a new metric δ_{lz} among measures that is based on the Hausdorff distance between their lift zonoids.

Definition 2.5 (Lift zonoid metric) *For $\mu, \nu \in \mathcal{M}_1$ let*

$$\delta_{lz}(\mu, \nu) = \delta_H(\widehat{Z}(\mu), \widehat{Z}(\nu)), \qquad \mu, \nu \in \mathcal{M}_1, \tag{2.46}$$

where δ_H is the Hausdorff distance. We call δ_{lz} the lift zonoid metric.

Proposition 2.23 (Metric) *δ_{lz} is a metric in $\mathcal{M}_1$.*

Proof. From Theorem 2.21 we see that $\delta_{lz}(\mu, \nu) = 0$ if and only if $\mu = \nu$. Symmetry and the triangle inequality follow from the corresponding properties of the Hausdorff distance. Q.E.D.

The support function of the lift zonoid of μ is, for any $(p_0, \boldsymbol{p}) \in \mathbb{R}^{d+1}$, the μ-integral over the function $f_{p_0, \boldsymbol{p}} = \max\{0, p_0 + \langle \boldsymbol{x}, \boldsymbol{p} \rangle\}, \boldsymbol{x} \in \mathbb{R}^d$. Define

$$\mathcal{F}_{lz} = \{f_{p_0, \boldsymbol{p}} \,:\, (p_0, \boldsymbol{p}) \in S^d\}.$$

The Hausdorff distance equals the supremum of the difference in support functions, hence

$$\delta_{lz}(\mu, \nu) = \sup_{f \in \mathcal{F}_{lz}} \left| \int_{\mathbb{R}^d} f \, d\mu - \int_{\mathbb{R}^d} f \, d\nu \right|. \tag{2.47}$$

Introducing the supremum norm $||.||_{\mathcal{F}_{lz}}$ on $\mathcal{M}_1$, this can be written as

$$\delta_{lz}(\mu, \nu) = ||\mu - \nu||_{\mathcal{F}_{lz}}.$$

We say that δ_{lz} is the measure metric induced on $\mathcal{M}_1$ by the class of functions $\mathcal{F}_{lz}$.

2.3.4 Linear transformations and projections

In this section the lift zonoid is demonstrated to be equivariant with regard to linear mappings of the measure. The lift zonoid of a linear image of a measure is a corresponding linear transform of the lift zonoid of the original measure.

Let $T_{\boldsymbol{A}}$ be a linear transformation $\mathbb{R}^d \to \mathbb{R}^k$, $T_{\boldsymbol{A}}(\boldsymbol{x}) = \boldsymbol{x}\boldsymbol{A}$ with some $d \times k$ matrix $\boldsymbol{A}$, and let $\mu^{\boldsymbol{A}} = \mu \circ T_{\boldsymbol{A}}^{-1}$ be the image measure of μ under $T_{\boldsymbol{A}}$. Then $\widehat{Z}(\mu) = \{\int_{\mathbb{R}^d} (1, \boldsymbol{x}) \, \nu(d\boldsymbol{x}) \,:\, \nu \leq \mu\}$, and therefore

$$\widehat{Z}(\mu^{\boldsymbol{A}}) = \left\{ \int (1, \boldsymbol{x}\boldsymbol{A}) \, \nu(d\boldsymbol{x}) \,:\, \nu \leq \mu \right\}.$$

With

$$\widehat{\boldsymbol{A}} = \begin{pmatrix} 1 & 0 & \dots & 0 \\ 0 & & & \\ \vdots & & \boldsymbol{A} & \\ 0 & & & \end{pmatrix}$$

we get $(1, \boldsymbol{x})\widehat{\boldsymbol{A}} = (1, \boldsymbol{x}\boldsymbol{A})$ and

$$\begin{aligned} \widehat{Z}(\mu^{\boldsymbol{A}}) &= \left\{ \int_{\mathbb{R}^d} (1, \boldsymbol{x})\ \nu(d\boldsymbol{x})\widehat{\boldsymbol{A}} \ : \ \nu \leq \mu \right\} \\ &= \{ \boldsymbol{z}\widehat{\boldsymbol{A}} \ : \ \boldsymbol{z} \in \widehat{Z}(\mu)\} \equiv \widehat{Z}(\mu)\widehat{\boldsymbol{A}} \ . \end{aligned}$$

We have shown:

Proposition 2.24 (Linear transformation) *For any* $\mu \in \mathcal{M}_1$ *and any linear transformation* $T_{\boldsymbol{A}}$*,*

$$\widehat{Z}(\mu^{\boldsymbol{A}}) = \widehat{Z}(\mu)\widehat{\boldsymbol{A}} \ .$$

For given μ and $0 \leq t \leq 1$, define the *t-cut* of the lift zonoid,

$$\widehat{Z}(\mu, t) \equiv \{(z_0, z_1, \dots, z_d) \in \widehat{Z}(\mu) : z_0 = t\}\,, \tag{2.48}$$

and the *t-slice* of the lift zonoid,

$$\widehat{Z}_t(\mu) \equiv pr_{1,\dots,d}\widehat{Z}(\mu, t)\,, \tag{2.49}$$

which is the projection of the t-cut on the last d coordinates. Obviously,

$$\widehat{Z}_0(\mu) = \boldsymbol{0}, \qquad \widehat{Z}_1(\mu) = \epsilon(\mu)\,. \tag{2.50}$$

In this notation the theorem says that

$$\widehat{Z}_t(\mu^{\boldsymbol{A}}) = \widehat{Z}_t(\mu)\boldsymbol{A}, \ \ 0 \leq t \leq 1\,. \tag{2.51}$$

We obtain from Proposition 2.24 the lift zonoids of arbitrary marginal measures. For any $J \subset \{1, 2, \dots, n\}$, $J \neq \emptyset$, consider the projection of points in $\mathbb{R}^d$ to coordinates with index in J. Let $\boldsymbol{A}_J$ be the corresponding projection matrix and $\boldsymbol{x}_J = \boldsymbol{x}\boldsymbol{A}_J$. Then $\mu_J = \mu \circ T_{\boldsymbol{A}_J}^{-1}$ is the marginal measure with respect to J. The extended matrix $\widehat{\boldsymbol{A}}_J$ projects points in $\mathbb{R}^{d+1}$ to their components with index in $\{0\} \cup J$. From Proposition 2.24 follows that the lift zonoid of a marginal measure is the corresponding projection of the lift zonoid:

Corollary 2.25 (Projection) *For* $\emptyset \neq J \subset \{1, 2, \dots, d\}$ *holds*

$$\widehat{Z}(\mu_J) = pr_J(\widehat{Z}(\mu)).$$

Finally we describe how the lift zonoid transforms under a shift of the measure. Consider a shift by a vector $\boldsymbol{c}$ in $\mathbb{R}^d$, $\boldsymbol{x} \mapsto \boldsymbol{x} + \boldsymbol{c}$, and the shifted measure $\mu^{\boldsymbol{c}}$, $\mu^{\boldsymbol{c}}(B) = \mu(B - \boldsymbol{c})$. The lift zonoid of $\mu^{\boldsymbol{c}}$ is derived from that of μ as follows.

Proposition 2.26 (Shift) *For any* $\mu \in \mathcal{M}_1$ *and* $\boldsymbol{c} \in \mathbb{R}^d$,

$$\widehat{Z}(\mu^{\boldsymbol{c}}) = \left\{(t, \boldsymbol{y}) : \boldsymbol{y} = \boldsymbol{x} + t\boldsymbol{c}, \ (t, \boldsymbol{x}) \in \widehat{Z}(\mu), \ 0 \leq t \leq 1\right\}. \tag{2.52}$$

Proof. For given $0 \leq t \leq 1$,

$$\begin{aligned}
\widehat{Z}(\mu^{\boldsymbol{c}}, t) &= \left\{(t, z_1, \dots, z_d) : g : \mathbb{R}^d \to [0,1], \int g(\boldsymbol{x} + \boldsymbol{c})\, \mu(d\boldsymbol{x}) = t, \right. \\
&\qquad \left. (z_1, \dots, z_d) = \int (\boldsymbol{x} + \boldsymbol{c})\, g(\boldsymbol{x} + \boldsymbol{c})\, \mu(d\boldsymbol{x}) \right\} \\
&= \left\{(t, z_1, \dots, z_d) : \tilde{g} : \mathbb{R}^d \to [0,1], \int \tilde{g}(\boldsymbol{x})\, \mu(d\boldsymbol{x}) = t, \right. \\
&\qquad \left. (z_1, \dots, z_d) = \int \boldsymbol{x}\, \tilde{g}(\boldsymbol{x})\, \mu(d\boldsymbol{x}) + t\boldsymbol{c} \right\},
\end{aligned}$$

hence

$$\widehat{Z}(\mu^{\boldsymbol{c}}, t) = \widehat{Z}(\mu, t) + (0, t\boldsymbol{c}), \tag{2.53}$$

and the proposition follows. Q.E.D.

Equation (2.53) can be rewritten as

$$\widehat{Z}_t(\mu^{\boldsymbol{c}}) = \widehat{Z}_t(\mu) + t\boldsymbol{c}, \ \ 0 \leq t \leq 1. \tag{2.54}$$

We reformulate and summarize the above results on shifts and linear transformations:

Corollary 2.27 (Affine transformation) *Let* $\boldsymbol{x} \to \boldsymbol{x}\boldsymbol{A} + \boldsymbol{c}$ *be an affine transformation* $\mathbb{R}^d \to \mathbb{R}^k$ *and* $\mu^{\boldsymbol{A},\boldsymbol{c}}$ *the image measure of some* $\mu \in \mathcal{M}_1$. *Then, for* $0 \leq t \leq 1$ *and* $\boldsymbol{x} \in \mathbb{R}^d$,

$$\boldsymbol{x}\boldsymbol{A} + t\boldsymbol{c} \in \widehat{Z}_t(\mu^{\boldsymbol{A},\boldsymbol{c}}) \quad \text{if and only if} \quad \boldsymbol{x} \in \widehat{Z}_t(\mu).$$

In other words,

$$\widehat{Z}_t(\mu^{\boldsymbol{A},\boldsymbol{c}}) = \widehat{Z}_t(\mu)\boldsymbol{A} + t\boldsymbol{c}, \quad 0 \leq t \leq 1.$$

2.3.5 Lift zonoids of spherical and elliptical distributions

Proposition 2.24 has some consequences for spherical and elliptical distributions. Their lift zonoids are composed of slices which are balls or ellipsoids in d-space, respectively.

A random vector $\boldsymbol{X}$ in $\mathbb{R}^d$ has a *spherical distribution* if for any orthogonal matrix $\boldsymbol{A}$ holds

$$\boldsymbol{X} =_{st} \boldsymbol{X}\boldsymbol{A},$$

where $=_{st}$ means equality in distribution. There exist several equivalent ways to characterize a spherical distribution. One of them relates $\boldsymbol{X}$ to the uniform distribution on the unit sphere. Let $\boldsymbol{U}$ denote a random vector that is uniformly distributed on the unit sphere, S^{d-1}. The random vector $\boldsymbol{X}$ has a spherical distribution if and only if $\boldsymbol{X} =_{st} R\,\boldsymbol{U}$, where R is some random variable on $[0,\infty[$ and independent of $\boldsymbol{U}$. Then the distribution of R is uniquely determined by the distribution of $\boldsymbol{X}$. The distribution function of R is called the *radial distribution function* and denoted by ψ. Thus, the spherical distributions can be parameterized by ψ. Write $\boldsymbol{X} \sim \mathcal{S}(\psi)$ for short.

An *elliptical* (also called *elliptically symmetric*) distribution is the affine image of a spherical distribution: The random vector $\boldsymbol{Y}$ in $\mathbb{R}^d$ has an elliptical distribution around $\boldsymbol{a}$ if

$$\boldsymbol{Y} =_{st} \boldsymbol{a} + \boldsymbol{X}\boldsymbol{B},$$

where $\boldsymbol{a} \in \mathbb{R}^d, \boldsymbol{B} \in \mathbb{R}^{k\times d}$ with rank k and $\boldsymbol{X}$ is spherical, $\boldsymbol{X} \sim \mathcal{S}(\psi)$. The parameters of this distribution are $\boldsymbol{a}, \boldsymbol{B}\boldsymbol{B}'$ and ψ. We write shortly $\boldsymbol{Y} \sim \mathcal{E}(\boldsymbol{a}, \boldsymbol{B}\boldsymbol{B}', \psi)$. For a detailed treatment of spherical and elliptical distributions, the reader is referred to Fang et al. (1990).

We give several examples concerning d-variate spherical and elliptical distributions that are most important in applications: The uniform distribution on a ball around the origin is a spherical distribution, while the uniform distribution on an arbitrary ellipsoid is an elliptical distribution. The uniform distribution on the unit sphere is another spherical distribution, which has many applications in the analysis of directional data (see, e.g. Watson (1984)) and in projection pursuit methods. The multivariate normal distribution, $N(\boldsymbol{a}, \Sigma)$ is elliptical; in the case of independence and homoscedasticity, $\Sigma = \beta^2 \boldsymbol{I}_d$, it is spherical. Many more examples are found in Fang et al. (1990).

As a spherical distribution is invariant against orthogonal transformations, it is clear from Equation (2.51) that the slices have to be balls:

Proposition 2.28 (Lift zonoid of a spherical distribution) *Let $\mu \in \mathcal{M}_1$ be spherical around the origin. Then, for any $0 < t < 1$, $\widehat{Z}_t(\mu)$ is a ball in $\mathbb{R}^d$ with center at the origin. Further, $\widehat{Z}_0(\mu) = \widehat{Z}_1(\mu) = \{\mathbf{0}\}$.*

Let $r_\psi(t)$ denote the radius of $\widehat{Z}_t(\mu)$ when $\mu \sim \mathcal{S}(\psi)$. Due to the projection theorem (Corollary 2.25) this radius can be calculated from the radius of the respective slice of a univariate marginal. Note that neither R nor $\boldsymbol{X}$ need have a density. But, if $\boldsymbol{X}$ has no atom at the origin, every marginal of $\boldsymbol{X}$ is continuously distributed. By spherical symmetry, all univariate marginals are the same. Their density is given by (Fang et al., 1990, Th. 2.10)

$$f_\psi(s) = C(d) \int_s^\infty r^{-d+2}(r^2 - s^2)^{\frac{d-1}{2}-1} d\psi(r), \qquad (2.55)$$

$$\text{with}^4 \quad C(d) = \frac{\Gamma(d/2)}{\Gamma((d-1)/2)\pi^{1/2}}.$$

Note that the marginal distribution is symmetric about zero and its lift zonoid is symmetric about the line at $z_0 = 1/2$; see Theorem 2.17. The radius $r_\psi(t)$ is the negative of the generalized Lorenz function of this marginal distribution at t. Let q_s denote the s-quantile of the univariate marginal. Then

$$r_\psi(t) = \begin{cases} \int_{q_t}^\infty s\, f_\psi(s)\, ds & \text{if } \frac{1}{2} \le t < 1,\ t = \frac{1}{2} + \int_0^{q_t} f_\psi(s)\, ds, \\ r_\psi(1-t) & \text{if } 0 < t < \frac{1}{2}. \end{cases} \qquad (2.56)$$

Now we turn to elliptical distributions. Any elliptical distribution is an affine transformation of a spherical distribution. From Corollary 2.27 we see that the slices of the lift zonoid are ellipsoids and their precise shape is as follows:

Proposition 2.29 (Lift zonoid of an elliptical distribution) *Let $\mu \in \mathcal{M}_1$ be elliptical, $\mu = \mathcal{E}(\boldsymbol{a}, \boldsymbol{BB}', \psi)$. Then, for any $0 < t < 1$, $\widehat{Z}_t(\mu)$ is an ellipsoid in $\mathbb{R}^d$ with center at $t\boldsymbol{a}$,*

$$\widehat{Z}_t(\mu) = \{\boldsymbol{x} \in \mathbb{R}^d : (\boldsymbol{x} - t\boldsymbol{a})(\boldsymbol{BB}')^{-1}(\boldsymbol{x} - t\boldsymbol{a})' \le r_\psi^2(t)\},$$

and $r_\psi(t)$ according to (2.56). Further, $\widehat{Z}_0(\mu) = \mathbf{0}$ and $\widehat{Z}_1(\mu) = \boldsymbol{a}$.

Example 2.10 (Multivariate normal distribution) The multivariate standard normal, $N(\mathbf{0}, I)$, is a spherical distribution. Its lift zonoid is composed of balls around the origin,

$$\widehat{Z}_t(N(\mathbf{0}, I)) = B(r_1(t)) = \{\boldsymbol{x} : \boldsymbol{x}\boldsymbol{x}' \le {r_1}^2(t)\} \qquad (2.57)$$

$$\text{with} \quad r_1(t) = \frac{1}{\sqrt{2\pi}} \exp\left(-\frac{u_t^2}{2}\right). \qquad (2.58)$$

The radius $r_1(t)$ is obtained as follows. The univariate projection of the t-slice $\widehat{Z}_t(N(\mathbf{0}, I))$ to, say, the first coordinate axis is the respective t-slice of a univariate standard normal. Thus, the radius of the ball is seen from Formula (2.42), which describes the lift zonoid of a univariate normal.

A general multivariate normal, $N(\boldsymbol{a}, \Sigma)$, with expectation $\boldsymbol{a}$ and covariance matrix Σ, is an affine transformation of a standard normal: $\boldsymbol{Y} \sim N(\boldsymbol{a}, \Sigma)$ if and only if $\boldsymbol{Y} =_{st} \boldsymbol{X}\boldsymbol{B} + \boldsymbol{a}$ with $\boldsymbol{B}\boldsymbol{B}' = \Sigma$ and $\boldsymbol{X} \sim N(\mathbf{0}, \boldsymbol{I})$. Thus, from Proposition 2.29 and Equation (2.57) obtain

$$\widehat{Z}_t(N(\boldsymbol{a}, \Sigma)) = \left\{\boldsymbol{x} \in \mathbb{R}^d : (\boldsymbol{x} - t\boldsymbol{a})\Sigma^{-1}(\boldsymbol{x} - t\boldsymbol{a})' \le {r_1}^2(t)\right\}. \qquad (2.59)$$

Example 2.11 (Uniform distribution on the unit sphere) If μ is *uniformly distributed on the unit sphere* in $\mathbb{R}^d$ then μ is spherical and the lift zonoid is built of slices which are balls, $B(r_2(t))$, around the origin having radius $r_2(t)$. We have $\psi(s) = 0$ if $0 \le s < 1$ and $\psi(s) = 1$ if $s \ge 1$. By (2.55) the univariate marginal density is

$$f_\psi(s) = C(d)(1 - s^2)^{\frac{d-1}{2}-1} \quad \text{if} \quad -1 < s < 1$$

and $f_\psi(s) = 0$ otherwise. From this and from (2.56) obtain that the radius is

$$\begin{aligned} r_2(t) &= \; = r_\psi(t) = \frac{1}{d-1} C(d)(1 - q_t^2)^{\frac{d-1}{2}} \quad \text{if} \quad \frac{1}{2} \le t < 1, \qquad (2.60) \\ r_2(t) &= r_2(1 - t) \quad \text{if} \quad 0 < t < \frac{1}{2}, \end{aligned}$$

Again, q_s denotes the s-quantile of the univariate marginal.

In particular, consider the unit spheres in dimensions two and three. If $d = 2$, we get $C(2) = 1/\pi$ and $q_t = \sin((t - 1/2)\pi)$ for $1/2 \le t < 1$. For a uniform distribution on the bivariate sphere, the slices are balls around the origin with radius

$$r_2(t) = \frac{1}{\pi} \sin(t\,\pi), \qquad 0 \le t \le 1.$$

If $d = 3$, we see that the univariate marginal is uniform on the interval $[-1, 1]$, $C(3) = 1/2$, and

$$r_2(t) = \frac{1}{4} - \left(t - \frac{1}{2}\right)^2, \qquad 0 \le t \le 1.$$

Example 2.12 (Uniform distribution on a ball or an ellipsoid) Let μ be *uniformly distributed on the unit ball* in $\mathbb{R}^d$. Again, μ is spherical and the slices of the lift zonoid are balls around the origin. Here $\psi(s) = s^d$ if $0 \le s < 1$ and $\psi(s) = 1$ if $s \ge 1$. From (2.55) and (2.56) is calculated that

$$f_\psi(s) = \frac{d}{d-1} C(d)\, (1 - s^2)^{(d-1)/2},$$

$$r_3(t) = r_\psi(t) = \frac{d}{(d-1)(d+1)} C(d)(1 - q_t^2)^{(d+1)/2} \quad \text{for } \frac{1}{2} \le t < 1, \quad (2.61)$$

and $r_3(t) = r_3(1-t)$ for $0 < t \le 1/2$. The quantile q_t, $1/2 \le t < 1$, is obtained by solving

$$t - \frac{1}{2} = \frac{d}{d-1} C(d) \int_0^{q_t} (1 - s^2)^{(d-1)/2}\, ds\,.$$

Especially, if $d = 2$, the last equations become

$$r_3(t) = \frac{2}{3\pi}(1 - q_t^2)^{3/2}, \quad t - \frac{1}{2} = \frac{1}{\pi}\left(q_t \sqrt{1 - q_t^2} + \arcsin q_t\right), \quad \frac{1}{2} \le t < 1\,.$$

If $d = 3$, the radius is given by

$$r_3(t) = \frac{3}{16}(1 - q_t^2)^2, \qquad t - \frac{1}{2} = \frac{3q_t}{4}\left(1 - \frac{q_t^3}{3}\right), \qquad \frac{1}{2} \le t < 1\,.$$

If the distribution μ is *uniform on a ball* in $\mathbb{R}^d$ with center $\boldsymbol{a}$ and radius β the slices of the lift zonoid are also balls. This follows from Corollary 2.27. The center of $\widehat{Z}_t(\mu)$ is $t\boldsymbol{a}$ and the radius is $\beta\, r_3(t)$.

The lift zonoid of a measure which is *uniformly distributed on an ellipsoid* in $\mathbb{R}^d$ is obtained from the latter lift zonoid by using Proposition 2.29.

2.4 Continuity and approximation

The embedding of d-variate measures into convex sets in the $(d+1)$-space is continuous. More precisely, the embedding of $\mathcal{M}_1$ into the class of centrally symmetric, convex compacts is continuous with respect to weak convergence of uniformly integrable measures in $\mathcal{M}_1$ and Hausdorff convergence of convex compacts, as will be shown in this section. A number of related continuity results will be established. Further we investigate the approximation of a measure by a weakly convergent sequences of measures from below and from above with regard to the lift zonoid order.

2.4.1 Convergence of lift zonoids

In this section weak continuity properties of the lift zonoid will be studied. A sequence of measures $(\mu^n)_{n\in\mathbb{N}}$ in $\mathcal{M}_1$ converges weakly to $\mu \in \mathcal{M}_1$, $\mu^n \xrightarrow{w} \mu$, if and only if $\int_{\mathbb{R}^d} f(x)\mu^n(dx) \to \int_{\mathbb{R}^d} f(x)\mu(dx)$ holds for any function $f : \mathbb{R}^d \to \mathbb{R}$ that is bounded and continuous. To state the main result the following notion of uniform integrability is needed:

Definition 2.6 (Uniform integrability) *A family of measures $(\mu^i)_{i\in I}$ in $\mathcal{M}_1$ is* uniformly integrable *if*

$$\lim_{\beta\to\infty} \sup_{i\in I} \int_{\|x\|\geq\beta} \|x\| \, \mu^i(dx) = 0 . \tag{2.62}$$

The main theorem of this section says that if a sequence $(\mu^n)_{n\in\mathbb{N}}$ of measures is uniformly integrable and converges weakly to μ in $\mathcal{M}_1$, the sequence of their lift zonoids converges, in the Hausdorff distance, to the lift zonoid of μ, that is, the lift zonoid distance between μ^n and μ goes to zero. In case the μ^n and μ have a density with respect to Lebesgue measure, the reverse implication holds, too.

Theorem 2.30 (Continuity of the lift zonoid distance) *Let μ and a sequence $(\mu^n)_{n\in\mathbb{N}}$ be in $\mathcal{M}_1$.*

(i) *If $(\mu^n)_{n\in\mathbb{N}}$ is uniformly integrable and $\mu^n \xrightarrow{w} \mu$, then $\delta_{lz}(\mu^n, \mu) \to 0$.*

(ii) *If, in addition, the μ^n and μ are L-continuous, the reverse implication is also true.*

We shall prove Theorem 2.30 below. Before, the uniform integrability is discussed and it is shown by an example that, given a weakly convergent sequence of measures, the sequence of their lift zonoids can, in the Hausdorff distance, converge to a convex compact which is no lift zonoid.

Example 2.13 (Limit of lift zonotopes) Consider the following sequence of probability distributions in $\mathbb{R}$: For any $n \in \mathbb{N}$ let μ^n have mass $(n-1)/n$ at zero and $1/n$ at n. Then $\epsilon(\mu^n) = 1$, hence $\mu^n \in \mathcal{M}_1$ for every n. The sequence $(\mu^n)_{n\in\mathbb{N}}$ converges weakly to δ_0, the Dirac measure at 0, whose lift zonoid $\widehat{Z}(\delta_0)$ equals the segment $[(0,0),(1,0)]$ in $\mathbb{R}^2$. The lift zonoid (2.31) of μ^n,

$$\widehat{Z}(\mu) = \frac{n-1}{n}\,[(0,0),(1,0)] + \frac{1}{n}\,[(0,0),(1,n)] \, ,$$

is a parallelogram with vertices $(0,0)$, $(\frac{1}{n},1)$, $(\frac{n-1}{n},0)$ and $(1,1)$. Obviously, the sequence of lift zonoids $\widehat{Z}(\mu^n)$ converges to the unit square, which differs from $\widehat{Z}(\delta_0)$. Now assume that the unit square is the lift zonoid of some measure ν. Since the square is contained in $\mathbb{R}^2_+$, ν must have nonnegative support, and since the point $(1,0)$ is a vertex, conclude $\alpha(\nu) = 1$ and $\epsilon(\nu) = 0$. Hence ν must be the Dirac measure at zero, $\nu = \delta_0$. But $\widehat{Z}(\delta_0) = [(0,0),(1,0)]$; contradiction. Therefore, the unit square, while being a zonoid, fails to be the lift zonoid of a measure. Note that in this example the sequence $(\mu^n)_{n\in\mathbb{N}}$ is weakly convergent but not uniformly integrable: For any β we have $\sup_n \int_{|x|\geq\beta} |x|\,\mu^n(dx) = 1$.

Recall that μ_p is the image of the measure μ under the linear transformation $x \mapsto \langle x, p\rangle$. It is well known (e.g., Billingsley (1979, Th. 29.4)) that a sequence $(\mu^n)_{n\in\mathbb{N}}$ converges weakly to μ in $\mathcal{M}_1$ if and only if the sequence $(\mu^n_p)_{n\in\mathbb{N}}$ converges weakly to μ_p for every $p \in S^{d-1}$ (or, equivalently, for every $p \in \mathbb{R}^d$).

Lemma 2.31 (Uniform integrability) *A sequence of measures $(\mu^n)_{n\in\mathbb{N}}$ in $\mathcal{M}_1$ is uniformly integrable if and only if the sequence $(\mu^n_p)_{n\in\mathbb{N}}$ is uniformly integrable for every $p \in S^{d-1}$.*

Proof. For any $p \in S^{d-1}$ and $\beta \in \mathbb{R}$ we get

$$\int_{|t|\geq\beta} |t|\,\mu^n_p(dt) = \int_{|\langle x,p\rangle|\geq\beta} |\langle x,p\rangle|\,\mu^n(dx) \leq \int_{\|x\|\geq\beta} |\langle x,p\rangle|\,\mu^n(dx), \tag{2.63}$$

since $|\langle x,p\rangle| \leq \|x\|$ for every x. We use the following formula by Helgason (1999, Lemma 7.2): For every $x \in \mathbb{R}^d$ and $k > 0$ holds

$$\int_{S^{d-1}} |\langle x,p\rangle|^k dp = \frac{2\pi^{\frac{d-1}{2}}\Gamma(\frac{k+1}{2})}{\Gamma(\frac{d+k}{2})}\,\|x\|^k. \tag{2.64}$$

Here dp means integration with respect to Lebesgue measure on the unit sphere.

(i) Assume that $(\mu^n)_{n\in\mathbb{N}}$ is uniformly integrable. Then

$$\begin{aligned}
0 &= \lim_{\beta\to\infty} \sup_{n\in\mathbb{N}} \int_{\|\boldsymbol{x}\|\geq\beta} \|\boldsymbol{x}\|\; \mu^n(d\boldsymbol{x}) \\
&= \lim_{\beta\to\infty} \sup_{n\in\mathbb{N}} \int_{\|\boldsymbol{x}\|\geq\beta} C(d) \int_{S^{d-1}} |\langle \boldsymbol{x}, \boldsymbol{p}\rangle|\; d\boldsymbol{p}\; \mu^n(d\boldsymbol{x}) \\
&= C(d) \lim_{\beta\to\infty} \sup_{n\in\mathbb{N}} \int_{S^{d-1}} \int_{\|\boldsymbol{x}\|\geq\beta} |\langle \boldsymbol{x}, \boldsymbol{p}\rangle|\; \mu^n(d\boldsymbol{x})\; d\boldsymbol{p} \\
&\geq C(d) \lim_{\beta\to\infty} \sup_{n\in\mathbb{N}} \int_{S^{d-1}} \int_{|t|\geq\beta} |t|\; \mu^n_{\boldsymbol{p}}(dt)\; d\boldsymbol{p} \;\geq\; 0\,,
\end{aligned}$$

where we have first employed (2.64) with $k = 1$ and then (2.63). The constant

$$C(d) = \frac{\Gamma(\frac{d+1}{2})}{2\pi^{\frac{d-1}{2}}}$$

is positive and depends on the dimension only. We conclude that

$$\lim_{\beta\to\infty} \sup_{n\in\mathbb{N}} \int_{|t|\geq\beta} |t|\; \mu^n_{\boldsymbol{p}}(dt) = 0 \qquad (2.65)$$

for every $\boldsymbol{p} \in S^{d-1}$, besides a subset of the sphere which has Lebesgue measure 0. For the remaining $\boldsymbol{p} \in S^d$, Equation (2.65) holds by continuity. Hence the sequence $(\mu^n_{\boldsymbol{p}})_{n\in\mathbb{N}}$ is uniformly integrable for every $\boldsymbol{p} \in S^d$.

(ii) To show the reverse, assume that (2.65) is satisfied for all $\boldsymbol{p} \in S^{d-1}$. Then, by inserting the canonical unit vectors $\boldsymbol{e}_j$ and notating $x_j = \langle \boldsymbol{x}, \boldsymbol{e}_j\rangle$, obtain

$$0 = \lim_{\beta\to\infty} \sup_{n\in\mathbb{N}} \int_{|x_j|\geq\beta} |x_j|\, \mu^n(d\boldsymbol{x})\,, \quad j = 1, \ldots, d\,.$$

There holds for every j

$$\begin{aligned}
&\int_{|x_j|\geq\beta} |x_j|\, \mu^n(d\boldsymbol{x}) \\
&\geq \int_{\max_k |x_k|\geq\beta} |x_j|\, \mu^n(d\boldsymbol{x}) - \sum_{k\neq j} \int_{|x_j|<\beta, |x_k|\geq\beta} |x_j|\, \mu^n(d\boldsymbol{x}) \\
&\geq \int_{\max_k |x_k|\geq\beta} |x_j|\, \mu^n(d\boldsymbol{x}) - \sum_{k\neq j} \int_{|x_j|<\beta, |x_k|\geq\beta} |x_k|\, \mu^n(d\boldsymbol{x})\,.
\end{aligned}$$

Taking $\sup_n$ and $\lim_{\beta\to\infty}$, the left hand side and the second term on the right hand side go to zero, hence

$$0 = \lim_{\beta\to\infty} \sup_{n\in\mathbb{N}} \int_{|x_j|\geq\beta} |x_j|\, \mu^n(d\boldsymbol{x}) \geq \lim_{\beta\to\infty} \sup_{n\in\mathbb{N}} \int_{\max_k |x_k|\geq\beta} |x_j|\, \mu^n(d\boldsymbol{x}) = 0\,.$$

Summing over j yields

$$
\begin{aligned}
0 &= \lim_{\beta\to\infty} \sup_{n\in\mathbb{N}} \int_{\max_k |x_k|\geq\beta} \sum_{j=1}^{d} |x_j|\,\mu^n(d\boldsymbol{x}) \\
&\geq \lim_{\beta\to\infty} \sup_{n\in\mathbb{N}} \int_{\max_k |x_k|\geq\beta} ||\boldsymbol{x}||\,\mu^n(d\boldsymbol{x}) \\
&= \lim_{\beta\to\infty} \sup_{n\in\mathbb{N}} \int_{||\boldsymbol{x}||\geq\beta} ||\boldsymbol{x}||\,\mu^n(d\boldsymbol{x})\,.
\end{aligned}
$$

Consequently, the sequence $(\mu^n)_{n\in\mathbb{N}}$ is uniformly integrable. Q.E.D.

As a second step towards the proof of Theorem 2.30 it will be demonstrated that the uniform integrability is a necessary restriction for the convergence of lift zonoids.

Proposition 2.32 (Convergence implies uniform integrability) *If* $\lim_{n\to\infty}\delta_{lz}(\mu^n,\mu)=0$ *then* $(\mu^n)_{n\in\mathbb{N}}$ *is uniformly integrable.*

Proof. Lift zonoids are compacts with nonnegative support functions. The Hausdorff convergence of lift zonoids, $\widehat{Z}(\mu^n)\to\widehat{Z}(\mu)$, implies therefore the existence of a compact K such that $\widehat{Z}(\mu^n)\subset K$ for all n. The support function of K is uniformly bounded by some number C for all $(p_0,\boldsymbol{p})\in S^d$. We choose $p_0=0$ and obtain from $\widehat{Z}(\mu^n)\subset K$ that

$$
\sup_{n\in\mathbb{N}} \int_{\mathbb{R}^d} \max\{0,\langle\boldsymbol{x},\boldsymbol{p}\rangle\}\mu^n(d\boldsymbol{x}) = \sup_{n\in\mathbb{N}} h(\widehat{Z}(\mu^n),0,\boldsymbol{p}) \leq h(K,0,\boldsymbol{p}) \leq C\,,
$$

and thus,

$$
\lim_{\beta\to\infty} \sup_{n\in\mathbb{N}} \int_{|\langle\boldsymbol{x},\boldsymbol{p}\rangle|\geq\beta} \max\{0,\langle\boldsymbol{x},\boldsymbol{p}\rangle\}\,\mu^n(d\boldsymbol{x}) = 0\,, \quad \boldsymbol{p}\in S^{d-1}\,. \tag{2.66}
$$

Let $\beta>0$. Then

$$
\begin{aligned}
&\int_{|t|\geq\beta} |t|\,\mu^n_{\boldsymbol{p}}(dt) \\
&= \int_{t\geq\beta} \max\{0,t\}\mu^n_{\boldsymbol{p}}(dt) + \int_{-t\geq\beta} \max\{0,-t\}\mu^n_{\boldsymbol{p}}(dt) \\
&= \int_{\langle\boldsymbol{x},\boldsymbol{p}\rangle\geq\beta} \max\{0,\langle\boldsymbol{x},\boldsymbol{p}\rangle\}\mu^n(d\boldsymbol{x}) + \int_{\langle\boldsymbol{x},-\boldsymbol{p}\rangle\geq\beta} \max\{0,\langle\boldsymbol{x},-\boldsymbol{p}\rangle\}\mu^n(d\boldsymbol{x})\,,
\end{aligned}
$$

$$\sup_n \int_{|t|\geq\beta} |t|\, \mu_{\boldsymbol{p}}^n(dt) \;\leq\; \sup_n \int_{\langle \boldsymbol{x},\boldsymbol{p}\rangle\geq\beta} \max\{0,\langle \boldsymbol{x},\boldsymbol{p}\rangle\}\mu^n(d\boldsymbol{x}) + \sup_n \int_{\langle \boldsymbol{x},-\boldsymbol{p}\rangle\geq\beta} \max\{0,\langle \boldsymbol{x},-\boldsymbol{p}\rangle\}\mu^n(d\boldsymbol{x})\,. \tag{2.67}$$

When β goes to infinity, the right hand side of (2.67) side converges to zero. That is, for every $\boldsymbol{p} \in S^{d-1}$ the sequence $\mu_{\boldsymbol{p}}^n$ is uniformly integrable. By Lemma 2.31 conclude that the sequence μ^n is uniformly integrable. Q.E.D.

Proof of Theorem 2.30 .
(i): Assume that $(\mu^n)_{n\in\mathbb{N}}$ is uniformly integrable and weakly convergent to μ. The support function of $\widehat{Z}(\mu^n)$ at $(p_0,\boldsymbol{p}) \in S^d$ is given by (Proposition 2.16)

$$\begin{aligned} h\left(\widehat{Z}(\mu^n),p_0,\boldsymbol{p}\right) &= \int_{\langle \boldsymbol{x},\boldsymbol{p}\rangle\geq -p_0} (p_0+\langle \boldsymbol{x},\boldsymbol{p}\rangle)\mu^n(d\boldsymbol{x}) \\ &= \int_{t\geq -p_0} (p_0+t)\mu_{\boldsymbol{p}}^n(dt) \\ &= \int_{t\geq -p_0,|t|<\beta} (p_0+t)\mu_{\boldsymbol{p}}^n(dt) + \int_{t\geq -p_0,|t|\geq\beta} (p_0+t)\mu_{\boldsymbol{p}}^n(dt). \end{aligned} \tag{2.68}$$

$$\lim_{n\to\infty} h\left(\widehat{Z}(\mu^n),p_0,\boldsymbol{p}\right) = \lim_{\beta\to\infty}\lim_{n\to\infty}\int_{t\geq -p_0,|t|<\beta} (p_0+t)\mu_{\boldsymbol{p}}^n(dt) + \lim_{\beta\to\infty}\lim_{n\to\infty}\int_{t\geq -p_0,|t|\geq\beta} (p_0+t)\mu_{\boldsymbol{p}}^n(dt). \tag{2.69}$$

For any β we have

$$\left|\lim_{n\to\infty}\int_{t\geq -p_0,|t|\geq\beta} (p_0+t)\mu_{\boldsymbol{p}}^n(dt)\right| \leq \sup_{n\in\mathbb{N}}\int_{|t|\geq\beta} (1+|t|)\mu_{\boldsymbol{p}}^n(dt)\,,$$

which, by uniform integrability of $(\mu_{\boldsymbol{p}}^n)_{n\in\mathbb{N}}$ (Lemma 2.31), goes to 0 when β goes to infinity. For any β, since $\mu_{\boldsymbol{p}}^n \xrightarrow{w} \mu_{\boldsymbol{p}}$,

$$\begin{aligned} \lim_{n\to\infty}\int_{t\geq -p_0,|t|<\beta} (p_0+t)\mu_{\boldsymbol{p}}^n(dt) &= \int_{t\geq -p_0,|t|<\beta} (p_0+t)\mu_{\boldsymbol{p}}(dt) \\ &= \int_{\langle \boldsymbol{x},\boldsymbol{p}\rangle\geq -p_0,|\langle \boldsymbol{x},\boldsymbol{p}\rangle|<\beta} (p_0+\langle \boldsymbol{x},\boldsymbol{p}\rangle)\mu(d\boldsymbol{x})\,. \end{aligned}$$

Letting β approach infinity we obtain that, for $(p_0, \boldsymbol{p}) \in S^d$,

$$\begin{aligned}\lim_{n\to\infty} h\left(\widehat{Z}(\mu^n), p_0, \boldsymbol{p}\right) &= \lim_{n\to\infty} \int_{\langle \boldsymbol{x},\boldsymbol{p}\rangle \geq -p_0} (p_0 + \langle \boldsymbol{x}, \boldsymbol{p}\rangle)\mu^n(d\boldsymbol{x}) \\ &= \int_{\langle \boldsymbol{x},\boldsymbol{p}\rangle \geq -p_0} (p_0 + \langle \boldsymbol{x}, \boldsymbol{p}\rangle)\mu(d\boldsymbol{x}) \\ &= h\left(\widehat{Z}(\mu), p_0, \boldsymbol{p}\right) .\end{aligned}$$

Moreover, by uniform integrability, the convergence is uniform in $(p_0, \boldsymbol{p}) \in S^d$. That means the support functions of $\widehat{Z}(\mu^n)$ converge uniformly on the sphere to the support function of $\widehat{Z}(\mu)$. We conclude that the sequence of lift zonoids $\widehat{Z}(\mu^n)$ converges, in the Hausdorff distance, to the lift zonoid $\widehat{Z}(\mu)$ of μ.

(ii): Assume that $\widehat{Z}(\mu^n) \xrightarrow{H} \widehat{Z}(\mu)$. Then we know from Proposition 2.32 that $(\mu^n)_{n\in\mathbb{N}}$ is uniformly integrable. The Hausdorff convergence of lift zonoids implies the convergence

$$\int_{\mathbb{R}^d} \max\{0, p_0 + \langle \boldsymbol{x}, \boldsymbol{p}\rangle\}\mu^n(d\boldsymbol{x}) \longrightarrow \int_{\mathbb{R}^d} \max\{0, p_0 + \langle \boldsymbol{x}, \boldsymbol{p}\rangle\}\mu(d\boldsymbol{x}) ,$$

uniformly in $(p_0, \boldsymbol{p}) \in S^d$. The directional derivative in direction $\boldsymbol{p}$ of the integrand is

$$\langle \nabla_{\boldsymbol{x}} \max\{0, p_0 + \langle \boldsymbol{x}, \boldsymbol{p}\rangle\}, \, \boldsymbol{p}\rangle = \begin{cases} \langle \boldsymbol{p}, \boldsymbol{p}\rangle = 1 - p_0^2 & \text{if } p_0 + \langle \boldsymbol{x}, \boldsymbol{p}\rangle > 0 , \\ 0 & \text{if } p_0 + \langle \boldsymbol{x}, \boldsymbol{p}\rangle < 0 , \end{cases}$$

for $(p_0, \boldsymbol{p}) \in S^d$. Note that this directional derivative is a uniform limit in $\boldsymbol{x}$. Therefore and by the L-continuity of μ^n, the integral $\int_{\mathbb{R}^d} \max\{0, p_0 + \langle \boldsymbol{x}, \boldsymbol{p}\rangle\}\mu^n(d\boldsymbol{x})$ has directional derivative $(1 - p_0^2) \int_{p_0+\langle \boldsymbol{x},\boldsymbol{p}\rangle \geq 0} \mu^n(d\boldsymbol{x})$ and the sequence of derivatives, divided by $(1 - p_0^2)$, converges,

$$\int_{p_0+\langle \boldsymbol{x},\boldsymbol{p}\rangle \geq 0} \mu^n(d\boldsymbol{x}) \longrightarrow \int_{p_0+\langle \boldsymbol{x},\boldsymbol{p}\rangle \geq 0} \mu(d\boldsymbol{x}) , \qquad (p_0, \boldsymbol{p}) \in S^d . \tag{2.70}$$

Obviously, the convergence (2.70) holds for any scalar multiple of $(p_0, \boldsymbol{p})$, too. Thus, we obtain

$$\int_{t\geq p_0} \mu_{\boldsymbol{p}}^n(dt) \longrightarrow \int_{t\geq p_0} \mu_{\boldsymbol{p}}(dt) , \qquad (p_0, \boldsymbol{p}) \in \mathbb{R}^{d+1} ,$$

which means weak convergence $\mu_{\boldsymbol{p}}^n \xrightarrow{w} \mu_{\boldsymbol{p}}$ for every $\boldsymbol{p} \in \mathbb{R}^d$, hence weak convergence $\mu^n \xrightarrow{w} \mu$. Q.E.D.

Corollary 2.33 (Compact support) *Assume that μ^n converges weakly to μ in $\mathcal{M}_1$ and that there is a compact which includes the support of μ^n for all n. Then* $\lim_{n\to\infty} \delta_{lz}(\mu^n, \mu) = 0$,

Proof. Under this assumption, the sequence μ^n is uniformly integrable. Q.E.D.

Corollary 2.34 (Lift zonoid superset) *Let μ^n weakly converge to μ in $\mathcal{M}_1$ and assume that there exists a measure $\nu \in \mathcal{M}_1$ with $\widehat{Z}(\mu^n) \subset \widehat{Z}(\nu)$ for all n. Then $\widehat{Z}(\mu^n)$ converges to $\widehat{Z}(\mu)$ in the Hausdorff distance.*

Proof. As in the proof of Proposition 2.32, the uniform integrability of the sequence $(\mu^n)_{n\in\mathbb{N}}$ follows from the inclusion $\widehat{Z}(\mu^n) \subset \widehat{Z}(\nu)$. Q.E.D.

Corollary 2.35 (Limiting lift zonoid) *Let μ^n weakly converge to μ in $\mathcal{M}_1$ and let $\widehat{Z}(\mu^n)$, in the Hausdorff sense, converge to a lift zonoid. Then* $\lim_{n\to\infty} \widehat{Z}(\mu^n) = \widehat{Z}(\mu)$.

Proof. Let $\widehat{Z}(\nu)$ be the limit zonoid. Since every lift zonoid is compact and has a nonnegative support function, one obtains that, with a big enough constant γ, $\widehat{Z}(\mu^n) \subset \widehat{Z}(\gamma\ \mu) = \gamma\ \widehat{Z}(\mu)$ for all n. The claim follows from the previous corollary. Q.E.D.

2.4.2 Monotone approximation of measures

The subsequent theorem says that a measure can be approximated by empirical measures which are smaller and by L-continuous measures which are larger, both with regard to the lift-zonoid order.

Theorem 2.36 (Monotone approximation) *Given $\mu \in \mathcal{M}_1$, there exists a sequence $(\nu^n)_{n\in\mathbb{N}}$ of empirical measures and a sequence $(\mu^n)_{n\in\mathbb{N}}$ of L-continuous measures such that both converge weakly to μ and*

$$\widehat{Z}(\nu^n) \subset \widehat{Z}(\mu) \subset \widehat{Z}(\mu^n) \quad \textit{for all } n.$$

Proof.
(i) For $n \in \mathbb{N}$ consider a partition $\{B_1^n, B_2^n, \dots, B_{k_n}^n\}$ of $\mathbb{R}^d$ with $\mu(B_i^n) \le 2^{-n}$. Let

$$y_i = \int_{B_i^n} x\mu(dx), \quad p_i = \mu(B_i^n), \quad i = 1, 2, \dots, k_n.$$

Then $\pi^n = \sum_{i=1}^{k_n} p_i \delta_{\boldsymbol{y}_i}$ is a discrete measure satisfying $\pi^n \preceq_{cx} \mu$ and $\pi^n \xrightarrow{w} \mu$. We shall approximate each π^n by an empirical measure from below. Define further

$$q_i = \frac{1}{n(k_n+1)}[n(k_n+1)p_i], \quad i = 1, 2, \ldots, k_n,$$

where $[\alpha]$ denotes the integer part of a given $\alpha \in \mathbb{R}$,

$$q_{k_n+1} = \sum_{i=1}^{k_n}(p_i - q_i), \quad \boldsymbol{y}_{k_n+1} = \sum_{i=1}^{k_n}(p_i - q_i)\boldsymbol{y}_i \, .$$

Let $\nu^n = \sum_{i=1}^{k_n+1} q_i \delta_{\boldsymbol{y}_i}$. Then ν^n is an empirical measure and $\nu^n \preceq_{cx} \pi^n$. For $n \to \infty$, ν^n converges weakly to μ. Note that μ is a *dilation* of ν^n, for every n. Hence by (8.13) in Section 8.4, $\widehat{Z}(\nu^n) \subset \widehat{Z}(\mu)$.

(ii) With respect to the approximation from above, proceed as follows: For $\boldsymbol{x} \in \mathbb{R}^d$ and $n \in \mathbb{N}$ let $\phi_{\boldsymbol{x}}^n$ be the multivariate normal distribution with expectation $\boldsymbol{x}$ and covariance matrix $n^{-1}\boldsymbol{I}_d$. $\boldsymbol{I}_d$ is the $d \times d$ unit matrix. Then $\phi_{\boldsymbol{x}}^n$ is larger in convex order (1.6) than the Dirac measure at $\boldsymbol{x}$ for every $\boldsymbol{x}$. The measure μ^n, $\mu^n(B) = \int_{\mathbb{R}^d} \phi_{\boldsymbol{x}}^n(B)\mu(d\boldsymbol{x})$ is, as a mixture, again larger in convex order than μ. Hence $\widehat{Z}(\mu) \subset \widehat{Z}(\mu^n)$. Further, μ^n is L-continuous. Because $\mu^n \xrightarrow{w} \mu$, the proof is complete. Q.E.D.

2.4.3 Volume of a lift zonoid

The volume of a lift zonoid is the expectation of the modulus of a random determinant:

Theorem 2.37 (Volume of a lift zonoid) *For a given $\mu \in \mathcal{M}_1$ consider independent random vectors $\boldsymbol{X}_0, \boldsymbol{X}_1, \ldots, \boldsymbol{X}_d$, defined on some probability space $(\Omega, \mathcal{A}, P)$, each distributed as $\frac{1}{\alpha(\mu)}\mu$. $\widehat{\boldsymbol{Q}}(\mu) = [(1,\boldsymbol{X}_0),(1,\boldsymbol{X}_1),\ldots,(1,\boldsymbol{X}_d)]$ denotes the $(d+1)\times(d+1)$ matrix having rows $(1,\boldsymbol{X}_0),(1,\boldsymbol{X}_1),\ldots,(1,\boldsymbol{X}_d)$. Then the $(d+1)$-dimensional volume of the lift zonoid $\widehat{Z}(\mu)$ is*

$$\mathrm{vol}_{d+1}\widehat{Z}(\mu) = \frac{1}{(d+1)!}(\alpha(\mu))^{d+1}\mathrm{E}(|\det \widehat{\boldsymbol{Q}}(\mu)|) \, . \tag{2.71}$$

Proof. Any lift zonoid, since it is compact, has a finite volume.
First, assume that μ is an empirical probability measure giving equal mass

to points $\boldsymbol{a}_1, \ldots, \boldsymbol{a}_n$. Then, by (2.11),

$$\mathrm{vol}_{d+1} \widehat{Z}(\mu) = \frac{1}{(d+1)!} \frac{1}{n^{d+1}} \sum_{i_0=1}^{n} \cdots \sum_{i_d=1}^{n} |\det[(1, \boldsymbol{a}_{i_0}), (1, \boldsymbol{a}_{i_1}), \ldots, (1, \boldsymbol{a}_{i_d})]| \,. \tag{2.72}$$

Further,

$$\mathrm{E}(|\det \widehat{\boldsymbol{Q}}(\mu)|) = \sum_{i_0=1}^{n} \cdots \sum_{i_d=1}^{n} |\det[(1, \boldsymbol{a}_{i_0}), \ldots, (1, \boldsymbol{a}_{i_d})]| \cdot P(\boldsymbol{X}_0 = \boldsymbol{a}_{i_0}, \ldots, \boldsymbol{X}_d = \boldsymbol{a}_{i_d}) \,.$$

Since $P(\boldsymbol{X}_0 = \boldsymbol{a}_{i_0}, \ldots, \boldsymbol{X}_d = \boldsymbol{a}_{i_d}) = n^{-(d+1)}$, Equation (2.71) is obtained for $\mu \in \mathcal{P}^n$. Now let $\mu \in \mathcal{P}_1$. According to Theorem 2.36, μ is approximated by a sequence of empirical measures ν^n such that $\widehat{Z}(\nu^n) \subset \widehat{Z}(\mu)$. From Corollary 2.34 we know that $\widehat{Z}(\nu^n)$ converges in the Hausdorff distance to $\widehat{Z}(\nu)$. Because the volume of a convex compact is continuous with respect to the Hausdorff distance, we have

$$\mathrm{vol}_{d+1} \widehat{Z}(\mu) = \lim_{n\to\infty} \mathrm{vol}_{d+1} \widehat{Z}(\nu^n) \,.$$

For every $n \in \mathbb{N}$ there exist independent random vectors $\boldsymbol{X}_0^{(n)}, \ldots, \boldsymbol{X}_d^{(n)}$ such that, for each $j = 0, \ldots, d$, $\boldsymbol{X}_j^{(n)} \sim \nu^n$ and $\lim_{n\to\infty} \boldsymbol{X}_j^{(n)} = \boldsymbol{X}_j$ pointwise. Let $\widehat{\boldsymbol{Q}}(\nu^n) = [(1, \boldsymbol{X}_0^{(n)}), \ldots, (1, \boldsymbol{X}_d^{(d)})]$. Then $\lim_n |\det \widehat{\boldsymbol{Q}}(\nu^n)| = \lim_n \sup_{m\le n} |\det \widehat{\boldsymbol{Q}}(\nu^n)| = |\det \widehat{\boldsymbol{Q}}(\nu)|$, and $\sup_{m\le n} |\det \widehat{\boldsymbol{Q}}(\nu^n)|$ is increasing. Therefore, by (2.72) and monotone convergence, we get

$$\begin{aligned} \lim_n \mathrm{vol}_{d+1} \widehat{Z}(\nu^n) &= \frac{1}{(d+1)!} \lim_n \mathrm{E}(|\det \widehat{\boldsymbol{Q}}(\nu^n)|) \\ &= \frac{1}{(d+1)!} \mathrm{E}(|\det \widehat{\boldsymbol{Q}}(\mu)|) \,. \end{aligned}$$

Finally, for $\mu \in \mathcal{M}_1$,

$$\widehat{Z}(\mu) = \alpha(\mu)\ \widehat{Z}\left(\frac{1}{\alpha(\mu)}\mu\right), \quad \mathrm{vol}_{d+1} \widehat{Z}(\mu) = (\alpha(\mu))^{d+1}\ \widehat{Z}\left(\frac{1}{\alpha(\mu)}\mu\right),$$

and hence (2.71). Q.E.D.

The volume of the zonoid of μ satisfies an analogous formula; see Theorem 2.10 above. The proof of Theorem 2.10 is left to the reader because it parallels that of Theorem 2.37.

2.5 Limit theorems

This section is concerned with the asymptotic behaviour of the lift zonoid. We draw an i.i.d. sample and consider the lift zonotope of the empirical distribution. As the empirical distribution is random, the lift zonotope of the sample, the *sample lift zonotope*, comes out to be a random set, *viz.* a finite sum of random segments. The question is how the random lift zonotope behaves in relation to the lift zonoid of the underlying probability distribution when the sample length goes to infinity.

In the sequel two limit theorems are proven, a law of large numbers and a central limit theorem. They say that the lift zonotope of the sample converges to the distribution's lift zonoid almost surely and that the convergence is governed by a Gaussian process on the sphere. The results are derived from limit theorems for Banach space valued random variables and random convex sets, respectively.

Let $\boldsymbol{X}_1, \dots, \boldsymbol{X}_n$ be an i.i.d. random sample, each $\boldsymbol{X}_i$ distributed as μ, $\mu \in \mathcal{P}_1$, and denote the random empirical distribution by $\widetilde{\mu}^n$. The lift zonoid of a single random observation $\boldsymbol{X}_i$ is the random segment $[(0, \mathbf{0}), (1, \boldsymbol{X}_i)]$, $i = 1, \dots, n$, and the the lift zonoid of the random empirical distribution is the arithmetic mean of n random segments,

$$\widehat{Z}(\widetilde{\mu}^n) = \frac{1}{n} \sum_{i=1}^{n} [(0, \mathbf{0}), (1, \boldsymbol{X}_i)] . \tag{2.73}$$

It is a random lift zonotope. The subsequent law of large numbers says that this random lift zonotope converges, in the Hausdorff distance, almost surely to the lift zonoid of μ. In other words, the empirical distribution converges, in the lift zonoid distance, almost surely to the true distribution.

Theorem 2.38 (Law of large numbers) *For any $\mu \in \mathcal{P}_1$ holds:*

$$\widehat{Z}(\widetilde{\mu}^n) \stackrel{H}{\longrightarrow} \widehat{Z}(\mu) \qquad a.s.$$

Proof. The theorem is a special case of a law of large numbers for random convex compact sets; see Artstein and Vitale (1975). Here are some details:

Because of the equivalence between lift zonoids and their support functions we have to demonstrate that

$$\sup_{(p_0, \boldsymbol{p}) \in S^d} |h(\widehat{Z}(\widetilde{\mu}^n), p_0, \boldsymbol{p}) - h(\widehat{Z}(\mu), p_0, \boldsymbol{p})| \longrightarrow 0 \quad a.s. \tag{2.74}$$

holds. Denote $Y_i(p_0, \boldsymbol{p}) = h([(0, \boldsymbol{0}), (1, \boldsymbol{X}_i)], p_0, \boldsymbol{p})$ and observe that

$$\begin{aligned} h(\widehat{Z}(\widetilde{\mu}^n), p_0, \boldsymbol{p}) &= \frac{1}{n}\sum_{i=1}^{n} Y_i(p_0, \boldsymbol{p}), \\ h(\widehat{Z}(\mu), p_0, \boldsymbol{p}) &= h(\mathrm{E}([(0, \boldsymbol{0}), (1, \boldsymbol{X}_1)]), p_0, \boldsymbol{p}) \\ &= \mathrm{E}(h([(0, \boldsymbol{0}), (1, \boldsymbol{X}_1)], p_0, \boldsymbol{p})) \\ &= \mathrm{E}(Y_1(p_0, \boldsymbol{p}))\,. \end{aligned}$$

Then (2.74) becomes:

$$\sup_{(p_0, \boldsymbol{p}) \in S^d} \left| \frac{1}{n}\sum_{i=1}^{n} Y_i(p_0, \boldsymbol{p}) - \mathrm{E}(Y_1(p_0, \boldsymbol{p})) \right| \longrightarrow 0 \quad a.s. \tag{2.75}$$

The random support functions $Y_1, \ldots, Y_n$ are i.i.d. random variables in the Banach space S^{d*} which is the space of continuous functions $S^d \to \mathbb{R}$ endowed with the supremum norm. For such random variables a strong law of large numbers is satisfied (see, e.g. van der Vaart and Wellner (1996) or Mourier (1956)), that is, (2.75) holds, hence the theorem. Q.E.D.

Recall that $\mathcal{P}_2$ denotes the set of probability distributions μ having finite second moments, $\int_{\mathbb{R}^d} ||\boldsymbol{x}||^2 \mu(d\boldsymbol{x}) < \infty$ and that $H(p_0, \boldsymbol{p})$ is the halfspace $\{\boldsymbol{x} \in \mathbb{R}^d \,:\, p_0 + \langle \boldsymbol{p}, \boldsymbol{x} \rangle \geq 0\}$. Recall further that δ_{lz} is the lift zonoid distance among measures and δ_H is the Hausdorff distance among convex sets. For distributions with finite second moments a central limit theorem is satisfied as follows.

Theorem 2.39 (Central Limit Theorem) *For any $\mu \in \mathcal{P}_2$ holds:*

$$\sqrt{n}\, \delta_{lz}(\widetilde{\mu}^n, \mu) = \sqrt{n}\, \delta_H\Big(\widehat{Z}(\widetilde{\mu}^n), \widehat{Z}(\mu)\Big) \stackrel{d}{\longrightarrow} \max_{(p_0, \boldsymbol{p}) \in S^d} |\mathbb{G}\,(p_0, \boldsymbol{p})|\,, \tag{2.76}$$

where $\mathbb{G} = \{\mathbb{G}\,(p_0, \boldsymbol{p})\}_{(p_0, \boldsymbol{p}) \in S^d}$ is a zero-mean continuous Gaussian process on the sphere that has the covariance function (2.77),

$$\Gamma_{\mathbb{G}}\,((p_0, \boldsymbol{p}), (q_0, \boldsymbol{q})) = \tag{2.77}$$

$$\int_{H(p_0, \boldsymbol{p}) \cap H(q_0, \boldsymbol{q})} (p_0 + \langle \boldsymbol{x}, \boldsymbol{p} \rangle)(q_0 + \langle \boldsymbol{x}, \boldsymbol{q} \rangle)\mu(d\boldsymbol{x})$$
$$- \int_{H(p_0, \boldsymbol{p})} (p_0 + \langle \boldsymbol{x}, \boldsymbol{p} \rangle)\mu(d\boldsymbol{x}) \int_{H(q_0, \boldsymbol{q})} (q_0 + \langle \boldsymbol{x}, \boldsymbol{q} \rangle)\mu(d\boldsymbol{x})\,.$$

Proof. Denote $C_i = [(0,\mathbf{0}),(1,\boldsymbol{X}_i)]$. Then

$$\begin{aligned}
\frac{1}{n}\sum_{i=1}^{n} C_i &= \widehat{Z}(\widetilde{\mu}^n),\\
\mathrm{E}(conv(C_1)) &= \mathrm{E}(C_1) \;=\; \widehat{Z}(\mu),\\
\|C_i\| &= \|(1,\boldsymbol{X}_i)\|,\\
\mathrm{E}(\|C_i\|^2) &\le 1+\mathrm{E}(\|\boldsymbol{X}_i\|^2) \;<\; \infty .
\end{aligned}$$

From Weil (1982), Theorem 3, conclude the convergence (2.76) to a zero mean continuous Gaussian process $\Gamma_{\mathbb{G}}$ with the covariance function

$$\begin{aligned}
&\Gamma_{\mathbb{G}}(\phi,\psi) = \\
&\mathrm{E}\left\{\int_{S^d} h(C_1,p_0,\boldsymbol{p})\,d\phi(p_0,\boldsymbol{p})\cdot\int_{S^d} h(C_1,p_0,\boldsymbol{p})\,d\psi(p_0,\boldsymbol{p})\right\}\\
&-\int_{S^d} h(\widehat{Z}(\mu),p_0,\boldsymbol{p})\,d\phi(p_0,\boldsymbol{p})\cdot\int_{S^d} h(\widehat{Z}(\mu),p_0,\boldsymbol{p})\,d\psi(p_0,\boldsymbol{p}),
\end{aligned}$$

where ϕ and ψ are elements of the dual space[5] of S^{d*}, that is, elements of S^d.

As the covariance function is bilinear, it is completely determined by its values at the one-point probability measures $\delta_{(p_0,\boldsymbol{p})}$ on the sphere, $(p_0,\boldsymbol{p}) \in S^d$. Thus, we can write

$$\begin{aligned}
\Gamma_{\mathbb{G}}((p_0,\boldsymbol{p}),(q_0,\boldsymbol{q})) &= \Gamma_{\mathbb{G}}(\delta_{p_0,\boldsymbol{p}},\delta_{q_0,\boldsymbol{q}})\\
&= \mathrm{E}\{h(C_1,p_0,\boldsymbol{p})\cdot h(C_1,q_0,\boldsymbol{q})\}\\
&\quad -h(\widehat{Z}(\mu),p_0,\boldsymbol{p})\cdot h(\widehat{Z}(\mu),q_0,\boldsymbol{q}).
\end{aligned}$$

With

$$h(C_1,p_0,\boldsymbol{p}) = \begin{cases} p_0+\langle \boldsymbol{X}_1,\boldsymbol{p}\rangle & \text{if } \boldsymbol{X}_1 \in H(p_0,\boldsymbol{p}),\\ 0 & \text{otherwise,}\end{cases}$$

and $h(\widehat{Z}(\mu),p_0,\boldsymbol{p}) = \int_{H(p_0,\boldsymbol{p})}(p_0+\langle\boldsymbol{p},\boldsymbol{x}\rangle\mu(d\boldsymbol{x}),$ obtain the covariance function (2.77). Q.E.D.

---

[5]The dual space of S^{d*} is, by reflexivity, isomorphic to S^d.

2.6 Representation of measures by a functional

In this section different from the rest of the book, we extend the view from measures on the Euclidean space to measures on a general metric space. Possibilities are discussed to represent them by a functional defined on some class of functions on the metric space. We use some standard notions, for which the reader is referred to, e.g., van der Vaart and Wellner (1996). It is also possible to define the lift zonoid of a measure in this more general setting.

Let $(\mathbb{D}, d)$ be a metric space, $\mathcal{D}$ its Borel σ-algebra and $\mathcal{M}$ the set of Borel measures on $(\mathbb{D}, \mathcal{D})$. How can a probability measure $\mu \in \mathcal{M}$ be represented in a visual way that is easy to handle and useful for statistical applications? The Borel σ-algebra $\mathcal{B}$, even for $\mathbb{D} = \mathbb{R}$, is not easily structured and difficult to overview. Since $\mathcal{B}$ is not visual at all, the same applies for the graph of μ.

A measure μ on $(\mathbb{D}, \mathcal{D})$ can be represented in many ways. One approach is to consider the integrals[6]

$$\mu f = \int_{\mathbb{D}} f d\mu, \quad f \in \mathcal{F}, \tag{2.78}$$

where $\mathcal{F}$ is a given class of measurable maps $(\mathbb{D}, d) \to (\overline{\mathbb{R}}, \overline{\mathcal{B}})$. In other words, the set function $\mu : \mathcal{D} \to [0,1]$, is replaced by the functional $\mu : \mathcal{F} \to \overline{\mathbb{R}}, f \mapsto \mu f$. Clearly, if $\mathcal{F}$ consists of the indicator functions of all Borel sets, the functional is identified with the set function μ, as there is a one-to-one correspondence between $\mathcal{D}$ and $\mathcal{F}$.

Now consider a subset $\mathcal{M}_0 \subset \mathcal{M}$ of measures. In order to represent the elements of $\mathcal{M}_0$ by (2.78), a minimal requirement is uniqueness: The functional must determine the underlying measure. Uniqueness is satisfied if and only if $\mathcal{F}$ *separates* $\mathcal{M}_0$, that is, for any μ and $\nu \in \mathcal{M}_0$,

$$\mu f = \nu f \quad \text{for all } f \in \mathcal{F} \quad \Longrightarrow \quad \mu = \nu. \tag{2.79}$$

Obviously, if $\mathcal{F}$ includes all indicator functions of sets $D \in \mathcal{D}$, $\mathcal{F}$ separates any $\mathcal{M}_0 \subset \mathcal{M}$. In general, $\mathcal{F}$ must be large enough to separate $\mathcal{M}_0$, but to be useful in a sampling context (see Section 2.6.1 below), it should not be too large.

In choosing an appropriate function, class $\mathcal{F}$ several aspects are relevant. One is the analytical tractability of the representation μ, which should be

[6] If for some $f \in \mathcal{F}$ the μ-integral does not exist, define μf as an *outer integral*; see, e.g., van der Vaart and Wellner (1996, p. 6).

easier to handle than the underlying μ. Another one is visuality: The graph of the functional $f \mapsto \mu f$ or another geometric object derived from μ can be suited to visualize μ and to reveal certain properties of it. A third aspect is statistical feasibility: A representation of probability measures should work well in a sampling situation.

In a given set $\mathcal{M}_0$ of measures on $(\mathbb{D}, \mathcal{D})$, a class $\mathcal{F}$ of measurable functions induces an ordering and a metric. Define, for μ and ν in $\mathcal{M}_0$,

$$\mu \preceq_{\mathcal{F}} \nu \quad \text{if} \quad \mu f \leq \nu f \quad \text{for all } f \in \mathcal{F}.$$

The relation $\preceq_{\mathcal{F}}$ is transitive and reflexive on any $\mathcal{M}_0$. It is also antisymmetric if $\mathcal{F}$ separates $\mathcal{M}_0$. Then, among the probability measures in $\mathcal{M}_0$, $\preceq_{\mathcal{F}}$ is a stochastic order[7] and called *integral order* generated by $\mathcal{F}$ or *stochastic dominance* with respect to $\mathcal{F}$. For the theory of such orderings, see, e.g., Mosler and Scarsini (1991), Müller (1997b) and Müller and Stoyan (2002).

To construct a distance on $\mathcal{M}_0$, consider

$$\delta_{\mathcal{F}}(\mu, \nu) = \sup_{f \in \mathcal{F}} |\mu f - \nu f| .$$

We set $|\mu f - \nu f| = \infty$ if μf or νf are not finite. Obviously, $\delta_{\mathcal{F}}(\mu, \nu) \geq 0$ and

$$\delta_{\mathcal{F}}(\mu, \pi) + \delta_{\mathcal{F}}(\pi, \nu) \geq \delta_{\mathcal{F}}(\mu, \nu) ,$$

for any μ, ν and π in $\mathcal{M}_0$. Hence $\delta_{\mathcal{F}}$ is a semi-metric on $\mathcal{M}_0$. If $\mathcal{F}$ separates $\mathcal{M}_0$, $\delta_{\mathcal{F}}$ is a metric. The metric $\delta_{\mathcal{F}}$ is called an *integral measure metric.* General properties of such metrics are investigated in Müller (1997a).

Moreover, the distance $\delta_{\mathcal{F}}$ defines a norm,

$$||\mu - \nu||_{\mathcal{F}} = \delta_{\mathcal{F}}(\mu, \nu),$$

for all $\mu, \nu \in L(\mathcal{M}_0)$, which is the linear space generated by $\mathcal{M}_0$. *Convergence in $\mathcal{F}$-norm* means convergence in $L(\mathcal{M}_0)$ with respect to the norm $|| \cdot ||_{\mathcal{F}}$.

We list a number of examples of classes $\mathcal{F}$ which can serve in a representation by (2.78). Each of them generates a stochastic order and a measure metric.

Example 2.14 (Uniform order) With $\mathcal{F} = \mathcal{F}_{\mathcal{D}}$, the class of indicator functions of Borel sets in $\mathbb{R}^d$, any measure is represented by itself. The ordering $\mu \preceq_{\mathcal{F}_{\mathcal{D}}} \nu$ means the uniform order $\mu \leq \nu$, that is, $\mu(B) \leq \nu(B)$ for every $B \in \mathcal{D}$. The metric is the maximum absolute difference, $\delta_{\mathcal{F}_{\mathcal{D}}}(\mu, \nu) = \max_{B \in \mathcal{D}} |\mu(B) - \nu(B)|$.

[7] A *stochastic order* is defined as a binary relation among probability distributions that is transitive, reflexive, and antisymmetric.

Example 2.15 (Cone order) Consider an ordered Polish space, with some simplicial cone[8] C and let $\mathcal{F}_C$ contain the indicator functions of sets $\boldsymbol{y} + C$, $\boldsymbol{y} \in \mathbb{D}$,

$$\mathcal{F}_C = \{f_{\boldsymbol{y}} : f_{\boldsymbol{y}} = 1_{\boldsymbol{y}+C}\,,\ \boldsymbol{y} \in \mathbb{D}\}\,.$$

Then the representation of a measure μ,

$$\mu f_{\boldsymbol{y}} = \mu(\boldsymbol{y} + C)\,,\ \boldsymbol{y} \in \mathbb{D}\,,$$

is a generalization of the usual distribution function.

Especially, with $\mathbb{D} = \mathbb{R}^d$ and $C = \mathbb{R}^d_-$, the usual distribution function is obtained. Then we write $\mathcal{F}_C = \mathcal{F}_{lo}$. $\mathcal{F}_C$ separates $\mathcal{M}$ according to (2.79) as C has a nonempty interior. The integral order generated by $\mathcal{F}_{lo}$ is the *lower orthant order* (see Dyckerhoff and Mosler (1997)) and the $\mathcal{F}_{lo}$-norm is the Kolmogorov-Smirnov norm used in statistical tests for homogeneity and goodness-of-fit. Exchanging $S = \mathbb{R}^d_-$ against another simplicial cone $S \subset \mathbb{R}^d$ yields tests that develop power on different sets of alternatives.

Example 2.16 (Halfspaces order) Let $\mathcal{H}$ be the set of closed halfspaces in $\mathbb{R}^d$. Consider the class $\mathcal{F}_{\mathcal{H}}$ of indicator functions of halfspaces

$$\mathcal{F}_{\mathcal{H}} = \{\, 1_{H_{p_0,\boldsymbol{p}}} \,:\, (p_0, \boldsymbol{p}) \in S^d\}\,,$$

where $H_{p_0,\boldsymbol{p}} = \{\boldsymbol{x} \in \mathbb{R}^d \,:\, p_0 + \langle \boldsymbol{x}, \boldsymbol{p}\rangle \geq 0\}$. A measure μ is then represented by

$$\mu(H_{p_0,\boldsymbol{p}}) \quad = \quad \mu_{\boldsymbol{p}}([-p_0, \infty[)\,, \quad (p_0, \boldsymbol{p}) \in S^d\,.$$

$\mathcal{F}_{\mathcal{H}}$ separates the set of all measures on $\mathcal{B}^d$; this is the classic Cramér-Wold theorem. The integral order generated by $\mathcal{F}_{\mathcal{H}}$ has been investigated in Muliere and Scarsini (1989) and applied to compare random cash flows or commodity bundles when the price vector is not fixed.

Example 2.17 (Lift zonoid order) Let $\mathbb{D}$ be a linear space and $\mathbb{D}^*$ denote its dual space. $\mathbb{D}^*$ consists of all linear continuous functionals $\boldsymbol{p} : \mathbb{D} \to \mathbb{R}$, $\boldsymbol{x} \mapsto \langle \boldsymbol{x}, \boldsymbol{p}\rangle$. Consider the class of functions

$$\mathcal{F}_{lz} = \{f_{p_0,\boldsymbol{p}} \,:\, p_0 \in \mathbb{R},\ \boldsymbol{p} \in \mathbb{D}^*\}\,, \tag{2.80}$$

$$\text{with} \quad f_{p_0,\boldsymbol{p}}(\boldsymbol{x}) = \max\{0, p_0 + \langle \boldsymbol{p}, \boldsymbol{x}\rangle\}\,.$$

[8]A *simplicial cone* is a cone that has a nonempty interior and is generated by a finite number of points.

The representation by $\mathcal{F}_{lz}$ is the central notion of this book. The function

$$(p_0, \boldsymbol{p}) \mapsto \mu f_{p_0,\boldsymbol{p}} = \int_{\mathbb{D}} \max\{0, p_0 + \langle \boldsymbol{p}, \boldsymbol{x}\rangle\}\mu(d\boldsymbol{x})$$

is positively homogeneous, convex and bounded. Therefore, it is the support function of a convex compact set. This set is the lift zonoid, which has been investigated here for the particular case when $\mathbb{D} = \mathbb{D}^* = \mathbb{R}^d$. Due to Theorem 2.21, the set $\mathcal{M}_1$ is separated by $\mathcal{F}_{lz}$.

Example 2.18 (Cone restricted lift zonoid order) Let C be a closed solid cone[9] in $\mathbb{D}^*$ and consider the class of functions

$$\mathcal{F}_{lzc} = \{f_{p_0,\boldsymbol{p}} : p_0 \in \mathbb{R},\ \boldsymbol{p} \in C\},$$

where $f_{p_0,\boldsymbol{p}}$ is defined as in the previous example. This representation uses a class of functions which is smaller than the class used in the lift zonoid representation. For $\mathbb{D} = \mathbb{R}^d$, $\mathcal{F}_{lzc}$ separates $\mathcal{M}_1$ as $\mathcal{F}_{lz}$ does, since the interior of C is nonempty. For details, see Koshevoy (200xb)

Further examples can be constructed along the lines of the integral orders investigated in Mosler and Scarsini (1991) and Müller (1997a).

2.6.1 Statistical representations

The representation of a probability measure on the Euclidean space by its distribution function satisfies limit theorems that are useful in statistical estimation and testing. For the empirical distribution of a random sample holds: (1) The empirical distribution function converges almost surely uniformly to its population analogue and (2) the convergence is governed by a Brownian bridge.

To be useful in a sampling context, the representation of probability measures by a class of functions $\mathcal{F}$ should also satisfy a uniform law of large numbers and a uniform central limit theorem. Therefore, an additional requirement will be that $\mathcal{F}$ is a Glivenko-Cantelli and a Donsker class for every $\mu \in \mathcal{M}_0$, as will be explained in this section.

For a sample of independent identically distributed random elements $\boldsymbol{X}_1, \boldsymbol{X}_2, \ldots, \boldsymbol{X}_n$ on $(\mathcal{D}, \mathbb{D})$ that are distributed as μ, consider the discrete random measure

$$\mathbb{P}_n^\mu(B) = \frac{1}{n}\sum_{i=1}^n \#\{i : \boldsymbol{X}_i \in B\} = \frac{1}{n}\sum_{i=1}^n \delta_{\boldsymbol{X}_i}(B),$$

[9] A closed solid cone is a closed cone that has a nonempty interior.

where $\# S$ is the number of elements in a set S and $\delta_{\boldsymbol{a}}$ is the Dirac measure at a point $\boldsymbol{a}$. The random measure induces a random map

$$f \longmapsto \mathbb{P}_n^\mu f = \int f \, d\mathbb{P}_n^\mu = \frac{1}{n} \sum_{i=1}^{n} f(\boldsymbol{X}_i), \quad f \in \mathcal{F}.$$

Then $\{\sqrt{n}(\mathbb{P}_n^\mu - \mu) f\}_{f \in \mathcal{F}}$ is the *$\mathcal{F}$-indexed empirical process,*

$$\sqrt{n}(\mathbb{P}_n^\mu - \mu) f = \frac{1}{\sqrt{n}} \sum_{i=1}^{n} \left(f(\boldsymbol{X}_i) - \int f d\mu \right), \quad f \in \mathcal{F}.$$

In the Euclidean case with $\mathcal{F} = \mathcal{F}_{lo}$, this is the usual empirical process in $\mathbb{R}^d$.

A class $\mathcal{F}$ is *Glivenko-Cantelli* for μ if

$$||\mathbb{P}_n^\mu - \mu||_{\mathcal{F}} = \sup_{f \in \mathcal{F}} |(\mathbb{P}_n^\mu - \mu) f| \to 0 \tag{2.81}$$

holds μ-almost surely. If (2.81) holds in probability μ, $\mathcal{F}$ is called *Glivenko-Cantelli in probability.*

The classic Glivenko-Cantelli theorem says that, regarding the usual empirical process, the class $\mathcal{F}_{lo}$ is Glivenko-Cantelli for every probability measure μ on $\mathbb{R}^d$.

The class $\mathcal{F}$ is named *Donsker* for μ if all $f \in \mathcal{F}$ are square-integrable and the $\mathcal{F}$-indexed empirical process converges, in $\mathcal{F}$-norm, weakly to a Gaussian process $\mathbb{G}$, that is,

$$||\sqrt{n}\,(\mathbb{P}_n^\mu - \mu) - \mathbb{G}\,||_{\mathcal{F}} \xrightarrow{w} 0. \tag{2.82}$$

Then the process $\mathbb{G}$ has zero mean and the covariance function amounts to

$$\mathrm{Cov}(\mathbb{G}\, f_1, \mathbb{G}\, f_2) = \mu((f_1 - \mu f_1)(f_2 - \mu f_2)) = \mu(f_1 \cdot f_2) - \mu f_1 \cdot \mu f_2. \tag{2.83}$$

With regard to the usual empirical process, the class $\mathcal{F}_{lo}$ is also Donsker for every $\mu \in \mathcal{P}$ which is square integrable.

Definition 2.7 (Statistical representation) *Let $\mathcal{M}_0$ be a set of Borel probability measures on $(\mathbb{D}, \mathcal{D})$. A class $\mathcal{F}$ of measurable maps $(\mathbb{D}, \mathcal{D}) \to (\overline{\mathbb{R}}, \overline{\mathcal{B}})$ is a* statistical representation *of $\mathcal{M}_0$ if*

(i) *$\mathcal{F}$ separates $\mathcal{M}_0$,*

(ii) *$\mathcal{F}$ is Glivenko-Cantelli for all $\mu \in \mathcal{M}_0$,*

(iii) *$\mathcal{F}$ is Donsker for those $\mu \in \mathcal{M}_0$ which are square integrable.*

The Definition 2.7 has been introduced in Koshevoy (200xb), where it is called the *representation* of a probability measure and where more details can be found.

Remark. To separate $\mathcal{M}_0$, $\mathcal{F}$ must be large enough, to be Glivenko-Cantelli and Donsker, it must be rather small. Any finite class $\mathcal{F}$ of square integrable functions is Donsker. For a detailed treatment of Glivenko-Cantelli and Donsker classes, the reader is referred to van der Vaart and Wellner (1996).

The subsequent discussion will be restricted to examples of measures on $\mathbb{R}^d$. However, many notions and results can be easily extended to a more general setting.

Examples

1. The class $\mathcal{F}_\mathcal{D}$ in Example 2.14 is no statistical representation since it is not Donsker.

2. The class $\mathcal{F}_{lo}$ in Example 2.15 is a statistical representation of all probability measures on $\mathbb{R}^d$. We reformulate this as follows:

   (a) A distribution function determines the underlying measure uniquely.

   (b)
   $$||\mathbb{P}_n^\mu - \mu||_{\mathcal{F}_{lo}} = \sup_{y \in \mathbb{R}^d} |\widehat{F}_n(y) - F_\mu(y)| \to 0 \qquad (2.84)$$
   holds μ-a.s., where F_μ is the distribution function of μ and $\widehat{F}_n$ is the empirical distribution function of a random sample from μ.

   (c) There holds (2.82) and the covariance function of the limiting Gaussian process is
   $$(x, y) \longmapsto \int_{(x \wedge y) + \mathbb{R}^d_-} d\mu - \int_{x + \mathbb{R}^d_-} d\mu \int_{y + \mathbb{R}^d_-} d\mu, \quad x, y \in \mathbb{R}^d,$$
   where $x \wedge y = (\min\{x_1, y_1\}, \ldots, \min\{x_d, y_d\})$.

3. Similarly, with any simplicial cone $C \subset \mathbb{R}^d$, the class $\mathcal{F}_C$ in Example 2.15 is a statistical representation of all Borel measures on $\mathbb{R}^d$.

4. The class $\mathcal{F}_\mathcal{H}$ in Example 2.16 is a statistical representation of all Borel measures on $\mathbb{R}^d$; see Koshevoy (200xb).

An important further example is the family $\mathcal{F}_{lz}$, which is discussed in the following section.

2.6.2 Lift zonoids and the empirical process

$\mathcal{F}_{lz}$ in Example 2.17 is a statistical representation of those Borel measures on $\mathbb{R}^d$ for which $\int_{\mathbb{R}^d} \|x\| \mu(dx)$ is finite. The class $\mathcal{F}_{lz}$ leads to the lift zonoid of a probability measure on $\mathbb{R}^d$. A probability measure $\mu \in \mathcal{P}_1$ is represented by the functional $f \mapsto \int f d\mu$, $f \in \mathcal{F}_{lz}$. Since $\mathcal{F}_{lz}$ is parameterized by elements of the sphere S^d, the functional representing μ can be identified with a function on the sphere. Moreover, this function extends to a convex and positively homogeneous function $\mathbb{R}^{d+1} \to \mathbb{R}$, which is the support function of a convex compact set in $\mathbb{R}^{d+1}$. As there is a one-to-one correspondence between convex compact sets in $\mathbb{R}^{d+1}$ and their support functions, every $\mu \in \mathcal{P}_1$ is represented by a convex compact in $\mathbb{R}^{d+1}$. This set is the lift zonoid of μ.

Theorem 2.40 (Lift zonoid representation) *For a probability measure $\mu \in \mathcal{P}_1$ holds:*
(i) $\widehat{Z}(\mu)$ *uniquely defines* μ,

(ii) $\widehat{Z}(\mathbb{P}_n^\mu) = \frac{1}{n}\sum_{i=1}^n [(0,\mathbf{0}),(1,\boldsymbol{X}_i)] \xrightarrow{H} \widehat{Z}(\mu) \qquad \mu\text{-}a.s.,$

(iii) $\sqrt{n}\ \delta_H\Big(\widehat{Z}(\mathbb{P}_n^\mu), \widehat{Z}(\mu)\Big) \xrightarrow{w} \max_{(p_0,\boldsymbol{p})\in S^d} |\mathbb{G} f_{p_0,\boldsymbol{p}}|,$

where $\mathbb{G} = \{\mathbb{G} f_{p_0,\boldsymbol{p}}\}_{(p_0,\boldsymbol{p})\in S^d}$ *is a zero-mean continuous Gaussian process on the sphere that has the covariance function* (2.77).

Proof. The theorem is a reformulation of Theorems 2.38 and 2.39. Q.E.D.

The class of functions $\mathcal{F}_{lzc}$ in Example 2.18 with $\mathbb{D} = \mathbb{R}^d$ is a statistical representation of all Borel measures on $\mathbb{R}^d$ that have finite first moments. The fact that the class is Donsker and Glivenko-Cantelli follows from the previous example since $\mathcal{F}_{lzc} \subset \mathcal{F}_{lz}$.

2.7 Notes

Zonotopes and zonoids have been used in various contexts in analysis and probability theory. Surveys of their properties are given in Choquet (1968), Bolker (1969), Schneider and Weil (1983), and Goodey and Weil (1993). Among other applications in probability and statistics, zonoids are widely used to describe random sets and point processes.

The definition of the lift zonoid and many of the above results have been first published in Koshevoy and Mosler (1998). In the original paper a flaw is found on page 382: $H_{\alpha(F)}$ should be H_1.

The notion of the statistical representation and most of the examples in Section 2.6 are taken from Koshevoy (200xb).

Dall'Aglio and Scarsini (2000a) provide a representation of the lift zonoid of a given measure as the range of a *nonatomic* vector measure; this simplifies some of the above proofs.

Every zonoid in $\mathbb{R}^d$ is, modulo a shift, the zonoid of a measure on $\mathbb{R}^d$. The zonoid of a measure has properties analogous to those shown for the lift zonoid in Theorem 2.19. The zonoid $Z(\mu)$ is also continuous on μ. But as different measures can have the same zonoid, the convergence of zonoids, in general, cannot imply the convergence of the underlying measures. The implication, however, is true for even measures (that is μ with $\mu(-B) = \mu(B)$ for all B) on the sphere; see Goodey and Weil (1993).

Concerning approximations in the lift zonoid order, a distribution in general can neither be approximated from above by discrete distributions nor from below by continuous distributions. As any zonoid, a lift zonoid can be approximated from outside by zonotopes; but these zonotopes need not be lift zonotopes. In fact, every lift zonoid has the origin as boundary point. If the origin is a smooth boundary point (that is, has a unique tangent hyperplane in $\mathbb{R}^{d+1}$) the lift zonoid cannot be approximated by lift zonotopes from outside. On the other hand, if a zonoid has no interior points it cannot be approximated from inside by lift zonoids of continuous distributions.

If a distribution has compact support, its lift zonoid contains the origin as a non-smooth boundary point. Then the lift zonoid can be approximated from outside by lift zonotopes. That means the distribution can be approximated in the lift zonoid order from above by discrete distributions. Also, if a given univariate distribution is not Dirac, the lift zonoid has a non-empty interior and can be approximated by lift zonoids of continuous distributions from inside.

3

Central regions

An important task of data analysis consists in identifying a part of the data which represents the typical features of the data generating process, while the remaining data points are seen as less typical or less probable regarding their hypothesized distribution. Often the data points are assumed to vary around a center and a *central region* is sought which includes the center and reflects the location and general shape of the data. Such a central region can also separate one or more outlying data points from the main body of the distribution. The border of a central region, called *contour set*, serves as a multivariate quantile.

In this way several central regions can be built, having different degrees of centrality. They form a nested family of sets around the center.

Central regions are defined for a probability measure as well. Since any data set can be regarded as an empirical probability measure, we shall stay with the more general case and investigate central regions of probability measures in the sequel.

For example, if the measure has a Lebesgue density, each density level set forms a central region, and the center is the mode. In general, a family of central regions can be seen as an analogue to the level sets of a density. Their position in space is to depict the location of the measure, their shape its dependence structure, and their closeness its dispersion. Central regions and density level sets have several aspects in common but others not. First, like density level sets, central regions have decreasing probability content since, by definition, they are nested; but, unlike density level sets, they have no maximum probability content per volume. Second, if a distribution has more than a single mode, this is immediately seen from its density level sets. On the other hand, most notions of central regions imply that each region is a connected set; those regions cannot reflect an eventual multimodality of the distribution and their use is essentially limited to the analysis of unimodal distributions. Under the assumption of a convex unimodal distribution[1], a family of central regions can be regarded as an estimate of the density level sets, and properties like consistency can be established.

Further, the family of all level sets of a density determines the probability measure uniquely. A natural postulate for central regions is that they should characterize the probability measure in the same way.

If the distribution is univariate and unimodal, a reasonable central region must be an interval. Take for example the intervals between two symmetric quantiles; they determine the measure uniquely and the smallest central region is a singleton, the usual median.

Tukey (1975) and Eddy (1985) suggested a multivariate analogue of the quantile function. Based on such quantiles, concepts of multivariate central regions have been introduced and discussed by Nolan (1992), Averous and Meste (1997a), Massé and Theodorescu (1994), Chaudhuri (1996) and Koltchinskii (1997). Also, related notions of multivariate location, skewness and kurtosis have been investigated by Averous and Meste (1990, 1994, 1997b).

The central regions are generalizations of univariate interquantile ranges and centered around certain multivariate medians. A simple early proposal is based on the Mahalanobis distance, hence on the first two moments of the measure: A central regions is an ellipsoid around the mean, consisting of all points having Mahalanobis distance from the mean that does not exceed a given value. Other central regions are defined by so called data depths, e.g., the halfspace depth or the Oja depth; see Section 4.2. The smallest central region defines a multivariate median: If the region is a singleton, like in the Mahalanobis case, the median is equal to this single point. If the smallest region contains more than one point, like the regions based on halfspace depth do, the median is usually defined as a proper gravity center of it. For surveys of multivariate medians and their relation to central regions, see Small (1990) and Niinimaa and Oja (1999).

In this chapter another concept of multivariate central regions is developed. They are based on the lift zonoid and therefore called *zonoid trimmed regions*. The zonoid trimmed regions are convex sets centered around the mean. They have nice mathematical properties, which are studied in detail. We also consider natural estimators of the trimmed regions and investigate

[1] A distribution is convex unimodal (Dharmadhikari and Joag-Dev, 1988) if it has a continuous L-density and the density level sets are convex.

their asymptotic behaviour. The zonoid trimmed regions give rise to a new concept of depth related to the mean. This will be the topic of the next chapter.

In the following Sections 3.1 and 3.2 we define zonoid trimmed regions for a probability measure, point out their relation with the lift zonoid, and demonstrate many useful geometric properties such as convexity, monotonicity, and affine equivariance. The largest region equals the measure's support while the smallest region contains the mean as single element. For empirical and for L–continuous measures the definition is particularly simple and well interpretable. The zonoid trimmed region $D_\alpha(\mu)$ of an L–continuous measure μ consists of the gravity centers (with respect to μ) of all sets having μ-measure α. Next, in Section 3.3 we discuss the trimming of univariate measures and contrast the zonoid trimming intervals with interquantile regions and with intervals bounded by expectiles. Some examples of multivariate zonoid trimming follow in Section 3.4. Then alternative notions of central regions and their shape in families of elliptical distributions are investigated (Section 3.5). In Section 3.6 two continuity theorems are established, stating that $D_\alpha(\mu)$ is continuous on α as well as on μ. Further a law of large numbers for zonoid trimmed regions is derived. Section 3.7 exhibits additional properties and formulae: strengthened uniqueness and monotonicity results and a formula for trimmed regions of a mixture. Section 3.8 contains a formula for the trimmed regions of an empirical measure and, as a corollary, a law of large numbers for order statistics. Remarks on the numerical computation of zonoid trimmed regions close the chapter (Section 3.9).

3.1 Zonoid trimmed regions

Let $\mathcal{P}_1$ denote the set of probability measures on $(\mathbb{R}^d, \mathcal{B}^d)$ that have finite first moments. Given a probability measure in $\mathcal{P}_1$ we define a family of central regions as follows.

Definition 3.1 (Zonoid trimmed regions) *For* $\mu \in \mathcal{P}_1$, $0 < \alpha \leq 1$, *call*

$$D_\alpha(\mu) = \left\{ \int_{\mathbb{R}^d} x\, g(x)\, \mu(dx) \ : \quad g : \mathbb{R}^d \to \left[0, \frac{1}{\alpha}\right] \text{ measurable and } \int_{\mathbb{R}^d} g(x)\, \mu(dx) = 1 \right\} \tag{3.1}$$

the zonoid α-trimmed region *of* μ. *For* $\alpha = 0$ *define*

$$D_0(\mu) = \mathbb{R}^d . \tag{3.2}$$

From the definition we see that the family of zoonoid trimmed regions is nested and the minimum, at $\alpha = 1$, contains just one point, the expectation of μ.

Theorem 3.1 (Monotonicity) *For $\mu \in \mathcal{P}_1$ holds:*

(i) $D_{\alpha_2}(\mu) \subset D_{\alpha_1}(\mu)$ *if* $0 \leq \alpha_1 < \alpha_2 \leq 1$,

(ii) $D_1(\mu) = \{\epsilon(\mu)\}$.

The following proposition provides the relation between an α-trimmed region and the lift zonoid and justifies thus the name *zonoid trimming*. Denote the hyperplane at $z_0 = \alpha$ by $G_\alpha = \{(z_0, z_1, \dots, z_d) : z_0 = \alpha\}$. Recall from (2.48) and (2.49) that $\widehat{Z}(\mu, \alpha) = \widehat{Z}(\mu) \cap G_\alpha$ and

$$\widehat{Z}_\alpha(\mu) = pr_{\{1,\dots,d\}}(\widehat{Z}(\mu, \alpha)),$$

which is the projection to the last d coordinates. From the definition (2.24) of the lift-zonoid obtain:

Proposition 3.2 (Relation to lift zonoid) *Let* $0 < \alpha \leq 1$. *Then*

$$D_\alpha(\mu) = \frac{1}{\alpha} \widehat{Z}_\alpha(\mu). \tag{3.3}$$

For an illustration of Proposition 3.2, see Example 3.1 and Figure 3.1 below.

In many statistical applications *affine equivariance* is a natural postulate on the statistical procedure. Consider a d-variate statistic like an empirical median or a central set. Affine equivariance means that, if the data points are subject to an affine transformation of $\mathbb{R}^d$ (in particular, are shifted, rescaled and/or rotated), the statistic transforms in the same way. The following theorem states that the zonoid trimmed regions are affine equivariant. This property is useful in many respects. Especially, the zonoid trimmed regions of a marginal distribution are easily determined from the trimmed regions of the joint distribution.

Theorem 3.3 (Affine equivariance) *Let $T_{\boldsymbol{A},\mathbf{c}} : \boldsymbol{x} \mapsto \boldsymbol{x}\boldsymbol{A} + \mathbf{c}$ with some $d \times k$ matrix $\boldsymbol{A}$ and $\mathbf{c} \in \mathbb{R}^k$, and let $\mu^{A,c} = \mu \circ T_{\boldsymbol{A},\mathbf{c}}^{-1}$ be the image measure of μ under $T_{\boldsymbol{A},\mathbf{c}}$. Then*

$$D_\alpha(\mu^{A,c}) = D_\alpha(\mu)\boldsymbol{A} + \mathbf{c} = \{\boldsymbol{x}\boldsymbol{A} + \mathbf{c} : \boldsymbol{x} \in D_\alpha(\mu)\}. \tag{3.4}$$

Proof. The result is immediately derived from Proposition 3.2 and Corollary 2.27. Q.E.D.

Corollary 3.4 (Marginals) *The zonoid trimmed regions of a marginal distribution of μ equal the proper projections of the zonoid trimmed regions of μ.*

For an empirical measure the definition of zonoid trimmed regions simplifies as follows: If μ is an empirical measure on the points $\boldsymbol{x}_1, \boldsymbol{x}_2, \ldots, \boldsymbol{x}_n \in \mathbb{R}^d$, we write $D_\alpha(\mu) = D_\alpha(\boldsymbol{x}_1, \ldots, \boldsymbol{x}_n)$. Setting $\lambda_i = g(\boldsymbol{x}_i)/n$ in the general definition of $D_\alpha(\mu)$, $0 < \alpha \leq 1$, we obtain

$$D_\alpha(\mu) = \left\{ \sum_{i=1}^{n} \lambda_i \boldsymbol{x}_i : \sum_{i=1}^{n} \lambda_i = 1,\ 0 \leq \lambda_i,\ \lambda_i \leq \frac{1}{\alpha} \text{ for all } i \right\}. \tag{3.5}$$

Note that for $0 < \alpha \leq \frac{1}{n}$, D_α in (3.5) equals the convex hull of the data, that is the convex hull, $conv(\mu)$, of the support of μ. It follows also that

$$D_*(\mu) \equiv \bigcup_{0<\alpha\leq 1} D_\alpha(\mu) = conv(\mu)\,. \tag{3.6}$$

If, for example, μ is the univariate empirical measure on the points 0 and 1, the zonoid trimmed regions are intervals:

$$D_\alpha(\mu) = \begin{cases} [0,1] & \text{if } \alpha \in \left]0, \frac{1}{2}\right]\,, \\ \left[1 - \frac{1}{2\alpha}, \frac{1}{2\alpha}\right] & \text{if } \alpha \in \left[\frac{1}{2}, 1\right]\,. \end{cases}$$

Further examples are postponed to Section 3.4.

Let $\mathcal{P}_{cont}$ denote the subclass of, with respect to Lebesgue measure, continuous distributions in $\mathcal{P}_1$. For $\mu \in \mathcal{P}_1$ and any Borel set U having positive μ–measure, consider the *μ-gravity center* of U,

$$\epsilon(\mu|U) = \frac{1}{\mu(U)} \int_U x\mu(dx)\,.$$

The point $\epsilon(\mu|U)$ is also called the *U-centroid* of μ.

When we apply our trimming notion to an L-continuous distribution we obtain the following simple and well interpretable result.

Theorem 3.5 (Sets of gravity centers) *Let* $\mu \in \mathcal{P}_{cont}$ *and* $\alpha > 0$. *Then*

(i) $D_\alpha(\mu) = \{\epsilon(\mu|U) \, : \, U \in \mathcal{B}^d, \, \mu(U) = \alpha\}$,

(ii) $D_\alpha(\mu) = conv\{\epsilon(\mu|H) \, : \, H$ *closed halfspace in* $\mathbb{R}^d$, $\mu(H) = \alpha\}$.

The theorem states that, for $0 < \alpha \leq 1$, the α-trimmed region of an L–continuous measure is equal to the set of U-centroids where U has probability α. In other words, a point $\boldsymbol{y}$ belongs to $D_\alpha(\mu)$ if and only if there is a Borel set having probability α and gravity center $\boldsymbol{y}$. Observe that, in the continuous distribution case, for every α there is some U with $\mu(U) = \alpha$. Hence, given α, the set of centroids is nonempty. Equivalently, $D_\alpha(\mu)$ is the convex hull of U-centroids where U is a closed halfspace having probability α.

Proof. (i): In the case of an L–continuous μ, extreme points of the lift zonoid are given by Theorem 2.4. With respect to (3.3), we see that every point of $D_\alpha(\mu)$ has the form $(\mu(U))^{-1} \int_U \boldsymbol{x}\mu(d\boldsymbol{x})$ with some U such that $\mu(U) = \alpha$. This proves (i).
(ii): By Theorem 2.4, extreme points of $\widehat{Z}(\mu)$ are generated by indicator functions of halfspaces. Due to the continuity of μ, for every $\alpha > 0$ and every direction in $\mathbb{R}^d$, there is a halfspace whose normal vector equals the chosen direction and which has μ-probability α. Extreme points of a cut at $z_0 = \alpha$ remain extreme points of its projection on the last d coordinates. That yields the proof of (ii). Q.E.D.

3.2 Properties of zonoid trimmed regions

Many properties of zonoid trimmed regions can, by the help of Proposition 3.2, be traced back to properties of the lift zonoid. The first property we derive is a uniqueness result: The family $\{D_\alpha(\mu)\}_{\alpha \in]0,1]}$ uniquely determines μ in $\mathcal{P}_1$.

Theorem 3.6 (Uniqueness) *Let* $\mu, \nu \in \mathcal{P}_1$, *and* $D_\alpha(\mu) = D_\alpha(\nu)$ *for all* $0 < \alpha \leq 1$. *Then* $\mu = \nu$.

Proof. From $D_\alpha(\mu) = D_\alpha(\nu)$ follows that $\widehat{Z}_\alpha(\mu) = \widehat{Z}_\alpha(\nu)$, $0 < \alpha \leq 1$. Further, there holds $\widehat{Z}_0(\mu) = \widehat{Z}_0(\nu) = \{\mathbf{0}\}$. Since $\widehat{Z}(\mu)$ and $\widehat{Z}(\nu) \subset \{(z_0, \boldsymbol{z}) \in \mathbb{R}^{d+1} : 0 \leq z_0 \leq 1\}$ we obtain that $\widehat{Z}(\mu) = \widehat{Z}(\nu)$ and, by Theorem 2.21, that $\mu = \nu$. Q.E.D.

Several geometric properties of zonoid trimmed regions are collected in the following theorem.

Theorem 3.7 (Convexity, compactness, symmetry) *Let* $\mu \in \mathcal{P}_1$.

(i) $D_\alpha(\mu)$ *is convex and compact if* $0 < \alpha \leq 1$.

(ii) *If the support of* μ *is in* $\mathbb{R}^d_+$, *then* $D_\alpha(\mu)$ *is contained in the d-dimensional rectangle between* $\mathbf{0}$ *and* $\frac{1}{\alpha}\epsilon(\mu)$.

(iii) $D_{\frac{1}{2}}(\mu)$ *is centrally symmetric with center* $\epsilon(\mu)$.

Proof. This follows from Propositions 3.2 and 2.15. Q.E.D.

Examples of zonoid trimmed regions will be given in Sections 3.3 and 3.4.

3.3 Univariate central regions

In this section the zonoid trimmed regions of a univariate measure are discussed and contrasted with central regions which rest on quantiles or expectiles. First we explain the relation between the zonoid trimmed regions and the Lorenz curve.

Let μ be a univariate probability measure with finite expectation. Then $\widehat{Z}(\mu)$ is the region between the generalized Lorenz curve and its dual, that is, the convex hull of the following points in $\mathbb{R}^2$ (see Theorem 2.17)

$$\left(t, \int_0^t Q_\mu(\beta)d\beta\right), \qquad 0 \leq t \leq 1, \quad \text{and}$$
$$\left(t, \int_{1-t}^1 Q_\mu(\beta)d\beta\right), \qquad 0 \leq t \leq 1,$$

where $Q_\mu(\beta) = \min\{x \in \mathbb{R} : F_\mu(x) \geq \beta\}$ signifies the β-quantile of μ, $0 < \beta \leq 1$. We use the convention that $\min \emptyset = \infty$. Thus, the zonoid trimmed region of a univariate distribution is

$$D_\alpha(\mu) = \left[\frac{1}{\alpha}\int_0^\alpha Q_\mu(\beta)d\beta, \frac{1}{\alpha}\int_{1-\alpha}^1 Q_\mu(\beta)d\beta\right]. \tag{3.7}$$

The lower bound of the interval is the average of all quantiles below the α-quantile, and the upper bound is the average of all quantiles above the $(1-\alpha)$-quantile. Thus, the univariate zonoid trimming comes out to be an *average quantile trimming*.

Example 3.1 (Univariate distribution type) Let $\mu_{a,b}$ belong to a distribution type as defined in Section 2.2.3 and have distribution function $F_{a,b}$.

The α-trimmed region of $\mu_{a,b}$ is given by

$$D_\alpha(\mu_{a,b}) = \tag{3.8}$$

$$\left\{x \in \mathbb{R} : \frac{1}{\alpha}E_{0,1}(Q_{0,1}(\alpha)) \leq \frac{x-a}{b} \leq \frac{1}{\alpha}\left[\epsilon(\mu_{0,1}) - E_{0,1}(Q_{0,1}(1-\alpha))\right], \right\},$$

$0 < \alpha \leq 1$. For the definitions of $F_{0,1}$, $Q_{0,1}$ and $E_{0,1}$, see Equations (2.38) and (2.39). This follows from Corollary 2.18. In particular, for the univariate normal (cf. Example 2.9),

$$D_\alpha(N(a,b^2)) = \left\{x \in \mathbb{R} : \left(\frac{x-a}{b}\right)^2 \leq \frac{1}{2\pi\alpha^2}e^{-u_\alpha^2}\right\}. \tag{3.9}$$

E.g., we choose $\alpha = 0.25$, $\alpha = 0.5$, and $\alpha = 0.75$ and calculate D_α for a univariate $N(1,1)$ distribution. With the standard normal quantiles $u_{0.75} = -u_{0.25} = 0.675$ and $u_{0.5} = 0$ we obtain the following intervals:

$$\begin{aligned} D_{0.25}(N(1,1)) &= [-0.2707, 2.2707], \\ D_{0.5}(N(1,1)) &= [0.2021, 1.7979], \\ D_{0.75}(N(1,1)) &= [0.5764, 1.4236]. \end{aligned}$$

They are depicted in Figure 3.1, which shows also the lift zonoid of $N(1,1)$. From Proposition 3.2 is known that the α-trimmed region relates to the α-cut of the lift zonoid. E.g., for $\alpha = 0.5$ the α-cut is a horizontal segment,

$$\widetilde{Z}(N(1,1), 0.5) = \{(0.5, x_1) : 0.1011 \leq x_1 0.8989\},$$

and the trimmed region is $1/\alpha$ times the interval $[0.1011, 0.8989]$.

Alternatively the interval between two symmetric quantiles can be considered as a central region. The *interquantile region* is the interval between the quantiles at α and $1-\alpha$, for some $\alpha \in [0, \frac{1}{2}]$,

$$D_\alpha^{qu}(\mu) = [Q_\mu(\alpha), Q_\mu(1-\alpha)].$$

$D_{0.5}^{qu}$ is the singleton whose only element is the median. Therefore the interquantile regions center around the median.

Newey and Powell (1987) define the *expectiles* of a random variable. An expectile relates to the mean in a similar way as a quantile relates to the median; see also Jones (1994) and Abdous and Remillard (1995). Intervals between expectiles play the same role with respect to the mean as interquantile intervals do with respect to the median. Such interexpectile intervals

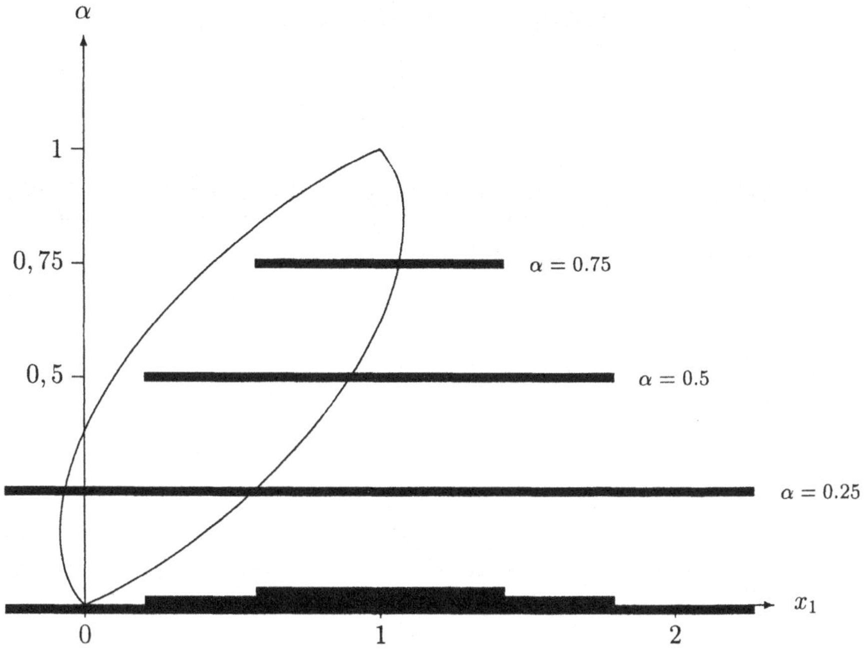

FIGURE 3.1: Lift zonoid and α-trimmed region of a univariate normal, $\mu = N(1,1)$, at $\alpha = 0.25$, $\alpha = 0.5$ and $\alpha = 0.75$.

have been studied in a more general setting by Averous and Meste (1990) and Breckling and Chambers (1988).

Newey and Powell (1987) define the τ-expectile of μ as the solution $\xi = \xi(\tau)$ of the equation

$$\frac{\tau}{1-\tau} = \frac{\int_{-\infty}^{\xi}(\xi - x)\mu(dx)}{\int_{\xi}^{\infty}(x-\xi)\mu(dx)}, \quad 0 < \tau < 1. \tag{3.10}$$

If, for example, μ is the empirical measure on the points 0 and 1, the interquantile intervals are $D_\alpha^{qu}(\mu) = [0,1]$ for $\alpha \in]\frac{1}{2}, 1]$, the interexpectile intervals are $[\tau, 1-\tau]$ for $0 < \tau \leq \frac{1}{2}$, and the zonoid trimmed intervals are

$$D_\alpha(\mu) = \begin{cases} [0,1] & \text{if } \alpha \in \left]0, \frac{1}{2}\right], \\ \left[1 - \frac{1}{2\alpha}, \frac{1}{2\alpha}\right] & \text{if } \alpha \in \left[\frac{1}{2}, 1\right]. \end{cases}$$

If μ is the continuous uniform distribution on $[0,1]$, obtain

$$D_\alpha(\mu) = D_{\frac{\alpha}{2}}^{qu}(\mu) = \left[\frac{\alpha}{2}, 1 - \frac{\alpha}{2}\right]$$

if $0 < \alpha \leq 1$, while the interexpectile intervals are of the form

$$\left[\frac{\sqrt{\tau}}{\sqrt{\tau}+\sqrt{1-\tau}}, 1 - \frac{\sqrt{\tau}}{\sqrt{\tau}+\sqrt{1-\tau}}\right].$$

The interquantile region $D_\alpha^{qu}(\mu)$ of a given univariate distribution μ is easily visualized. It equals the closure of the difference of two level sets of the distribution function; see Figure 3.2.

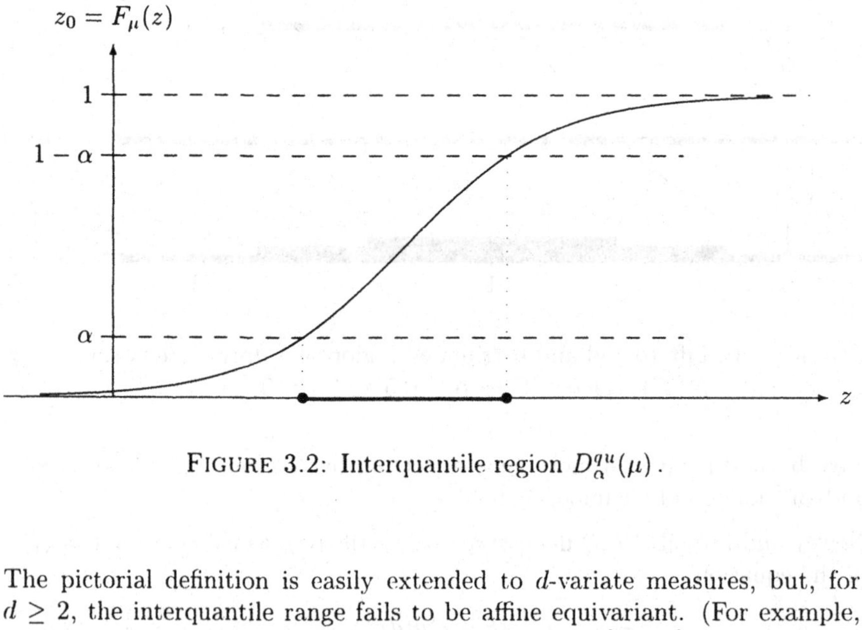

FIGURE 3.2: Interquantile region $D_\alpha^{qu}(\mu)$.

The pictorial definition is easily extended to d-variate measures, but, for $d \geq 2$, the interquantile range fails to be affine equivariant. (For example, consider a uniform distribution on the square $[-1,1]^2 \subset \mathbb{R}^2$ and transform it by $\boldsymbol{x} \mapsto -\boldsymbol{x}$. The distribution is invariant against the transform, but the central region is not.)

3.4 Examples of multivariate zonoid trimmed regions

In this section the multivariate notion of zonoid trimmed regions is illustrated by simple examples, and zonoid central regions are given for important special distributions.

Example 3.2 (Two-point distribution in $\mathbb{R}^d$) As a first example consider a two-point distribution in $\mathbb{R}^d$. Assume that $\mu(\{\boldsymbol{x}\}) = p$ and $\mu(\{\boldsymbol{y}\}) = 1 - p$ for two given points $\boldsymbol{x}, \boldsymbol{y} \in \mathbb{R}^d$ and some $0 \le p \le \frac{1}{2}$. Then, by (3.6), $D_*(\mu)$ is the segment $[\boldsymbol{x}, \boldsymbol{y}]$ between $\boldsymbol{x}$ and $\boldsymbol{y}$, and $D_1 = \{\boldsymbol{\epsilon}(\mu)\} = \{p\boldsymbol{x} + (1-p)\boldsymbol{y}\}$. In view of the affine equivariance (Theorem 3.3) it is convenient to map the segment onto the unit interval in $\mathbb{R}$ and determine the central regions of a univariate two-point distribution $\tilde{\mu}$ supported by 0 and 1, with $\tilde{\mu}(\{0\}) = p$. Then $Q_{\tilde{\mu}}(\beta) = 0$ if $0 < \beta \le p$, $Q_{\tilde{\mu}}(\beta) = 1$ if $p < \beta \le 1$. Equation (3.7) yields

$$D_\alpha(\tilde{\mu}) = \begin{cases} [0,1] & \text{if } 0 < \alpha \le p\,, \\ \left[\frac{\alpha - p}{\alpha}, 1\right] & \text{if } p < \alpha < 1 - p\,, \\ \left[\frac{\alpha - p}{\alpha}, \frac{1-p}{\alpha}\right] & \text{if } 1 - p \le \alpha \le 1\,. \end{cases}$$

Transforming back, we obtain the zonoid trimmed regions of μ:

$$D_\alpha(\mu) = \begin{cases} [\boldsymbol{x}, \boldsymbol{y}] & \text{if } 0 < \alpha \le p\,, \\ \left[\boldsymbol{x} + \frac{\alpha - p}{\alpha}(\boldsymbol{y} - \boldsymbol{x}), \boldsymbol{y}\right] & \text{if } p < \alpha < 1 - p\,, \\ \left[\boldsymbol{x} + \frac{\alpha - p}{\alpha}(\boldsymbol{y} - \boldsymbol{x}), \boldsymbol{x} + \frac{1-p}{\alpha}(\boldsymbol{y} - \boldsymbol{x})\right] & \text{if } 1 - p \le \alpha \le 1\,. \end{cases} \tag{3.11}$$

Example 3.3 (Empirical distribution in $\mathbb{R}^2$) For a given empirical distribution in $\mathbb{R}^d$, the zonoid trimmed regions can, in principle, be calculated on the basis of their defining equation (3.5). For details of the computation see Section 4.5. Figure 1.9 in chapter 1 exhibits, for different values of α, the trimmed regions of an empirical measure on ten points in the plane.

The next three examples treat trimmed regions of elliptical distributions. For a spherical or elliptical distribution on $\mathbb{R}^d$, the α-trimmed regions are balls or ellipsoids, respectively. As for any $\alpha > 0$ an α-trimmed region equals the α-slice of the lift zonoid multiplied by $1/\alpha$, the regions can be calculated from Propositions 2.28 and 2.29.

Example 3.4 (Multivariate normal distribution) Let μ be the multivariate standard normal distribution, $\mu = N(\mathbf{0}, I)$. Then (see Example 2.10 in Section 2.3.5 above)

$$D_\alpha(N(\mathbf{0}, I)) = B\left(\frac{r_1(\alpha)}{\alpha}\right) \quad \text{with} \quad r_1(\alpha) = \frac{1}{\sqrt{2\pi}} \exp\left(-\frac{u_\alpha^2}{2}\right), \quad \alpha \in]0,1]\,.$$

Again, u_α denotes the α-quantile of the univariate standard normal distribution. In view of Theorem 3.3 the zonoid trimmed regions of a general multivariate normal $N(\boldsymbol{a}, \Sigma)$, where Σ is positive semidefinite, are given by

$$D_\alpha(\mathrm{N}(\boldsymbol{a}, \Sigma)) = \left\{\boldsymbol{x} : (\boldsymbol{x} - \boldsymbol{a})\Sigma^{-1}(\boldsymbol{x} - \boldsymbol{a})' \le \left(\frac{r_1(\alpha)}{\alpha}\right)^2\right\}. \tag{3.12}$$

Figure 3.3 exhibits zonoid regions for $n = 500$ and $n = 50$ data that have been simulated from a bivariate normal distribution with standard normal marginals and correlation $\rho = 0.5$.

Example 3.5 (Uniform distribution on the unit sphere) Let μ be the uniform distribution on the boundary of the unit ball in $\mathbb{R}^d$. Then, for any α, $0 < \alpha \leq 1$, the α-trimmed region is obtained from the α-slice by multiplication with $1/\alpha$. The α-slice is a ball which has already been calculated in Section 2.3.5. Thus,

$$D_\alpha(\mu) = B\left(\frac{r_2(\alpha)}{\alpha}\right)$$

with $r_2(\alpha)$ as in Example 2.11. Especially, if μ is uniform on the spheres in $\mathbb{R}^2$ or $\mathbb{R}^3$,

$$r_2(\alpha) = \begin{cases} \frac{\sin \alpha\, \pi}{\pi} & \text{if } d = 2\,, \\ \frac{1}{4} - \left(\alpha - \frac{1}{2}\right)^2 & \text{if } d = 3\,. \end{cases}$$

Example 3.6 (Uniform distribution on a ball) We consider the uniform distribution μ on a given ball in $\mathbb{R}^d$ having center $\boldsymbol{a}$ and radius β. Then, $D_1(\mu) = \{\boldsymbol{a}\}$, $D_0(\mu) = \mathbb{R}^d$. For $0 < \alpha < 1$, the trimmed regions are balls around $\boldsymbol{a}$,

$$D_\alpha(\mu) = B\left(\boldsymbol{a}, \frac{r_3(\alpha)}{\alpha}\right),$$

with $r_3(\alpha)$ given by (2.61). This follows from Example 2.12. There holds $r_3(\alpha) = r_3(1-\alpha)$ for all $\alpha \in]0,1[$. Specifically, for a uniform distribution on the *unit disc* in $\mathbb{R}^2$ we obtain

$$r_3(\alpha) = \frac{2}{3\pi}\,(1-q_\alpha^2)^{3/2}, \quad \alpha - \frac{1}{2} = \frac{1}{\pi}\left(q_\alpha\sqrt{1-q_\alpha^2} + \arcsin q_\alpha\right), \quad \frac{1}{2} \leq \alpha < 1\,.$$

A uniform distribution on the *unit ball* in $\mathbb{R}^3$ has trimmed regions with

$$r_3(\alpha) = \frac{3}{16}\,(1-q_\alpha^2)^2, \qquad \alpha - \frac{1}{2} = \frac{3q_\alpha}{4}\left(1 - \frac{q_\alpha^3}{3}\right), \quad \frac{1}{2} \leq \alpha < 1\,.$$

If μ is *uniform on an ellipsoid* or on the boundary of an ellipsoid, the zonoid central regions are easily derived from the preceding examples by a proper affine transformation.

Example 3.7 (Bivariate exponential distribution) Figure 3.4 shows zonoid α-trimmed regions for $n = 500$ and $n = 50$ observations simulated from a Marshall–Olkin distribution. $\boldsymbol{X} = (X_1, X_2)$ is Marshall–Olkin distributed if $X_1 = \min\{Z_1, Z_3\}$, $X_2 = \min\{Z_2, Z_3\}$, and Z_1, Z_2, Z_3 have independent exponential distributions. Here $Z_j \sim Exp(1/2)$ for $j = 1, 2, 3$.

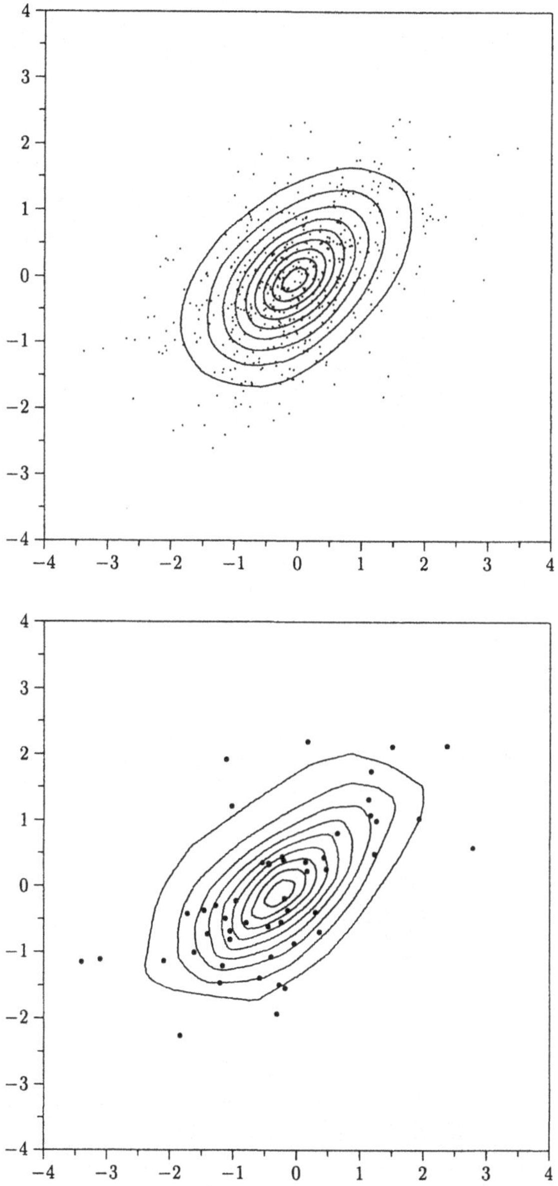

FIGURE 3.3: Zonoid trimmed regions ($\alpha = 0.1, \ldots, 0.9$) for $n = 500$ and $n = 50$ bivariate normal observations, (X_1, X_2) with $\mu_1 = \mu_2 = 0$, $\rho = 1/2, \sigma_1 = 1 = \sigma_2 = 1)$.

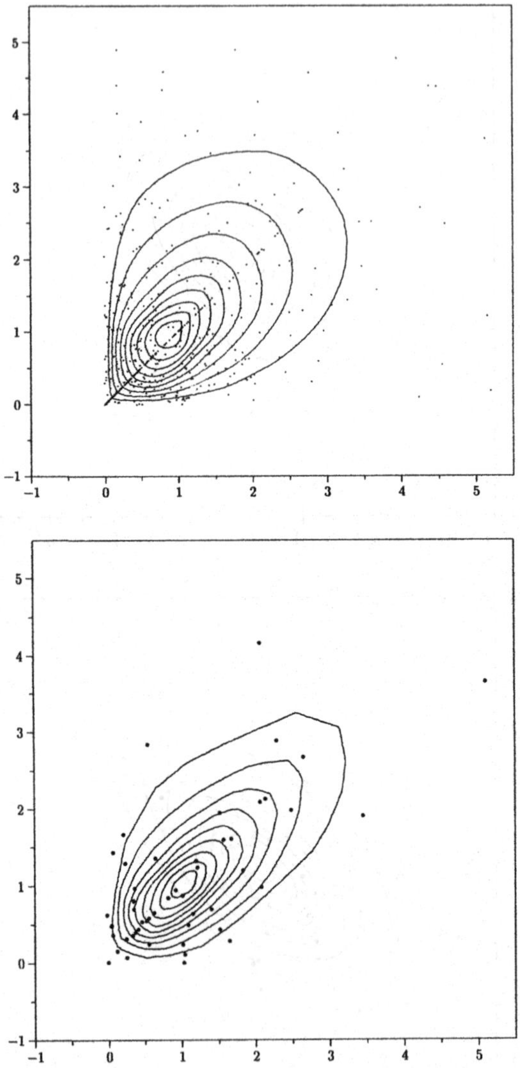

FIGURE 3.4: Zonoid trimmed regions ($\alpha = 0.1, \ldots, 0.9$) for $n = 500$ and $n = 50$ dependent exponential observations (Marshall–Olkin distribution), $X_1 = \min\{Z_1, Z_3\}$, $X_2 = \min\{Z_2, Z_3\}$, where Z_1, Z_2, Z_3 have independent exponential distributions, $Z_j \sim Exp(1/2)$ for $j = 1, 2, 3$.

3.5 Notions of multivariate central regions

Given a d-variate probability measure μ, any family of nested sets in $\mathbb{R}^d$ can be considered as a family of central regions. The intersection of all sets in the family determines the median set of μ.

Definition 3.2 (Family of central regions, R-median) *Let $\mathcal{P}_0 \subset \mathcal{P}_1$. A family of central regions is a mapping*

$$\mathcal{R} : \mu \mapsto \{R_\alpha(\mu)\}_{\alpha_0 \leq \alpha \leq \alpha_1}, \quad \mu \in \mathcal{P}_0, \quad \textit{with}$$

$$R_\alpha(\mu) \subset R_{\alpha'}(\mu) \quad \textit{if } \alpha_0 \leq \alpha' \leq \alpha \leq \alpha_1 .$$

$$\mathcal{R}\text{−Med}(\mu) = \bigcap_{\alpha_0 \leq \alpha \leq \alpha_1} R_\alpha(\mu)$$

is the $\mathcal{R}$-median set, and the gravity center of $\mathcal{R}$−Med(μ) is the $\mathcal{R}$-median.

Widely used central regions are:
1. Likelihood central regions. If μ is continuous with respect to Lebesgue measure, the nonempty level sets of the density form a family of central regions, called the *likelihood central regions.* Their median set is the set of largest modes.

2. Mahalanobis central regions. Let $\mathcal{P}_2$ denote the set of d-variate probability measures that have finite moments up to order 2. For every $\mu \in \mathcal{P}_2$ that has a positive definite covariance matrix Σ_μ,

$$d_{Mah}(\boldsymbol{x}, \boldsymbol{y}) = (\boldsymbol{x} - \boldsymbol{y})\Sigma_\mu^{-1}(\boldsymbol{x} - \boldsymbol{y})'$$

is the Mahalanobis distance (Mahalanobis, 1936) between $\boldsymbol{x}$ and $\boldsymbol{y}$ in $\mathbb{R}^d$. It gives rise to the *Mahalanobis central regions,*

$$D_\alpha^{Mah}(\mu) = \{\boldsymbol{x} : (\boldsymbol{x} - \epsilon(\mu))\Sigma_\mu^{-1}(\boldsymbol{x} - \epsilon(\mu))' \leq \frac{1-\alpha}{\alpha}\}, \quad 0 < \alpha \leq 1 . \quad (3.13)$$

They are ellipsoids centered at the expectation vector. The *Mahalanobis median* is the mean. Observe that in the nondegenerate normal case, if Σ has full rank, the zonoid trimmed regions of a multivariate normal distribution are parallel to the Mahalanobis central regions.

Figure 3.5 shows 50 observations that have been simulated from a bivariate normal distribution with standard marginals and correlation $\rho = 0.5$. The figure exhibits the Mahalanobis regions for $\alpha = 0.1, 0.2, \ldots, 0.9$. Figure 3.6

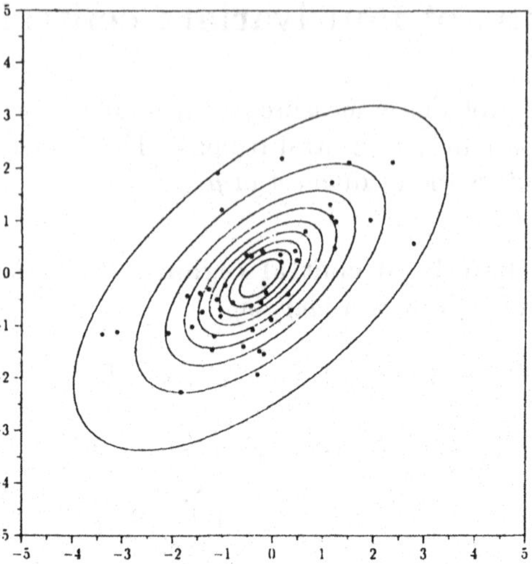

FIGURE 3.5: Mahalanobis trimmed regions ($\alpha = 0.1, \ldots, 0.9$) for $n = 50$ bivariate normal observations, (X_1, X_2) with $\mu_1 = \mu_2 = 0$, $\rho = 1/2, \sigma_1 = 1 = \sigma_2 = 1$).

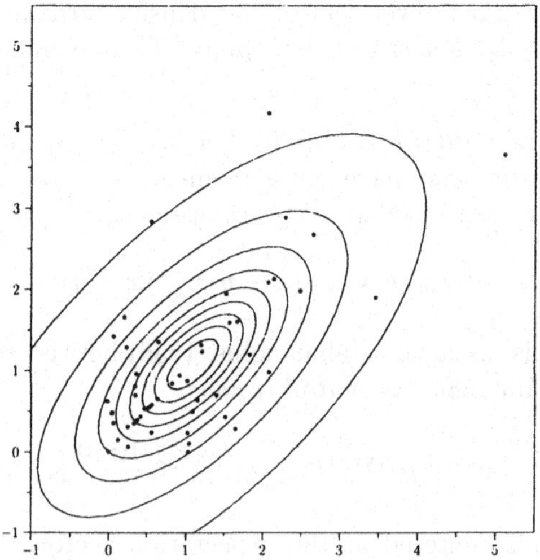

FIGURE 3.6: Mahalanobis trimmed regions ($\alpha = 0.1, \ldots, 0.9$) for $n = 50$ dependent exponential observations (Marshall–Olkin distribution), $X_1 = \min\{Z_1, Z_3\}$, $X_2 = \min\{Z_2, Z_3\}$, where Z_1, Z_2, Z_3 have independent exponential distributions, $Z_j \sim Exp(1/2)$ for $j = 1, 2, 3$.

presents Mahalanobis regions for 50 observations simulated from a Marshall-Olkin distribution; cf. Example 3.7.

3. Central regions based on data depths. A *data depth* is a function that indicates, in some sense, how deep a point is located with respect to a given multivariate data cloud (or to a given probability distribution) in $\mathbb{R}^d$. The depth defines a center of the cloud, that is the set of deepest points, and measures how far a point lies from the center. Special, widely investigated data depths are the halfspace depth (also called Tukey depth), the Oja depth and the simplicial depth. Each data depth generates a family of central regions for a given distribution: A trimmed region is the set of all points whose depth is not less than some value. The zonoid trimmed regions are also generated by a data depth, which we call the *zonoid depth.* The zonoid depth, contrasted with other notions af data depth, will be the topic of Chapter 4.

Obviously, a data depth is affine invariant if and only if the central regions which it generates are affine equivariant. The zonoid depth and all above listed notions of depth are affine invariant. Therefore the central regions which they generate are affine equivariant.

An affine (resp. orthogonal) equivariant central region must have a particular form if the distribution is elliptically (resp. spherically) symmetric. For a definition of elliptical and spherical distributions, see Section 2.3.5 above.

Proposition 3.8 (Spherical and elliptical distributions)
(i) *If μ is spherical and $\mathcal{R}$ is orthogonal equivariant, then $\mathcal{R}(\mu)$ must be a family of balls around the origin.*

(ii) *If μ is elliptical around $\boldsymbol{a}$ and $\mathcal{R}$ is affine equivariant, then $\mathcal{R}(\mu)$ must be a family of ellipsoids of proper dimension with center $\boldsymbol{a}$.*

(iii) *If μ is nondegenerate elliptical (i.e., elliptical with $k = d$) and $\mathcal{R}$ is affine equivariant, then the sets in $\mathcal{R}(\mu)$ parallel the Mahalanobis central sets.*

Proof. (i): Sphericity of a measure μ means that μ is invariant against orthogonal transformations. It follows that, if $\mathcal{R}$ is orthogonal equivariant, a set in $\mathcal{R}(\mu)$ must stay invariant under orthogonal transformations, that is, it must be a ball around the origin.
(ii) If μ is elliptical around $\boldsymbol{a}$, $\mu \in \mathcal{E}(\boldsymbol{a}, BB', \phi)$, it is, by $\boldsymbol{x} \mapsto \boldsymbol{a} + \boldsymbol{x}B$, the affine image of a spherical measure. As $\mathcal{R}$ is affine equivariant, it is also orthogonal equivariant and every set in $\mathcal{R}(\mu)$ is the affine image of a ball around the origin, that is, an ellipsoid of proper dimension, having center $\boldsymbol{a}$.
(iii) Obvious. Q.E.D.

In particular, when applied to a multivariate normal distribution, any affine equivariant notion of central regions yields a family of sets which parallel the Mahalanobis central regions.

3.6 Continuity of the trimmed regions and law of large numbers

In this section we demonstrate two continuity theorems: The zonoid trimmed regions are, with respect to the Hausdorff distance, continuous on α as well as on μ. As a consequence, a law of large numbers for zonoid trimmed regions is derived.

Theorem 3.9 (Continuous on α) *Let $\mu \in \mathcal{P}_1$, $\alpha_n \to \alpha$ and $\alpha > 0$. Then*

$$D_{\alpha_n}(\mu) \xrightarrow{H} D_\alpha(\mu).$$

Proof. Let $\alpha_n \to \alpha > 0$. Then, due to the convexity and compactness of $\widehat{Z}(\mu)$, obtain $\text{proj}_{\alpha_n} \widehat{Z}(\mu) \xrightarrow{H} \text{proj}_\alpha \widehat{Z}(\mu)$. Therefore, $\lim_n D_{\alpha_n}(\mu) = \lim_n \frac{1}{\alpha_n} \text{proj}_{\alpha_n} \widehat{Z}(\mu) = \frac{1}{\alpha} \lim_n \text{proj}_{\alpha_n} \widehat{Z}(\mu) = \frac{1}{\alpha} \text{proj}_\alpha \widehat{Z}(\mu) = D_\alpha(\mu)$. Q.E.D.

Next we demonstrate continuity of the central regions with respect to the underlying measure.

Theorem 3.10 (Continuous on μ) *Assume that $\mu^n \xrightarrow{w} \mu$ in $\mathcal{P}_1$ and one of the following four restrictions is satisfied:*

- *There exists a compact set that contains the supports of all μ^n.*
- *The sequence μ^n is uniformly integrable.*
- *There exists some $\nu \in \mathcal{P}_1$ such that $\widehat{Z}(\mu^n) \subset \widehat{Z}(\nu)$ for all n.*
- *There exists some $\nu \in \mathcal{P}_1$ such that $\widehat{Z}(\mu^n) \xrightarrow{H} \widehat{Z}(\nu)$ for all n.*

Then, in the Hausdorff distance,

$$D_\alpha(\mu) = \lim_n D_\alpha(\mu^n), \quad \text{for } 0 < \alpha \leq 1.$$

Proof. Let $\mu^n \xrightarrow{w} \mu$ and $\lim_n \epsilon(\mu^n) = \epsilon(\mu)$. Then, either by Theorem 2.30 or one of its corollaries, $\widehat{Z}(\mu^n) \xrightarrow{H} \widehat{Z}(\mu)$ and $\widehat{Z}(\mu^n, \alpha) \xrightarrow{H} \widehat{Z}(\mu, \alpha)$ holds for any $0 < \alpha \leq 1$. We conclude $D_\alpha(\mu^n) \xrightarrow{H} D_\alpha(\mu)$ for $0 < \alpha \leq 1$. Q.E.D.

We apply the Theorem 3.10 to obtain a law of large numbers for zonoid trimmed regions. Let $\boldsymbol{X}_1, \ldots, \boldsymbol{X}_n$ be i.i.d. random vectors in $\mathbb{R}^d$ that are distributed as μ, and let $\widetilde{\mu}^n$ denote their random empirical measure. By the Glivenko-Cantelli theorem we know that μ-a.s. $\lim_{n\to\infty} \widetilde{\mu}^n = \mu$, and by the usual law of large numbers we obtain convergence of expectations, $\epsilon(\widetilde{\mu}^n)$ converges μ-a.s. to $\epsilon(\mu)$. Because of this, we conclude from Theorem 3.10 :

Corollary 3.11 (Law of large numbers)

$$\lim_{n\to\infty} D_\alpha(\widetilde{\mu}^n) = D_\alpha(\mu) \quad \textit{holds } \mu\textit{-a.s.}, \quad 0 < \alpha \leq 1,$$

Further consequences of the continuity of zonoid trimmed regions are found in the subsequent section.

3.7 Further properties of zonoid trimmed regions

In this section we present additional properties of the zonoid trimming and refinements of the previous results.

By envoking continuity on α we can demonstrate that for a general measure in $\mathcal{P}_1$ the set $D_*(\mu)$ equals the convex hull of the support and we are able to weaken the uniqueness theorem. It is shown that the monotonicity of zonoid trimmed regions is strict. Further, the zonoid trimmed regions of a mixed distribution are determined, which are unions of certain mixtures of trimmed regions.

First, we use the Theorem 3.9 to derive a uniqueness result that is stronger than Theorem 3.6 :

Theorem 3.12 (Uniqueness by parameter subset) *Let $\mu, \nu \in \mathcal{P}_1$.*

(i) *If $D_\alpha(\mu) = D_\alpha(\nu)$ for all α in some dense subset of $\left]0, \frac{1}{2}\right]$, then $\mu = \nu$.*

(ii) *If $D_\alpha(\mu) = D_\alpha(\nu)$ for all α in some dense subset of $\left[\frac{1}{2}, 1\right[$, then $\mu = \nu$.*

Proof. Due to Theorem 3.6, $D_\alpha(\mu) = D_\alpha(\nu)$ for $0 < \alpha \le 1$ implies $\mu = \nu$. From the central symmetry of the lift zonoid (Proposition 2.15) we obtain that the set $\widehat{Z}_\alpha(\mu) - \alpha\epsilon(\mu)$ is centrally symmetric to the set $\widehat{Z}_{1-\alpha}(\mu) - (1-\alpha)\epsilon(\mu)$. Therefore it is sufficient to have $D_\alpha(\mu) = D_\alpha(\nu)$ for all α in either the lower or the upper half of the unit interval, and in view of Theorem 3.9 a dense subset is enough. Q.E.D.

Now, consider the union of α-trimmed regions with $\alpha > 0$. Previously we have seen (cf. (3.6)) that it equals the convex hull of the support if the measure is empirical. By envoking the second continuity theorem (Theorem 3.10) it is possible to prove this for a general probability measure as well.

Theorem 3.13 (Convex hull of support) *For any $\mu \in \mathcal{P}_1$ holds*

$$\bigcup_{0<\alpha\le 1} D_\alpha(\mu) = conv(\mu).$$

Proof. Denote $D_*(\mu) = \bigcup_{0<\alpha\le 1} D_\alpha(\mu)$. Let $\mu^n \xrightarrow{w} \mu$ be a sequence of empirical measures according to Theorem 2.36. As $\widehat{Z}(\mu^n) \subset \widehat{Z}(\mu)$, there holds $D_\alpha(\mu^n) \subset D_\alpha(\mu)$ for all $\alpha > 0$, and therefore $D_*(\mu^n) \subset D_*(\mu)$. We conclude that $\lim_n D_*(\mu^n) \subset D_*(\mu)$. Theorem 3.10 yields that $D_*(\mu) \subset \lim_n D_*(\mu^n)$, hence $D_*(\mu) = \lim_n D_*(\mu^n)$. Let $\mathcal{H}^d$ be the set of closed halfspaces in $\mathbb{R}^d$. Due to Equation (3.19), $D_*(\mu^n) = \bigcap\{H \in \mathcal{H}^d : \mu^n(H) = 1\}$ holds. When n approaches infinity we get $D_*(\mu) = \lim_n D_*(\mu^n) = \bigcap\{H \in \mathcal{H}^d : \mu(H) = 1\} = conv(\mu)$. Q.E.D.

We know already that the zonoid trimmed regions are (weakly) decreasing on α. Now it is demonstrated that the decrease is strict as soon as the regions differ from $conv(\mu)$.

Theorem 3.14 (Decreasing on parameter)
Let $\mu \in \mathcal{P}_1$, $0 < \alpha_1 < \alpha_2 \le 1$. Then

(i) $D_{\alpha_2}(\mu) \subset D_{\alpha_1}(\mu) \subset conv(\mu)$,

(ii) $D_{\alpha_2}(\mu) = D_{\alpha_1}(\mu)$ *only if* $D_{\alpha_1}(\mu) = conv(\mu)$.

Proof. Recall that $G_\alpha = \{z \in \mathbb{R}^{d+1} : z_0 = \alpha\}$ and $\widehat{Z}(\mu,\alpha) = \widehat{Z}(\mu) \cap G_\alpha$. For $0 \le \alpha_1 < \alpha_2$, we know from Theorems 3.1 and 3.13 that $D_{\alpha_2}(\mu) \subset D_{\alpha_1}(\mu) \subset D_*(\alpha) = conv(\mu)$, hence (i). Assume $D_{\alpha_2}(\mu) = D_{\alpha_1}(\mu)$. Then

$$\widehat{Z}(\mu,\alpha_1) = \frac{\alpha_1}{\alpha_2}\,\widehat{Z}(\mu,\alpha_2). \qquad (3.14)$$

As $\widehat{Z}(\mu)$ is convex and $\mathbf{0} \in \widehat{Z}(\mu)$, (3.14) implies that, for any $0 < \alpha \le \alpha_1$,

$$\widehat{Z}(\mu, \alpha) \subset \frac{\alpha}{\alpha_2} \widehat{Z}(\mu, \alpha_2)\,, \tag{3.15}$$

hence $D_\alpha(\mu) \subset D_{\alpha_2}(\mu)$. Conclude that

$$conv(\mu) = \bigcup_{\alpha > 0} D_\alpha(\mu) \subset D_{\alpha_2}(\mu) = D_{\alpha_1}(\mu) \subset \bigcup_{\alpha>0} D_\alpha(\mu)\,,$$

which proves (ii). Q.E.D.

Moreover, given two zonoid trimmed regions $D_{\alpha_2} \subset D_{\alpha_1}$, if some boundary point of the larger region D_{α_1} is included in the smaller region D_{α_2}, the larger region must equal the convex hull of the support, $D_{\alpha_1} = conv(\mu)$; for illustration, see Figure 1.9 in chapter 1.

Finally we prove that the α-trimmed region of a mixture of measures is the union of certain mixtures of trimmed regions:

Theorem 3.15 (Mixture) *Let μ, $\nu \in \mathcal{P}_1$, $0 \le \beta \le 1$. Then, for $0 < \alpha \le 1$,*

$$D_\alpha(\beta\mu + (1-\beta)\nu) =$$
$$\bigcup\left\{\frac{\beta\gamma}{\alpha} D_\gamma(\mu) + \frac{(1-\beta)\gamma'}{\alpha} D_{\gamma'}(\nu) \;:\; \beta\gamma + (1-\beta)\gamma' = \alpha\right\}. \tag{3.16}$$

Proof. By positive linearity of the lift zonoid (Theorem 2.19) we know that $\widehat{Z}(\beta\mu + (1-\beta)\nu) = \beta\widehat{Z}(\mu) + (1-\beta)\widehat{Z}(\nu)$. Therefore,

$$\widehat{Z}(\beta\mu + (1-\beta)\nu, \alpha) = \bigcup\left\{\beta\widehat{Z}(\mu,\gamma) + (1-\beta)\widehat{Z}(\nu,\gamma') \;:\; \beta\gamma + (1-\beta)\gamma' = \alpha\right\}.$$

Thus,

$$\begin{aligned}
&\frac{1}{\alpha}\widehat{Z}_\alpha(\beta\mu + (1-\beta)\nu) \\
&= \bigcup\left\{\frac{\beta}{\alpha}\widehat{Z}_\gamma(\mu) + \frac{1-\beta}{\alpha}\widehat{Z}_{\gamma'}(\nu) \;:\; \beta\gamma + (1-\beta)\gamma' = \alpha\right\} \\
&= \bigcup\left\{\frac{\beta\gamma}{\alpha} D_\gamma(\mu) + \frac{(1-\beta)\gamma'}{\alpha} D_{\gamma'}(\nu) \;:\; \beta\gamma + (1-\beta)\gamma' = \alpha\right\},
\end{aligned}$$

that is (3.16). Q.E.D.

3.8 Trimming of empirical measures

In this section a useful formula is proven which describes the zonoid trimmed regions of an empirical measure. We illustrate the formula by applying it to a law of large numbers for univariate zonoid trimming and order statistics.

Let μ be an empirical probability measure on n given points in $\mathbb{R}^d$, $\boldsymbol{x}_1, \ldots, \boldsymbol{x}_n$. The lift zonoid of μ is a sum of line segments

$$\widehat{Z}(\mu) = \sum_{i=1}^{n} \left[\mathbf{0}, \left(\frac{1}{n}, \frac{\boldsymbol{x}_i}{n}\right)\right]. \tag{3.17}$$

Theorem 3.16 (Trimmed regions of an empirical distribution) . *Let $\alpha \in \left[\frac{k}{n}, \frac{k+1}{n}\right]$, $k = 1, \ldots, n-1$. Then*

$$D_\alpha(\mu) = conv\left\{\frac{1}{\alpha n}\sum_{j=1}^{k} \boldsymbol{x}_{i_j} + \left(1 - \frac{k}{\alpha n}\right)\boldsymbol{x}_{i_{k+1}} \; : \; \{i_1, \ldots, i_{k+1}\} \subset N\right\} \tag{3.18}$$

where $N = \{1, \ldots, n\}$. For $\alpha \in \left]0, \frac{1}{n}\right]$,

$$D_\alpha(\mu) = conv\{\boldsymbol{x}_1, \ldots, \boldsymbol{x}_n\} = \bigcap\{H \in \mathcal{H}^d : \mu(H) = 1\}, \tag{3.19}$$

where $\mathcal{H}^d$ is the set of closed halfspaces in $\mathbb{R}^d$.

Proof. The lift zonoid of an empirical measure is a convex polytope,

$$\begin{aligned} \widehat{Z}(\mu) &= conv\left\{\sum_{i=1}^{n} \delta_i \left(\frac{1}{n}, \frac{1}{n}\boldsymbol{x}_i\right) : \delta_i \in \{0,1\}, i \in N\right\} \quad (3.20) \\ &= conv\left(\bigcup_{k=0}^{n} V_k\right). \end{aligned}$$

where $V_k = \{\sum_{j\in J} \left(\frac{1}{n}, \frac{1}{n}\boldsymbol{x}_{i_j}\right) : J \subset N, |J| = k\}$ for $k = 1, 2, \ldots, n$, $V_0 = \{(0, \mathbf{0})\}$. The V_k are pairwise disjoint sets. Every extreme point $\boldsymbol{z}^*$ of

$$\widehat{Z}(\mu, \alpha) = \{(z_0, z_1, \ldots, z_d) \in \widehat{Z}(\mu) : z_0 = \alpha\}.$$

is either an extreme point of $\widehat{Z}(\mu)$ or the intersection of an edge (i.e. one-dimensioned face) of $\widehat{Z}(\mu)$ with the hyperplane at $z_0 = \alpha$. Note that every edge of the lift zonoid has either the origin as vertex or is a shift of an edge that has the origin as vertex. More precisely, every edge of $\widehat{Z}(\mu)$ can be written (Shephard, 1974) as

$$\left[(0,\mathbf{0}), \left(\frac{1}{n}, \frac{1}{n}\boldsymbol{x}_s\right)\right] + \sum_{i\in J}\left(\frac{1}{n}, \frac{1}{n}\boldsymbol{x}_i\right) \tag{3.21}$$

with some $J \subset N \setminus \{s\}$ and $s \in N$. We proceed in three steps.

First we prove (3.18) for $\alpha = \frac{k}{n}$, $k = 1, \dots, n$. Obviously, $conv(V_k) \subset \widehat{Z}(\mu, \frac{k}{n})$ holds. Let $\boldsymbol{z}^*$ be an extreme point of $\widehat{Z}(\mu, \frac{k}{n})$. Then $\boldsymbol{z}^*$ either is an extreme point of $\widehat{Z}(\mu)$, hence in $\boldsymbol{z}^* \in V_k$, or belongs to an edge of $\widehat{Z}(\mu)$. In the latter case (3.21) implies that $\boldsymbol{z}^* \in V_k$. We conclude that $conv(V_k) = \widehat{Z}(\mu, \frac{k}{n})$. Then

$$D_{\frac{k}{n}}(\mu) = \frac{n}{k}\,\widehat{Z}_{\frac{k}{n}}(\mu) = \frac{n}{k}\; conv\left\{\sum_{j=1}^{k}\frac{1}{n}\boldsymbol{x}_{i_j} \;:\; \{i_1, \dots, i_k\} \subset N\right\}.$$

That yields Equation (3.18) for $\alpha = \frac{k}{n}$.

Second, we establish (3.18) for $\alpha \in]\frac{k}{n}, \frac{k+1}{n}[$, $k = 1, \dots, n-1$. Again let $\boldsymbol{z}^*$ be an extreme point of $\widehat{Z}(\mu, \frac{k}{n})$. As $\boldsymbol{z}^*$ cannot be an extreme point of $\widehat{Z}(\mu)$, (3.21) yields

$$\boldsymbol{z}^* = \left(\alpha - \frac{k}{n}\right)\left(1, \boldsymbol{x}_{i_{k+1}}\right) + \sum_{j=1}^{k}\frac{1}{n}\left(1, \boldsymbol{x}_{i_j}\right). \tag{3.22}$$

$\widehat{Z}(\mu, \alpha)$ is the convex hull of all $\boldsymbol{z}^*$ given by (3.22). Therefore,

$$\frac{1}{\alpha}\widehat{Z}_\alpha(\mu) = conv\left\{\sum_{j=1}^{k}\frac{1}{n\alpha}\boldsymbol{x}_{i_j} + \left(1 - \frac{k}{n\alpha}\right)\boldsymbol{x}_{i_{k+1}} \;:\; \{i_1, \dots, i_{k+1}\} \subset N\right\}.$$

That yields Equation (3.18) also for $\alpha \in]\frac{k}{n}, \frac{k+1}{n}[$.

It remains to demonstrate (3.19), the second equality of which is obvious. For $\alpha = 1/n$, Equation 3.18 specializes to $D_{1/n} = conv\{\boldsymbol{x}_1, \dots, \boldsymbol{x}_n\} = conv(\mu)$. From this conclude that, for $0 < \alpha \le 1/n$,

$$conv(\mu) = D_{\frac{1}{n}}(\mu) \subset D_*(\mu) = conv(\mu)\,,$$

which completes the proof. Q.E.D.

As an application of the previous Theorem 3.16 we present a law of large numbers for univariate zonoid trimming and order statistics.

Let $X_1, \ldots, X_n$ be given i.i.d. random variables in $\mathbb{R}$ that are distributed as μ, and let $X_{(1)} \le \ldots \le X_{(n)}$ denote their order statistics. Consider the random empirical distribution is $\widetilde{\mu}^n$. Its zonoid α-trimmed region $D_\alpha(\widetilde{\mu}^n)$ is given by Equation (3.18) and equals the interval

$$\left[\frac{1}{\alpha n}\sum_{j=1}^{[\alpha n]} X_{(j)} + \frac{\{\alpha n\}}{\alpha n} X_{([\alpha n]+1)}, \frac{1}{\alpha n}\sum_{j=n-[\alpha n]+1}^{n} X_{(j)} + \frac{\{\alpha n\}}{\alpha n} X_{(n-[\alpha n])}\right]. \tag{3.23}$$

Note that, if $\alpha \in [\frac{k}{n}, \frac{k+1}{n}]$, $[\alpha n] = k$ and $\{\alpha n\} = [\alpha n] - k$, where as usual $\{\beta\}$ and $[\beta]$ are the fraction and the integer parts of a real number β, respectively. There holds

$$\lim_{n\to\infty} D_\alpha(\widetilde{\mu}^n) = \lim_{n\to\infty}\left[\frac{1}{\alpha n}\sum_{j=1}^{[\alpha n]} X_{(j)}, \frac{1}{\alpha n}\sum_{j=n-[\alpha n]+1}^{n} X_{(j)}\right]. \tag{3.24}$$

Thus, in the univariate case, from Corollary 3.11 a law of large numbers for order statistics is derived. Equation (3.24) means that

$$\left[\frac{1}{\alpha n}\sum_{j=1}^{[\alpha n]} X_{(j)}, \frac{1}{\alpha n}\sum_{j=n-[\alpha n]+1}^{n} X_{(j)}\right] \longrightarrow \left[\frac{1}{\alpha}\int_0^\alpha Q_\mu(\beta)d\beta, \frac{1}{\alpha}\int_{1-\alpha}^1 Q_\mu(\beta)d\beta\right]$$

μ-a.s. if $0 < \alpha \le 1$. From this we obtain:

Corollary 3.17 (Law of large numbers for order statistics) *For i.i.d. random vectors $X_1, X_2, X_3, \ldots,$ in $\mathbb{R}^d$ holds:*

$$\frac{1}{[\alpha n]}\sum_{j=1}^{[\alpha n]} X_{(j)} \longrightarrow \frac{1}{\alpha}\int_0^\alpha Q_\mu(\beta)d\beta \quad \mu\text{-a.s.} \quad \textit{and} \tag{3.25}$$

$$\frac{1}{[\alpha n]}\sum_{j=n-[\alpha n]+1}^{n} X_{(j)} \longrightarrow \frac{1}{\alpha}\int_{1-\alpha}^1 Q_\mu(\beta)d\beta \quad \mu\text{-a.s.} \tag{3.26}$$

3.9 Computation of zonoid trimmed regions

In view of the possible applications of zonoid trimmed regions it is important to have special algorithms to calculate one or more trimmed regions of a data set.

Dyckerhoff et al. (1996) have built an efficient algorithm to compute the zonoid depth of a point in a given data set; see Section 4.5 in the next chapter. In principle, if an efficient algorithm is available to compute the depth of a point, it is an easy task to compute approximate trimmed regions by evaluating the depth of many points. However, this can be time consuming. Thus, a direct approach should be superior.

Dyckerhoff (2000) has developed an algorithm to calculate the zonoid trimmed regions of a given data set in two dimensions. The algorithm is based on the support function of the trimmed region, which is a convex set. The set of extreme points equals the set of unique maximizers of the support function. It is computed by letting $\boldsymbol{p}$ vary in S^1 and calculating all points that are unique maximizers of the support function in direction $\boldsymbol{p}$.

In a first step the algorithm computes all lines through two different points of the data set and the angles between the normal vectors of the lines and the positive first coordinate axis. In a next step these angles are sorted in increasing order. Each of the angles defines two extreme points of the zonoid trimmed region. It should be noted that extreme points associated with different angles are not necessarily different. By going through the sequence of angles all the extreme points of the zonoid trimmed region are constructed in a counterclockwise order. The complexity of this algorithm is determined by the sorting of angles in the first step. Since $\binom{n}{2}$ angles have to be sorted the complexity is $O(n^2 \log n)$. Once the sorting of the angles is completed, the trimmed region can be calculated with complexity $O(n^2)$. If two or more zonoid trimmed regions have to be computed of the same data with different α, the initialization step has to be executed only once, which results in a considerable reduction of computation time.

3.10 Notes

The notion of zonoid trimmed regions has been proposed in Koshevoy and Mosler (1997b). The discrete case and its computational aspects are developed in Dyckerhoff et al. (1996). Chapter 3 refers essentially to these two papers. Observe that the definition of $D_0(\mu)$ given above differs from that given in Koshevoy and Mosler (1997b): What I call $D_*(\mu)$ here corresponds to $D_0(\mu)$ there.

The problem of convergence of sample contour sets to their population analogue is addressed in He and Wang (1997).

Most medians proposed in the literature can be regarded as minimal central regions or gravity centers of such regions. In particular, the Tukey median is central to the halfspace trimmed regions. The Oja median and the simplicial median are central to the regions that are generated by the Oja depth and the simplicial depth, respectively.

To compute the Tukey median, an exact algorithm for dimension $d = 2$ is given in Rousseeuw and Ruts (1998) and an approximate one for higher dimensions in Rousseeuw and Struyf (1998). For the bivariate Oja median, Niinimaa et al. (1992) provide an algorithm; see also Niinimaa (1992).

For the central regions generated by halfspace depth, Rousseeuw and Ruts (1996) present an algorithm that computes the halfspace depth in the bivariate case; see also Johnson et al. (1998).

The complement of a central region is an *outlying region*; its elements can be seen as *outliers*. Cramer (2002) studies the identification of outliers by central regions and data depth.

The robustness of central regions is an important issue in data analysis. In terms of worst case behaviour, the breakdown point (Donoho and Gasko, 1992) of a statistical estimate is a natural indicator of its robustness. Any notion of central regions cannot have a higher breakdown than the pertaining median. In this sense, the zonoid trimmed regions are not robust. They have breakdown zero since their median is the mean.

Other central regions have better breakdown properties. The breakdown of the Tukey median is investigated in Donoho and Gasko (1992) and Jeyaratnam (1991), where also a robustified version of the Tukey median is proposed. Chen (1995) discusses the breakdown of the simplicial median. For the breakdown point of the Oja median and others, see Niinimaa et al. (1990), while their influence functions are treated in Niinimaa and Oja (1995).

4

Data depth

A data depth is a function that indicates, in some sense, how deep a point is located with respect to a given data cloud (or to a given probability distribution) in d-space. The depth defines a center of the cloud, that is the set of deepest points, and measures how far away a point is located from the center. Various notions of data depth can be employed in procedures of multivariate data analysis, such as cluster analysis and the detection of outlying data. In multivariate statistical inference they are also used to construct rank tests for homogeneity against scale and location alternatives.

Each data depth defines a family of nested central regions. Such a central region is the set of all points whose depth is not less than some value. The *median set* consists of all points that have maximum depth. If the median set is a singleton, its single element is the *median* with respect to the depth. If not, the gravity center of the median set is taken as the median.

Given a data cloud and a depth notion, one can calculate the depth of each data point with respect to the whole cloud. Then the points can be ordered according to their depth as it decreases from the center outwards. Thus, every depth is a kind of mid rank which allows to order multivariate data by their centrality.

Several different notions of data depth have been proposed in the literature: Mahalanobis depth based on the distance of the same name (Mahalanobis, 1936), halfspace depth (Hotelling, 1929; Tukey, 1975), Oja depth related to the Oja median (Oja, 1983), simplicial depth (Liu, 1990), majority depth (Singh, 1991), and others. For earlier contributions see the following surveys: Barnett (1976) about general concepts of ordering multivariate data, Huber (1972) and Rousseeuw and Leroy (1987, Ch. 7) about relations of data depths with robust statistics, and Small (1990) about depths and medians. Under a more general view, several types of data depths are investigated in Zuo and Serfling (2000b,c). Mizera (1998) develops a differential calculus for depth functions.

Recently, functionals of depths have been employed as descriptive statistics to depict the location, dispersion, asymmetry, and tail behaviour of multivariate data (Liu et al., 1999; Zuo and Serfling, 2000a,d) and simple graphical procedures for comparing distributions have been based on them. Inference has been done by bootstrapping such statistics; see Yeh and Singh (1997) for confidence intervals and Liu and Singh (1997) for tests. Brown et al. (1992a), Liu and Singh (1993), Dyckerhoff (1998), and others have used data depth for testing hypotheses. In particular, depth based rank tests for homogeneity against location and scale alternatives are found in Dyckerhoff (1998); see also Chapter 5 below.

This chapter presents the zonoid depth and several other data depths which can be used for different tasks, in particular for

- characterizing a distribution by means of a depth function of the pertaining trimmed regions,
- defining a multivariate median,
- constructing multivariate rank tests,
- clustering and classifying data.

In order to tackle these tasks, different properties of depths are required. In the first task we have to consider the depth of all points in order to define trimmed regions that characterize the whole distribution. In the second task only one parameter of the distribution has to be identified. In the third task data from a random sample have to be ranked in some way, while in the fourth task the metric properties of the data are important.

Different depths measure different aspects of centrality and vary in important aspects: their ability to identify the underlying distribution, the median they produce, their invariance and continuity properties, their robustness, and their computational feasibility.

The first aspect is the sensitivity of the depth to the underlying distribution: Have two distinct distributions depth functions that can be distinguished? For example, the *Mahalanobis depth* (which is reciprocal to the Mahalanobis distance from the mean increased by one) depends on the first two moments of the distribution alone. Therefore, two distributions that have the same mean vector and covariance matrix have also the same Mahalanobis depth function.

Second comes invariance: If both, the point and the distribution, are subject to the same transformation, say affine or orthogonal, does the value of the depth remain constant? In many applications the statistical procedure should be invariant against affine transformations or, at least, rigid Euclidean motions. This applies particularly if the data, in each dimension, are measured on an interval scale, and, in addition, linear combinations of the variables have a meaningful interpretation, as it is the case in factor analysis. The Mahalanobis depth and most of the other data depths considered below are affine invariant, the remaining depths are rigid motion invariant, and some are invariant against more general transformations of the data. Several data depths (halfspace depth, majority depth, and others) are combinatorial invariant, which means that they reflect nothing besides the combinatorial structure of the data.

Another important aspect is continuity: Continuity means that the depth varies only slowly with small perturbations of the data. We distinguish continuity in two respects, *viz.* on the point and on the distribution. The Mahalanobis depth is continuous in both respects. Continuity on the point implies that the depth has a finite maximum and that the trimmed regions are closed.

Other properties concern the computability and the robustness of the depth. Concerning computability, the various depth notions differ in the availability of efficient algorithms to calculate the depth of a point or, which is an even more involved task, to calculate a *depth contour*, which is the border of a level set; see Chapter 3. The Mahalanobis depth contours are ellipsoids, which are particularly easy to calculate. Some applications are computationally more intensive than others. In bootstrap tests for homogeneity, e.g., the depth has to be calculated a large number of times, while the point estimation of a median affords not so many calculations. Therefore, in certain applications we are restricted to the use of only a few, easily computable depth notions, while in others we have a larger choice among different depths to use.

Each depth yields a family of trimmed regions including the median. In general, different depths produce different trimmed regions. Depending on the depth they are based on, these regions and, in particular, medians are more or less robust against outlying data. For example, the median produced by the Mahalanobis depth is the mean vector, which, of course, is not robust and has breakdown zero. Since every trimmed region includes the deepest point (= median) as an element, the breakdown point of a trimmed region cannot be higher than the breakdown point of the median. The breakdown of various data depths has been analyzed by Donoho and Gasko (1992) and Koshevoy (1999a); see also the endnotes of Chapter 3. In applications where the robustness of the statistical procedure is an issue it is advisable to employ a depth whose median has a sufficiently large breakdown point.

When a depth is used for constructing a multivariate rank test (see Chapter 5), the richness of values the depth attains is important. If the depth has only few values or if it is constant on a large set, too many ties will arise, which impair the power of the test. In cluster analysis (see Section 7.5), if clusters are built from trimmed regions, the shape of the trimmed regions plays some role.

In the sequel a data depth, named *zonoid depth,* is developed which has particularly nice properties: The zonoid depth determines the distribution in a unique way. It assumes its unique maximum (equal to one) at the mean, decreases on rays extending from the mean and vanishes outside the convex hull of the support. The zonoid depth increases if the distribution is dilated. Moreover, it is continuous on the point as well as on the distribution and can be efficiently calculated.

In terms of robustness the zonoid depth is not competitive: As the zonoid depth median is the mean, the asymptotic breakdown point of this median equals zero and the same holds for any zonoid trimmed region. The zonoid depth shares this nonrobustness with the Mahalanobis depth. Other depths, like the halfspace depth, show better robustness properties.

In Section 4.1 the zonoid depth is defined and the relations of its sample and population versions to the zonoid trimmed regions are considered. In Section 4.2 the principal properties of the zonoid depth are investigated. Section 4.3 surveys several popular data depths and relates their properties to the zonoid depth. Section 4.4 covers the combinatorial invariance of depths. In Section 4.5 an efficient algorithm is presented that calculates the zonoid depth of an arbitrary point in $\mathbb{R}^d$ with respect to a given empirical distribution of d-variate data.

4.1 Zonoid depth

A notion of data depth is introduced, called zonoid data depth or, shortly, zonoid depth and its properties are investigated. Recall that, given a measure $\mu \in \mathcal{P}_1$ and a measurable function $g : \mathbb{R}^d \to \mathbb{R}_+$, $\nu(B) = \int_B g(x)\mu(dx), B \in \mathcal{B}^d$, defines a measure, shortly $\nu = g\mu$. Then ν is μ-continuous with density g.

Definition 4.1 (Zonoid depth) *For given $\mu \in \mathcal{P}_1$ and $\boldsymbol{y} \in \mathbb{R}^d$, define the* zonoid depth *of $\boldsymbol{y}$ with respect to μ by*

$$d_Z(\boldsymbol{y}|\mu) \tag{4.1}$$
$$= \begin{cases} \sup\{\frac{1}{\|g\|_\infty} : g \geq 0 \text{ measurable}, g\mu \in \mathcal{A}(\mu, \boldsymbol{y})\} & \text{if } \mathcal{A}(\mu, \boldsymbol{y}) \neq \emptyset\,, \\ 0 & \text{otherwise}, \end{cases}$$

where

$$\mathcal{A}(\mu, \boldsymbol{y}) = \{\nu \in \mathcal{P}_1 : \nu \text{ is } \mu\text{-continuous and } \epsilon(\nu) = \boldsymbol{y}\}.$$

Note that $\mathcal{A}(\mu, \boldsymbol{y})$ is the subset of probability distributions ν that have expectation $\epsilon(\nu) = \boldsymbol{y}$ and possess a density g with respect to μ, $\nu = g\mu$.

Relation to zonoid trimmed regions

If μ is an empirical distribution on the points $\boldsymbol{x}_1, \ldots, \boldsymbol{x}_n$, write

$$d_Z(\boldsymbol{y}|\mu) = d_Z(\boldsymbol{y}|\boldsymbol{x}_1, \ldots, \boldsymbol{x}_n).$$

This is called the *sample version* of the zonoid depth in contrast to the *population version*, where μ is a general probability measure. In the empirical distribution case the class $\mathcal{A}(\mu, \boldsymbol{y})$ consists of all discrete distributions supported by $\boldsymbol{x}_1, \ldots, \boldsymbol{x}_n$. We obtain $g(\boldsymbol{x}_i) = n\lambda_i$, $\lambda_i = \nu(\{\boldsymbol{x}_i\})$, and $\sum_i \lambda_i \boldsymbol{x}_i = \boldsymbol{y}$. Note that $\mathcal{A}(\mu, \boldsymbol{y})$ is empty if and only if $\boldsymbol{y} \notin conv\{\boldsymbol{x}_1, \ldots, \boldsymbol{x}_n\}$; in this case, $d_Z(\boldsymbol{y}|\boldsymbol{x}_1, \ldots, \boldsymbol{x}_n) = 0$. For $\boldsymbol{y} \in conv\{\boldsymbol{x}_1, \ldots, \boldsymbol{x}_n\}$ holds

$$\begin{aligned} & d_Z(\boldsymbol{y}|\boldsymbol{x}_1, \ldots, \boldsymbol{x}_n) \\ = \ & \sup\left\{ \frac{1}{n \max\limits_{1 \leq i \leq n} \lambda_i} : \boldsymbol{y} = \sum_{i=1}^n \lambda_i \boldsymbol{x}_i,\ \sum_{i=1}^n \lambda_i = 1,\ \lambda_i \geq 0\ \forall i \right\} \\ = \ & \sup\left\{ \alpha : n \max_{1 \leq i \leq n} \lambda_i = \frac{1}{\alpha}, \boldsymbol{y} = \sum_{i=1}^n \lambda_i \boldsymbol{x}_i,\ \sum_{i=1}^n \lambda_i = 1,\ \lambda_i \geq 0\ \forall i \right\} \\ = \ & \sup\left\{ \alpha : \alpha\lambda_i \leq \frac{1}{n}, \boldsymbol{y} = \sum_{i=1}^n \lambda_i \boldsymbol{x}_i,\ \sum_{i=1}^n \lambda_i = 1,\ \lambda_i \geq 0\ \forall i \right\} \qquad (4.2) \\ = \ & \sup\left\{ \alpha : \boldsymbol{y} \in D_\alpha(\mu) \right\}, \end{aligned}$$

where $D_\alpha(\mu)$ is the zonoid trimmed region; see Definition 3.1.

If μ is L-continuous with density f, it follows by Theorem 3.5 that

$$d_Z(y|\mu) = \begin{cases} \sup\left\{\alpha : \frac{1}{\alpha}\int_U x\mu(dx) = y,\ \mu(U) = \alpha\right\} & \text{if this set is nonempty,} \\ 0 & \text{otherwise.} \end{cases}$$

Observe that in this case the supremum in (4.1) is attained at a density g which is constant on some set U and vanishes outside. Then $\alpha = 1/||g||_\infty = \mu(U)$.

The last relation between the zonoid depth and the zonoid trimmed regions is valid for general probability distributions as well:

Proposition 4.1 (Trimmed regions and zonoid depth) *For* $\mu \in \mathcal{P}_1$ *holds:*

$$d_Z(y|\mu) = \sup\{\alpha : y \in D_\alpha(\mu)\}, \quad y \in \mathbb{R}^d, \tag{4.3}$$

$$D_\alpha(\mu) = \{y \in \mathbb{R}^d : d_Z(y|\mu) \geq \alpha\}, \quad 0 \leq \alpha \leq 1, \tag{4.4}$$

$$D_\alpha(\mu) = \left\{y \in \mathbb{R}^d : g\mu \in \mathcal{A}(\mu, y), ||g||_\infty \leq \frac{1}{\alpha}\right\}, \quad 0 < \alpha \leq 1. \tag{4.5}$$

Proof. According to Definition 3.1, $y \in D_\alpha(\mu)$ for some $\alpha > 0$ if and only if there exists a measurable function $g : \mathbb{R}^d \to \mathbb{R}_+$ such that

$$y = \int_{\mathbb{R}^d} x\, g(x) d\mu(x), \quad \int_{\mathbb{R}^d} g(x) d\mu(x) = 1, \quad \text{and} \quad 0 \leq g(x) \leq \frac{1}{\alpha} \text{ for all } x. \tag{4.6}$$

(4.6) says that g is a density with $||g||_\infty \leq \frac{1}{\alpha}$ and $\epsilon(g\mu) = y$. Hence Equation (4.5) holds. Note further that $\mathcal{A}(\mu, y)$ is not empty if and only if $y \in D_\alpha(\mu)$ for some $\alpha > 0$. In this case,

$$\begin{aligned} d_Z(y|\mu) &= \sup\left\{\alpha : g\mu \in \mathcal{A}(\mu, y), ||g||_\infty = \frac{1}{\alpha}\right\} \\ &= \sup\left\{\alpha : g\mu \in \mathcal{A}(\mu, y), ||g||_\infty \leq \frac{1}{\alpha}\right\} \\ &= \sup\{\alpha : y \in D_\alpha(\mu)\}. \end{aligned}$$

Otherwise, there holds $d_Z(y|\mu) = 0$ and $\alpha = 0$. We conclude (4.3). Finally (4.4) follows from (4.3). Q.E.D.

Several properties of the zonoid data depth are immediately seen from Proposition 4.1: The depth of $\boldsymbol{y}$ equals zero if $\boldsymbol{y}$ lies outside $cl \bigcup_{\alpha>0} D_\alpha = conv(\mu)$; it equals one if $\boldsymbol{y}$ is the expectation.

Recall (Proposition 3.2) that D_α equals $\frac{1}{\alpha}$ times the α-slice, $\widehat{Z}_\alpha(\mu)$, of the lift zonoid, that is, the projection of the d-dimensional cut of the lift zonoid at $z_0 = \alpha$. We conclude:

Corollary 4.2 *The zonoid depth of* $\boldsymbol{y}$ *is the maximal value* α *at which* $\alpha\, \boldsymbol{y} \in \widehat{Z}_\alpha(\mu)$.

Every point outside the convex hull of the distribution's support has zonoid depth zero. The following corollary says that every point of the relative interior of this convex hull lies on the relative border $r\partial D_\alpha$ for exactly one $\alpha > 0$ which equals its depth.

Corollary 4.3 (Relative border of trimmed regions)

(i) *For any* $\boldsymbol{y}$ *in the relative interior of* $conv(\mu)$ *holds* $d_Z(\boldsymbol{y}|\mu) > 0$ *and*

$$\boldsymbol{y} \in r\partial D_\alpha(\mu) \quad \Longleftrightarrow \quad \alpha = d_Z(\boldsymbol{y}|\mu)\,. \tag{4.7}$$

(ii) *For any* $\boldsymbol{y} \in r\partial conv(\mu)$,

$$\boldsymbol{y} \in r\partial D_\alpha(\mu) \quad \Longleftrightarrow \quad 0 \le \alpha \le d_Z(\boldsymbol{y}|\mu). \tag{4.8}$$

Proof. The corollary follows from Proposition 4.1 in connection with Theorems 3.14 and 3.13. Q.E.D.

4.2 Properties of the zonoid depth

The zonoid depth has attractive properties, which we shall investigate now. The first property is affine invariance, the second is a projection property. For $\mu \in \mathcal{P}_1$, consider a general affine transform, $T_{\boldsymbol{A},\boldsymbol{c}} : \boldsymbol{x} \mapsto \boldsymbol{x}\boldsymbol{A}+\boldsymbol{c}$ with some $d \times k$ matrix $\boldsymbol{A}$ and $\boldsymbol{c} \in \mathbb{R}^k$. Let $\mu^{\boldsymbol{A},\boldsymbol{c}} = \mu \circ T^{-1}_{\boldsymbol{A},\boldsymbol{c}}$ denote the image measure of μ under $T_{\boldsymbol{A},\boldsymbol{c}}$. Then $\mu_{\boldsymbol{p}} = \mu^{\boldsymbol{p}',0}$ is the univariate marginal distribution when μ is projected into some direction $\boldsymbol{p} \in S^{d-1}$.

Theorem 4.4 (Affine invariance) *For* $\mu \in \mathcal{P}_1$ *and any* $d \times k$ *matrix* $\boldsymbol{A}$ *and* $\boldsymbol{c} \in \mathbb{R}^k$ *holds*

$$\begin{aligned} d_Z(\boldsymbol{y}\boldsymbol{A}+\boldsymbol{c}|\mu^{\boldsymbol{A},\boldsymbol{c}}) &\geq d_Z(\boldsymbol{y}|\mu) \quad \textit{and} \\ d_Z(\boldsymbol{y}\boldsymbol{A}+\boldsymbol{c}|\mu^{\boldsymbol{A},\boldsymbol{c}}) &= d_Z(\boldsymbol{y}|\mu) \quad \textit{if } \boldsymbol{A} \textit{ has full rank } d\,. \end{aligned}$$

Proof. From Proposition 4.1 and Theorem 3.3 obtain that

$$\begin{aligned} d_Z(\boldsymbol{y}\boldsymbol{A}+\boldsymbol{c}|\mu^{\boldsymbol{A},\boldsymbol{c}}) &= \sup\{\alpha : \boldsymbol{y}\boldsymbol{A}+\boldsymbol{c} \in D_\alpha(\mu^{\boldsymbol{A},\boldsymbol{c}})\} \\ &= \sup\{\alpha : \boldsymbol{y}\boldsymbol{A}+\boldsymbol{c} \in D_\alpha(\mu)\boldsymbol{A}+\boldsymbol{c}\} \\ &= \sup\{\alpha : \boldsymbol{y}\boldsymbol{A} \in D_\alpha(\mu)\boldsymbol{A}\} \\ &\geq \sup\{\alpha : \boldsymbol{y} \in D_\alpha(\mu)\}. \end{aligned}$$

The last inequality holds with "=" if the matrix $\boldsymbol{A}$ possesses full rank d. Q.E.D.

The next two theorems describe the main properties of $d_Z(\boldsymbol{y}|\mu)$ as a function of $\boldsymbol{y}$ and μ, respectively.

Theorem 4.5 (Dependence on $\boldsymbol{y}$) *For $\mu \in \mathcal{P}_1$, the zonoid depth is*

(i) **zero at infinity:**

$$\lim_{M\to\infty} \sup_{\|\boldsymbol{y}\|\geq M} d_Z(\boldsymbol{y}|\mu) = 0,$$

(ii) **continuous on $\boldsymbol{y}$:**

$$\lim_{\boldsymbol{y}_n\to\boldsymbol{y}} d_Z(\boldsymbol{y}_n|\mu) = d_Z(\boldsymbol{y}|\mu)$$

if $\boldsymbol{y}_n \in conv(\mu)$ for all n and $\boldsymbol{y}_n \to \boldsymbol{y}$,

(iii) **unity only at expectation:**

$$d_Z(\boldsymbol{\epsilon}(\mu)|\mu) = 1 > d_Z(\boldsymbol{y}|\mu) \quad \textit{for all } \boldsymbol{y} \neq \boldsymbol{\epsilon}(\mu),$$

(iv) **decreasing on rays:**

$$d_Z(\boldsymbol{\epsilon}(\mu)+\gamma_1\boldsymbol{p}|\mu) \geq d_Z(\boldsymbol{\epsilon}(\mu)+\gamma_2\boldsymbol{p}|\mu)$$

if $\boldsymbol{p} \in S^{d-1}$ and $0 \leq \gamma_1 \leq \gamma_2$.

Theorem 4.5 says that the depth is *maximum at expectation*, *vanishes at infinity*, and *decreases on rays* extending from the deepest point. The maximum zonoid depth, attained at expectation, is one. Thus, the zonoid depth generates a unique median, the expectation. Moreover, it vanishes uniformly at infinity, and the restriction of the zonoid depth to the convex hull of the support is continuous. In particular, if the distribution is centrally symmetric[1], the zonoid depth takes its maximum at the center of symmetry.

[1]A probability measure μ is *centrally symmetric* about $\boldsymbol{c}$ if a random vector $\boldsymbol{X}$ which is distributed by μ satisfies $\boldsymbol{X}-\boldsymbol{c} =_{st} \boldsymbol{c}-\boldsymbol{X}$. μ is *angularly symmetric* about $\boldsymbol{c}$ if the distribution of $(\boldsymbol{X}-\boldsymbol{c})/\|\boldsymbol{X}-\boldsymbol{c}\|$ is centrally symmetric about $\boldsymbol{0}$. Note that every centrally symmetric distribution is also angular symmetric, but the reverse is not true.

Proof of Theorem 4.5. (i): Let $(y_n)_{n\in\mathbb{N}}$ be a an unbounded sequence in $\mathbb{R}^d$. Assume that there exists some α such that $d_Z(y_n|\mu) \geq \alpha > 0$ holds for all n. Then $y_n \in D_\alpha(\mu)$ for all n, which contradicts the compactness of $D_\alpha(\mu)$. Therefore, $\sup_n d_Z(y_n|\mu) = 0$.

(ii): Let $\mu \in \mathcal{P}_1$, $y_n \in conv(\mu)$ for $n \in \mathbb{N}$, and $y_n \to y$. Then $(d_Z(y_n|\mu))_{n\in\mathbb{N}}$ is a bounded sequence. In order to prove that $\lim_n d_Z(y_n|\mu) = d_Z(y|\mu)$, we shall show that, whenever a subsequence $(d_Z(y_{n_j}|\mu))_{j\in\mathbb{N}}$ converges, then

$$\lim_j d_Z(y_{n_j}|\mu) = d_Z(y|\mu). \tag{4.9}$$

Let $(d_Z(y_{n_j}|\mu))_{j\in\mathbb{N}}$ be a convergent subsequence and assume that y is in the relative interior of the convex hull of the support, $y \in rint\,conv(\mu)$. Then w.l.o.g. $y_{n_j} \in rint\,conv(\mu)$ for all j and, as $D_\alpha(\mu)$ is continuous on α (Theorem 3.9), $\lim_j D_{d_Z(y_{n_j}|\mu)}(\mu) = D_{\lim_j d_Z(y_{n_j}|\mu)}(\mu)$ in the Hausdorff distance. Since $y_{n_j} \to y$ and $y_{n_j} \in r\partial D_{d_Z(y_{n_j}|\mu)}(\mu)$ for all j, we get $y \in r\partial D_{\lim_j d_Z(y_{n_j}|\mu)}(\mu)$. (4.9) follows from (4.7).

If $y \in r\partial\, conv(\mu)$, conclude from (4.8) that $\lim_j d_Z(y_{n_j}|\mu) \leq d_Z(y|\mu)$. Assume that $\lim_j d_Z(y_{n_j}|\mu) < d_Z(y|\mu)$ and let $\gamma = (\lim_j d_Z(y_{n_j}|\mu) + d_Z(y|\mu))/2$. Then $D_\gamma(\mu) \supset D_{d_Z(y|\mu)}(\mu)$, and for all j that are larger than some j_0 holds $0 < d_Z(y_{n_j}|\mu) < \gamma$, hence $y_{n_j} \notin D_\gamma(\mu)$. That, again with Theorem 3.9, contradicts the convergence $y_{n_j} \to y$.

(iii): We have $D_1(\mu) = \partial D_1(\mu) = \{\epsilon(\mu)\}$. If $conv(\mu)$ is a singleton, the claim is obvious. If not, the relative interior of $conv(\mu)$ is non-empty and contains $\epsilon(\mu)$; then the claim follows from Corollary 4.3 (i).

(iv): As all D_α are convex and contain $\epsilon(\mu)$, they are starshaped about $\epsilon(\mu)$. Further, they decrease on α (Theorem 3.14). This yields the monotonicity of the depth on rays that extend from $\epsilon(\mu)$. Q.E.D.

Theorem 4.6 states that the dependence on the distribution μ is also continuous.

Theorem 4.6 (Continuous on μ) *Let $\mu \in \mathcal{P}_1$, $y \in rint\,conv(\mu)$, and $\mu, \mu^n \in \mathcal{P}_1, n \in \mathbb{N}$, with $\mu^n \overset{w}{\to} \mu$. Assume that the sequence μ^n is uniformly integrable. Then $d_Z(y|\mu^n) \to d_Z(y|\mu)$.*

Proof. Let $y \in rint\,conv(\mu)$. Then $d_Z(y|\mu) > 0$ holds by Corollary 4.3. The sequence $d_Z(y|\mu^n)$ is bounded. There exists a subsequence of $d_Z(y|\mu^n)$, say $d_Z(y|\mu^{n(k)})$, such that $0 < d_Z(y|\mu^{n(k)})$ for all k and $\lim_k d_Z(y|\mu^{n(k)}) = d_Z(y|\mu) > 0$. According to Theorem 3.10,

$\lim_j D_{d_Z(y|\mu^{n(k)})}(\mu^j) = D_{d_Z(y|\mu^{n(k)})}(\mu)$ for every $k \in \mathbb{N}$. Taking $\lim_k$ on both sides we obtain

$$\begin{aligned}\lim_k D_{d_Z(y|\mu^{n(k)})}(\mu^{n(k)}) &= \lim_k \lim_j D_{d_Z(y|\mu^{n(k)})}(\mu^j) \\ &= \lim_k D_{d_Z(y|\mu^{n(k)})}(\mu) \\ &= D_{\lim_k d_Z(y|\mu^{n(k)})}(\mu)\,. \end{aligned} \tag{4.10}$$

The last equation holds by Theorem 3.9. As $y \in rint\, conv(\mu)$, obtain $y \in r\partial D_{d_Z(y|\mu^{n(k)})}(\mu^{n(k)})$ for all k due to (4.7). Therefore, by the Hausdorff convergence (4.10), $y \in r\partial D_{\lim_k d_Z(y|\mu^{n(k)})}(\mu)$, and, again by (4.7),

$$\lim_k d_Z(y|\mu^{n(k)}) = d_Z(y|\mu).$$

As the last equation holds for every convergent subsequence $(\mu^{n(k)})$ of (μ^n) and the sequence $d_Z(y|\mu^n)$ is bounded, we conclude that $\lim_n d_Z(y|\mu^n) = d_Z(y|\mu)$. Q.E.D.

The zonoid depth has another nice property: The depth of a point y with respect to the distribution μ is determined by the depths of the projections $\langle p, y\rangle$ with respect to the marginals μ_p. It is concluded from Theorem 4.4 that

$$d_Z(y|\mu) \le \inf_{p\in S^{d-1}} d_Z(\langle p, y\rangle|\mu_p)\,, \quad y \in \mathbb{R}^d\,.$$

Moreover, it can be shown (Dyckerhoff, 2002) that the inequality must be an equality:

Theorem 4.7 (Projection infimum) *For $\mu \in \mathcal{P}_1$,*

$$d_Z(y|\mu) = \inf_{p\in S^{d-1}} d_Z(\langle p, y\rangle|\mu_p)\,, \quad y \in \mathbb{R}^d\,. \tag{4.11}$$

If a depth, like d_Z, satisfies (4.11), one says that it satisfies the *projection property*. A depth that has this property can be numerically approximated by choosing a number of directions and calculating the pertaining univariate depths and their minimum over all directions. This proves to be particularly useful when d is large.

Now we study the behavior of the zonoid depth when the data are subject to a dilation.

Definition 4.2 (Dilation) *For μ and $\nu \in \mathcal{P}_1$, ν is a* dilation *of μ if there exists a Markov kernel $(\boldsymbol{x}, B) \mapsto M(\boldsymbol{x}, B)$ such that*

$$\nu(B) = \int_{\mathbb{R}^d} M(\boldsymbol{x}, B)\, \mu(d\boldsymbol{x}), \quad B \in \mathcal{B}^d,$$
$$\boldsymbol{x} = \int_{\mathbb{R}^d} \boldsymbol{y}\, M(\boldsymbol{x}, d\boldsymbol{y}), \quad \boldsymbol{x} \in \mathbb{R}^d.$$

A dilation can be characterized in many ways. The dilation order is the same as the convex order. A more detailed discussion of this order and its connections with the lift zonoid are given in Section 8.4. Here we mention only the equivalent characterization by random vectors with an error term: If μ and ν are the distributions of two random vectors $\boldsymbol{X}$ and $\boldsymbol{Y}$, then ν is a dilation of μ if and only if there exists some $\boldsymbol{U}$ such that $\mathrm{E}[\boldsymbol{U}|\boldsymbol{X}] = 0$ and $\boldsymbol{Y}$ is distributed like $\boldsymbol{X} + \boldsymbol{U}$. The vector $\boldsymbol{U}$ can be interpreted as noise added to the random vector $\boldsymbol{X}$.

If both μ and ν are empirical distributions on the points $\boldsymbol{x}_1, \ldots, \boldsymbol{x}_n$ and $\boldsymbol{y}_1, \ldots, \boldsymbol{y}_m$, respectively, ν is a dilation of μ if and only if each $\boldsymbol{x}_i$ is a weighted mean of the $\boldsymbol{y}$'s,

$$\boldsymbol{x}_i = \sum_{j=1}^{n} \pi_{ij}\, \boldsymbol{y}_j \quad \text{for every } i,$$

where $[\pi_{ij}]$ forms a doubly stochastic matrix. That means μ is regained from its dilation ν by averaging. Briefly, a dilation is the opposite of an average. As ν, in this sense, is more dispersed than μ, the zonoid depth of a given point should be larger with ν than with μ, which will now be demonstrated.

Theorem 4.8 (Increasing on dilation) *Let $\mu, \nu \in \mathcal{P}_1$ and ν be a dilation of μ. Then $d_Z(\boldsymbol{y}|\mu) \leq d_Z(\boldsymbol{y}|\nu)$.*

Theorem 4.8 says that the zonoid depth at any given point increases if the data is subject to some additive noise.

Proof. Assume that ν is a dilation of μ. Dilation implies inclusion of the lift zonoids; see Theorem 8.13. Therefore, $D_\alpha(\mu) \subset D_\alpha(\nu)$ for every $\alpha \in [0,1]$ or, equivalently, $d_Z(\boldsymbol{y}|\mu) \leq d_Z(\boldsymbol{y}|\nu)$ for every $\boldsymbol{y} \in \mathbb{R}^d$. Q.E.D.

Finally we state a uniqueness result:

Theorem 4.9 (Uniqueness of μ) *For $\mu, \nu \in \mathcal{P}_1$,*

$$d_Z(\boldsymbol{y}|\mu) = d_Z(\boldsymbol{y}|\nu) \quad \textit{for all } \boldsymbol{y} \in \mathit{conv}(\mu) \quad \Longrightarrow \quad \mu = \nu.$$

Proof. As, for $\alpha > 0$, $D_\alpha(\mu) = \{y : d_Z(y|\mu) \geq \alpha\} \subset conv(\mu)$, conclude from the premise that $D_\alpha(\mu) = D_\alpha(\nu)$ holds for all $\alpha > 0$. By Theorem 3.6 follows $\mu = \nu$. Q.E.D.

4.3 Different notions of data depth

Various notions of data depth have been proposed in different contexts of applications. Most of them have been defined *ad hoc* to solve specific tasks in data analysis. They differ in their deepest points, in the shape of their depth contours, their invariance properties, their computability, and their robustness against outliers. To meet different goals and needs of applications, differences in these respects are useful and desired.

In general, a ***depth*** is an arbitrary real function, defined at any point in $\mathbb{R}^d$ and at distributions in some set of d-variate probability measures. But a number of properties or postulates are intrinsic or, at least, desirable to hold for any meaningful notion of data depth: A data depth should decrease monotonically on rays from a deepest point and vanish at infinity. If there is a center of – properly defined – symmetry, the depth should take its maximum at this center. These postulates will be investigated in Section 5.1 below. Further, it is desirable that the depth is a continuous function of the point as well as of the distribution. A depth function should monotonically increase with dilations of the distribution and different distributions should have different depth functions. We have seen in the previous section that the zonoid depth satisfies all these postulates.

The last property, uniqueness, is particularly important if the depth is used in inference as a nonparametric statistic. Further, if the depth contours are employed as multivariate quantiles, the depth contours of a sample should converge to the depth contours of the underlying population.

Given a depth, the *α-trimmed region* includes all points whose depth is at least α. Some properties of the depth have immediate consequences for the depth trimmed regions: If the depth is affine invariant, its trimmed regions are affine equivariant. If the depth decreases on rays from a median, the trimmed regions are starshaped about the median and, hence, connected. If the depth is upper continuous (on the point y), the trimmed regions are closed. If it is upper continuous and vanishes at $||y|| \to \infty$, the α-trimmed regions are compact for any $\alpha > 0$. For details on the mutual relations between the depth and the depth trimmed regions, see Section 5.1.

In this section several data depths will be shortly surveyed, some of which are well known, and their properties with those of the zonoid depth will be contrasted. Some invariance properties and computational issues are postponed to Sections 4.4 and 4.5 and to Chapter 5.

All the depths below are defined for μ in $\mathcal{P}_1$ or some some subset thereof. In the sequel we always assume that $conv(\mu)$ has full affine dimension d. Otherwise the analysis has to be restricted to a proper subspace.

We start with the $\mathbb{L}_2$-depth, $d_{\mathbb{L}_2}$, which is based on the mean outlyingness of a point, as measured by the Euclidean distance.

$\mathbb{L}_2$-depth. If $\mu \in \mathcal{P}_1$ and $\boldsymbol{X}$ are distributed as μ define

$$d_{\mathbb{L}_2}(\boldsymbol{y}|\mu) = (1 + \mathrm{E}||\boldsymbol{y} - \boldsymbol{X}||)^{-1} . \tag{4.12}$$

This is the *population version* of the $\mathbb{L}_2$-depth. In general, the *sample version* of a depth is the population version taken at an empirical distribution. Hence

$$d_{\mathbb{L}_2}(\boldsymbol{y}|\boldsymbol{x}_1, \ldots, \boldsymbol{x}_n) = \left(1 + \frac{1}{n}\sum_{i=1}^{n} ||\boldsymbol{y} - \boldsymbol{x}_i||\right)^{-1} \tag{4.13}$$

is the sample version of the $\mathbb{L}_2$-depth. Obviously, the $\mathbb{L}_2$-depth depends continuously on $\boldsymbol{y}$, vanishes at infinity, and is maximum at any *spatial median*[2]. If the distribution is centrally symmetric, the center is a spatial median, hence the maximum is attained at the center. Monotonicity with respect to the deepest point as well as convexity and compactness of the trimmed regions derive immediately from the triangle inequality. Like the zonoid depth, the $\mathbb{L}_2$-depth is, for uniformly integrable sequences of measures, continuous on μ.

Unlike the zonoid depth, the $\mathbb{L}_2$-depth does not increase on dilations of μ, but decreases: As $||\boldsymbol{y} - \boldsymbol{x}||$ is convex in $\boldsymbol{x}$, the expectation $\mathrm{E}[||\boldsymbol{y} - \boldsymbol{X}||]$ increases with a dilation of μ. Hence (4.12) decreases.

The $\mathbb{L}_2$-depth is also invariant against rigid Euclidean motions, but not affine invariant. An affine invariant version is constructed as follows: Given a positive definite $d \times d$ matrix M define the *Mahalanobis norm* (Mahalanobis, 1936; Rao, 1988),

$$||\boldsymbol{y}||_M = \sqrt{\boldsymbol{y}M^{-1}\boldsymbol{y}'}, \quad \boldsymbol{y} \in \mathbb{R}^d . \tag{4.14}$$

Let $\boldsymbol{D_X}$ be a positive definite $d \times d$ matrix that depends continuously on (weak convergence of) the distribution and measures the dispersion of $\boldsymbol{X}$ in an affine equivariant way. The latter means that

$$\boldsymbol{D}_{\boldsymbol{X}\boldsymbol{A}+\boldsymbol{b}} = \boldsymbol{A}'\boldsymbol{D_X}\boldsymbol{A} \quad \text{holds for any } \boldsymbol{A} \text{ of full rank and any } \boldsymbol{b}. \tag{4.15}$$

[2]By definition, a spatial median is a point $\boldsymbol{y} \in \mathbb{R}^d$ which minimizes $\mathrm{E}||\boldsymbol{y} - \boldsymbol{X}||$.

Then an *affine invariant* $\mathbb{L}_2$*-depth* is given by

$$\left(1 + \mathrm{E}\|y - X\|_{D_X^{-1}}\right)^{-1}. \tag{4.16}$$

A most prominent example for D_X is the covariance matrix Σ_X of X. (Note that the covariance matrix is positive definite, as the convex hull of the support, $conv(\mu)$, is assumed to have full dimension.)

If robustness of the depth is desired, proper choices for D_X that satisfy (4.15) are the *minimum volume ellipsoid* (MVE) estimator, the *minimum covariance determinant* (MCD) estimator and similar robust covariance estimators; see Rousseeuw and Leroy (1987) and Lopuhaä and Rousseeuw (1991).

Zuo and Serfling (2000a) have investigated the affine invariant $\mathbb{L}_2$-depth with $D_X = \Sigma_X$. Besides invariance, it has similar properties as the $\mathbb{L}_2$-depth.

Mahalanobis depth. Based on the Mahalanobis norm a useful class of affine invariant depths can be constructed,

$$d_{Mah}(y|\mu) = \left(1 + \|y - c_X\|^2_{D_X}\right)^{-1}. \tag{4.17}$$

Here, c_X is a vector that measures the location of X in a continuous and affine equivariant way and, as before, D_X is a matrix that satisfies (4.15) and depends continuously on the distribution.

In particular, if c_X and D_X are the expectation and the covariance matrix, $\mathrm{E}[X] = \epsilon(\mu)$ and Σ_μ, we obtain the widely used *Mahalanobis depth*,

$$d_{Mah}(y|\mu) = \left(1 + (y - \epsilon(\mu))\Sigma_\mu^{-1}(y - \epsilon(\mu))'\right)^{-1}. \tag{4.18}$$

Its sample version is

$$d_{Mah}(y|x_1, \ldots, x_n) = \left(1 + (y - \overline{x})\Sigma_x^{-1}(y - \overline{x})'\right)^{-1}, \tag{4.19}$$

where $\overline{x} = n^{-1}\sum x_i$ and $\Sigma_x = n^{-1}\sum(x_i - \overline{x})'(x_i - \overline{x})$. In contrast to the Mahalanobis depth (4.18) resp. (4.19) we call (4.17) the *generalized Mahalanobis depth.*

Like the zonoid depth, the Mahalanobis depth is affine invariant and continuous on y and μ. It has range $]0, 1]$, vanishing for $\|y\| \to \infty$ and taking its unique maximum, equal to 1, at the expectation. Therefore, if μ is centrally symmetric, the Mahalanobis depth is maximum at the center. For any μ the trimmed regions are ellipsoids around the expectation, which are given by (3.13) and the depth decreases on rays from the deepest point. If μ is multivariate normal the same ellipsoids are zonoid trimmed regions, but for different α; see Example 3.4. It follows that the Mahalanobis depth of a given μ is a strictly increasing transform of the zonoid depth of the normal distribution $N(\epsilon(\mu), \Sigma_\mu)$. It is also easily seen for the Mahalanobis depth that the α-contour set of a sample from μ converges almost surely to the α-contour set of μ, for any $\alpha > 0$.

Concerning uniqueness however, the Mahalanobis depth, unlike the zonoid depth, fails in identifying the underlying distribution. As only the first two moments are used, any two distributions which have the same first two moments cannot be distinguished by their Mahalanobis depth functions. However, within the family of nondegenerate d-variate normal distributions – or, more general, within any affine family of nondegenerate d-variate distributions having finite second moments – a single contour set of the Mahalanobis depth suffices to identify the distribution.

The generalized Mahalanobis depth (4.17) has similar properties and does neither determine the distribution in a nonparametric setting.

Halfspace depth. Hodges (1955), building on ideas of Hotelling (1929), introduces the halfspace depth as follows:

$$d_H(y|\mu) = \inf\{\mu(H) : H \text{ a closed halfspace}, \quad y \in H\}. \tag{4.20}$$

This is the population version of the halfspace depth. The sample version is the same defined on the empirical distribution of the data.

The halfspace depth is affine invariant. Outside the convex hull of the support of μ the depth attains its minimal value, which is zero. The maximum value of the halfspace depth depends on the distribution. If μ has an L-density, the halfspace depth depends continuously on y; otherwise the dependence on y is noncontinuous and there can be more than one point where the maximum is attained. The set of all such points is mentioned as the *halfspace median set* and its gravity center as the *Tukey median* (Tukey, 1975).

If μ is angular symmetric, the halfspace depth is maximal at the center and the Tukey median coincides with the center. In general, the depth decreases monotonically on rays from any deepest point.

As a function of μ the halfspace depth is obviously noncontinuous. It determines the distribution in a unique way if the distribution is either discrete (Struyf and Rousseeuw, 1999; Koshevoy, 200xc) or continuous with compact support (Koshevoy, 200xa). The halfspace depth in a sample from μ, that is, in its empirical distribution function, converges almost surely to the halfspace depth in μ (Donoho and Gasko, 1992). The same holds for the depth contours (Massé and Theodorescu, 1994; He and Wang, 1997; Zuo and Serfling, 2000d) if the distribution is elliptic. The halfspace trimmed regions are connected and convex.

The next two depth notions involve simplices in $\mathbb{R}^d$.

Simplicial depth. Liu (1990) defines the *simplicial depth* as follows:

$$d_{Sim}(\boldsymbol{y}|\mu) = P\left(\boldsymbol{y} \in conv(\{\boldsymbol{X}_1, \ldots, \boldsymbol{X}_{d+1}\})\right), \tag{4.21}$$

where $\boldsymbol{X}_1, \ldots, \boldsymbol{X}_{d+1}$ are i.i.d. by μ. The sample version reads as

$$d_{Sim}(\boldsymbol{y}|\boldsymbol{x}_1, \ldots, \boldsymbol{x}_n) = \frac{1}{\binom{n}{d+1}} \#\Big\{\{i_1, \ldots, i_{d+1}\} : \boldsymbol{y} \in conv(\{\boldsymbol{x}_{i_1}, \ldots, \boldsymbol{x}_{i_{d+1}}\})\Big\}. \tag{4.22}$$

The simplicial depth is affine invariant. It takes values from 0 at infinity to some positive value, which depends on the distribution. In general, the point of maximum simplicial depth is not unique: the *simplicial median* is defined as the gravity center of these points. The sample simplicial depth converges almost surely uniformly in $\boldsymbol{y}$ to its population version (Liu, 1990; Dümbgen, 1992). The simplicial depth has positive breakdown (Chen, 1995).

If the distribution is L-continuous, the simplicial depth behaves very well: It varies continuously on $\boldsymbol{y}$ (Liu, 1990, Th. 2), is maximum at a center of angular symmetry, and decreases monotonously from a deepest point. The *simplicial trimmed regions* of an L-continuous distribution are connected and compact (Liu, 1990). For a continuous elliptical distribution the sample α-trimmed regions converge almost surely to their population counterparts, which are ellipsoids (Zuo and Serfling, 2000d, Cor. 4.2).

However, if the distribution is discrete, each of these properties can fail; for counterexamples see, e.g., Zuo and Serfling (2000a). The simplicial depth characterizes an empirical measure if the supporting points are in general position[3]; see Koshevoy (1997). But as the corresponding central regions can be non-convex, they seem to be less suited for statistical purposes.

Oja depth. For any points $\boldsymbol{v}_1, \ldots, \boldsymbol{v}_{d+1} \in \mathbb{R}^d$, the convex hull $conv(\{\boldsymbol{v}_1, \ldots, \boldsymbol{v}_{d+1}\})$ is a simplex and the vertices of the convex hull are contained in the set $\{\boldsymbol{v}_1, \ldots, \boldsymbol{v}_{d+1}\}$. Its d-dimensional volume equals

$$\mathrm{vol}_d(conv\{\boldsymbol{v}_1, \ldots, \boldsymbol{v}_{d+1}\}) = \frac{1}{d!} \left|\det((1, \boldsymbol{v}_1), \ldots, (1, \boldsymbol{v}_{d+1}))\right|.$$

Oja (1983) introduced a location measure for an empirical distribution on $\boldsymbol{x}_1, \ldots, \boldsymbol{x}_n$ as follows: The *Oja median* minimizes the average volume of

[3] A subset R of $\mathbb{R}^d$ is *in general position* if their convex hull has affine dimension d, that is no more than d of its points are lying on the same hyperplane. Equivalently, each subset of $d+1$ points in R spans a simplex with d-dimensional volume greater than 0.

simplices

$$\frac{1}{\binom{n}{d}} \sum_{1 \le i_1 < \cdots < i_d \le n} \text{vol}_d(conv\{\boldsymbol{y}, \boldsymbol{x}_1, \dots, \boldsymbol{x}_d\})$$

$$= \frac{(n-d)!}{n!} \sum_{1 \le i_1 < \cdots < i_d \le n} |\det((1, \boldsymbol{y}), (1, \boldsymbol{x}_{i_1}), \dots, (1, \boldsymbol{x}_{i_d}))| \, .$$

The *Oja depth* function is defined by Zuo and Serfling (2000a),

$$d_{Oja}(\boldsymbol{y}|\mu) = \left(1 + \frac{\text{E}\,(\text{vol}_d(conv\{\boldsymbol{y}, \boldsymbol{X}_1, \dots, \boldsymbol{X}_d\}))}{\sqrt{\det \boldsymbol{D}_{\boldsymbol{X}}}}\right)^{-1},$$

where $\boldsymbol{X}_1, \dots, \boldsymbol{X}_d$ are random vectors independently distributed as μ, E denotes the expectation, and $\boldsymbol{D}_{\boldsymbol{X}}$ has been defined above. Note that this definition of the Oja depth is affine invariant. In particular, we can choose $\boldsymbol{D}_{\boldsymbol{X}} = \Sigma_\mu$. Then the sample version of the Oja depth is

$$d_{Oja}(\boldsymbol{y}|\boldsymbol{x}_1, \dots, \boldsymbol{x}_n) \tag{4.23}$$
$$= \left(1 + \frac{\frac{(n-d)!}{n!} \sum_{1 \le i_1 < \cdots < i_d \le n} |\det((1, \boldsymbol{y}), (1, \boldsymbol{x}_{i_1}), \dots, (1, \boldsymbol{x}_{i_d}))|}{\sqrt{\det \boldsymbol{\Sigma}_{\boldsymbol{x}}}}\right)^{-1}.$$

The Oja depth is continuous on $\boldsymbol{y}$, zero at infinity and maximum at the Oja median, which is not unique. It decreases monotonically on rays from all points in the median set. Consequently, the *Oja trimmed regions* are starshaped and compact. In the case of a centrally symmetric distribution the median set includes the center. The Oja depth determines the distribution uniquely among those measures which have compact support of full dimension (Koshevoy, 200xa).

Majority depth. Define the *major side*, $\text{Maj}(\boldsymbol{x}_1, \dots, \boldsymbol{x}_d)$, of d points in $\mathbb{R}^d$ as follows: If $\boldsymbol{x}_1, \dots, \boldsymbol{x}_d$ are linearly independent, consider the hyperplane generated by them; it delimits two closed halfspaces. If one of the halfspaces has larger μ-mass than the other, $\text{Maj}(\boldsymbol{x}_1, \dots, \boldsymbol{x}_d)$ equals that halfspace. In all other cases, $\text{Maj}(\boldsymbol{x}_1, \dots, \boldsymbol{x}_d) = \mathbb{R}^d$.

Singh's *majority depth* (Liu and Singh, 1993) is defined by

$$d_{Maj}(\boldsymbol{y}|\mu) = P\,(\boldsymbol{y} \in \text{Maj}(\boldsymbol{X}_1, \dots, \boldsymbol{X}_d))\ , \tag{4.24}$$

where $\boldsymbol{X}_1, \dots, \boldsymbol{X}_d$ are i.i.d. by μ. The majority depth satisfies the following properties (Liu and Singh, 1993): Affine invariance, maximality at a point of angular symmetry, monotonicity on rays from the *majority median* (which is not unique), convergence of sample trimmed regions to their population counterparts. Note further that $d_{Maj}(\mathbf{0}|\mu) = 1$ if μ is rotation symmetric on the sphere. However, the majority depth is noncontinuous, and the trimmed regions are closed but not compact as the depth can be nonzero at infinity.

In Chapter 6 the depth of a hyperplane is defined and studied (Koshevoy and Mosler, 1999a). It is connected with an extension of majority depth. Related to the majority depth is also the *dual majority depth* (Koshevoy, 1999b). It will be defined and discussed in Section 4.4 below.

Peeling depth. There exist several variants of peeling in the literature. Here, we mention only the simplest, called *convex-hull peeling* (Barnett, 1976). Related notions have been proposed by Bebbington (1978), Titterington (1978), and others. For an empirical distribution on $\boldsymbol{x}_1, \ldots, \boldsymbol{x}_n$, the convex-hull peeling depth is constructed by a sequential procedure as follows:

Consider first the convex hull of $\boldsymbol{y}$ and all $\boldsymbol{x}$'s, $C_1 = conv\{\boldsymbol{y}, \boldsymbol{x}_1, \ldots, \boldsymbol{x}_n\}$, and the set of its extreme points, E_1. Calculate $C_k = C_{k-1} \setminus E_{k-1}$ in steps $k = 2, 3, \ldots$. Continue until $\boldsymbol{y} \in E_k$. Define the *convex-hull peeling depth*, shortly *CHP-depth*, of $\boldsymbol{y}$ by

$$d_{CHP}(\boldsymbol{y}|\boldsymbol{x}_1, \ldots, \boldsymbol{x}_n) = \begin{cases} \min\{k : \boldsymbol{y} \in E_k\} & \text{if such a } k \text{ exists,} \\ 0 & \text{otherwise.} \end{cases} \tag{4.25}$$

The convex-hull peeling depth is affine invariant. As this depth assumes only integer values, it is not continuous as a real function. The depth is not robust: Its breakdown point is generally bounded above by $\frac{1}{p+1} + \frac{1}{n}$. No population version exists and the asymptotic behaviour of the depth is mostly unknown. For some properties of peeling depth, see Donoho and Gasko (1992).

Projection depth. The *projection depth* has been proposed by Zuo and Serfling (2000a).

$$d_{proj}(\boldsymbol{y}|\mu) = \left(1 + \sup_{\boldsymbol{p} \in S^{d-1}} \frac{|\langle \boldsymbol{p}, \boldsymbol{y}\rangle - \mathrm{med}(\langle \boldsymbol{p}, \boldsymbol{X}\rangle)|}{\mathrm{Dmed}(\langle \boldsymbol{p}, \boldsymbol{X}\rangle)}\right)^{-1}, \tag{4.26}$$

where $\mathrm{med}(U)$ is the usual median of a univariate random variable U and $\mathrm{Dmed}(U) = \mathrm{med}(|U - \mathrm{med}(U)|$ is the median absolute deviation from the median. The projection depth has good properties, which are discussed in detail by Zuo and Serfling (2000a). For breakdown properties of the employed location and scatter statistics, see Zuo (2000).

4.4 Combinatorial invariance

In the previous section a number of depth notions have been surveyed. All of them are either affine invariant or can be made affine invariant by slight modifications. Some of them are invariant to much more general transformations of the d-space, namely, their sample versions depend only on the combinatorial structure of the data.

Observe that the halfspace depth, the simplicial depth, the majority depth and the peeling depth have certain features in common. In case $d = 1$ they yield the same median, which is the usual one. The usual univariate median depends only on the relative position of the data points and does not reflect any distances or other metrical features of the data. In dimension one the relative position of points is given by their natural order, which is invariant against any strictly increasing transformation of the line.

In dimensions two and more, there exists no natural ordering of points. But the relative position of points can be described in a similar way, stating that a point lies on one or the other 'side' of the remaining points.

Combinatorial structures have been extensively studied in the literature. Comprehensive references are Björner et al. (1999) and Bokowski (1993). In this section the invariance of depths with respect to the combinatorial equivalence of data matrices is discussed. We present a result by Koshevoy (1999b), which states that the sample versions of the above mentioned four depths (and a fifth one) depend only on the combinatorial structure of the data $[\boldsymbol{y}, \boldsymbol{x}_1, \ldots, \boldsymbol{x}_n]$, that is, they are *combinatorial invariant.* Combinatorial structure is defined through Radon partitions as follows.

A Radon partition of a data matrix divides the set of row indices, $K = \{1, \ldots, k\}$, into three disjoint subsets K_+, K_-, and K_0 so that the convex hulls of the pertaining points intersect,

$$conv\{\boldsymbol{s}_j : j \in K_+\} \cap conv\{\boldsymbol{s}_j : j \in K_-\} \neq \emptyset\,.$$

More formally, let $\boldsymbol{S} = [\boldsymbol{s}_1, \ldots, \boldsymbol{s}_k]$ be a $d \times k$ data matrix. A *Radon partition* of $\boldsymbol{S}$ is a map τ,

$$\begin{aligned} &\tau : K \to \{-1, 0, 1\} \\ \text{with} \quad &conv\{\boldsymbol{s}_j \,:\, \tau(j) = -1\} \cap conv\{\boldsymbol{s}_j \,:\, \tau(j) = 1\} \neq \emptyset\,. \end{aligned}$$

It is clear that for any Radon partition of a data matrix $\boldsymbol{S}$ there exists a similar one with K_+ and K_- interchanged. (These two partitions will not be distinguished in the sequel.) The set $K_+ \cup K_-$ is mentioned as the *support* of τ. A Radon partition of $\boldsymbol{S}$ is *minimal* if no other Radon partition of $\boldsymbol{S}$ has smaller support.

Radon partitions are also applied to a finite set of points in $\mathbb{R}^d$. A *Radon partition of a finite set* $S \subset \mathbb{R}^d$ is a Radon partition of any data matrix built from S, modulo permutations of the index set K. See also Definition 4.3 below.

Example 4.1 (Four points in the plane) Two data sets in $\mathbb{R}^2$, each consisting of four points in general position, and their minimal Radon partitions are exhibited in Figure 4.1. All points of the left-hand data are extremal in their convex hull, while one point of the right-hand data is lying in the interior of the convex hull of the other three. Obviously, up to permutations of K and interchanges of K_+ and K_-, these are the only minimal Radon partitions of the example data. Each of the two minimal Radon partitions is supported by the whole set of four points.

Moreover, if a given four-point set in $\mathbb{R}^2$ is in general position, it can be classified into one of two categories corresponding to the two example sets. These two categories are mentioned as *combinatorial structures*; each is described by a Radon partition. For four-point sets in $\mathbb{R}^2$, there exist exactly two different minimal Radon partitions (besides permutations of K and interchanges of K_+ and K_-) and, therefore, exactly two different combinatorial structures.

In general, a minimal Radon partition in the d-space is supported by at most $d+2$ of its points. If the rows of the data matrix are in general position, the support contains exactly $d+2$ of them. If $d+1$ or less points in $\mathbb{R}^d$ are in general position, they have no Radon partition.

For example, three points in $\mathbb{R}^2$ possess one Radon partition if they are collinear, while they possess no Radon partition if they are in general position. Four collinear points in the plane have six different Radon partitions, four of which are minimal, and each minimal partition is supported by three points. Note that modulo rotations of the plane only two of the four are different.

Definition 4.3 (Combinatorial structure, combinatorial invariance) (i) *The* combinatorial structure *of a data matrix* $\boldsymbol{S}$ *is the set of all minimal Radon partitions of* $\boldsymbol{S}$. *It is denoted by* $C(\boldsymbol{S})$.

(ii) *Two* $d \times k$ *data matrices* $\boldsymbol{S} = [\boldsymbol{s}_1, \ldots, \boldsymbol{s}_k]$ *and* $\boldsymbol{R} = [\boldsymbol{r}_1, \ldots, \boldsymbol{r}_k]$ *are* combinatorial equivalent *if there exists a permutation* $\sigma : K \to K$ *such that* $C(\boldsymbol{S}) = C(\boldsymbol{R}_\sigma)$. *Here* $\boldsymbol{R}_\sigma = [\boldsymbol{r}_{\sigma(1)}, \ldots, \boldsymbol{r}_{\sigma(k)}]$ *is obtained from* $\boldsymbol{R}$ *by permuting the rows* $\boldsymbol{r}_j$.

(iii) *A real-valued statistic* $t : (\boldsymbol{s}_1, \ldots, \boldsymbol{s}_k) \mapsto t(\boldsymbol{s}_1, \ldots, \boldsymbol{s}_k)$ *is* combinatorial invariant *if* $t(\boldsymbol{s}_1, \ldots, \boldsymbol{s}_k) = t(\boldsymbol{r}_1, \ldots, \boldsymbol{r}_k)$ *whenever the matrices* $\boldsymbol{S}$ *and* $\boldsymbol{R}$ *are combinatorial equivalent.*

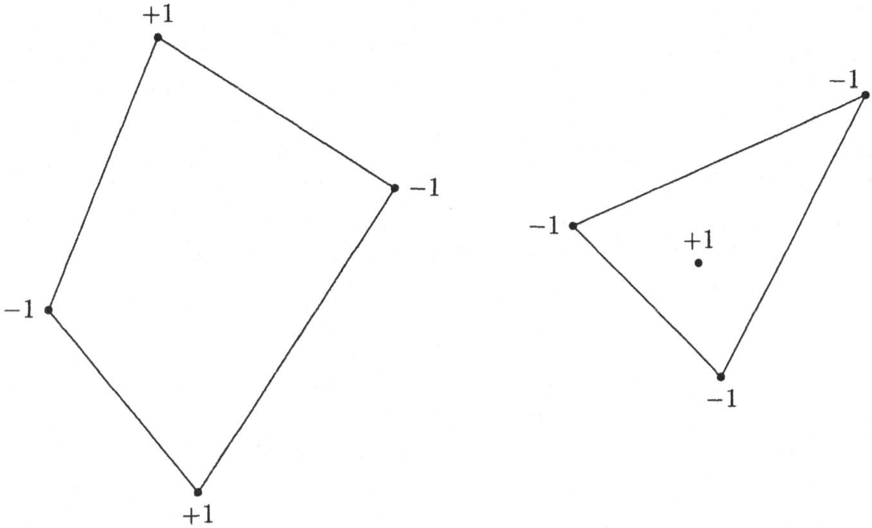

FIGURE 4.1: The two minimal Radon partitions of sets in $\mathbb{R}^2$ that consist of four points.

Two data matrices $\boldsymbol{S}$ and $\boldsymbol{R}$ are combinatorial equivalent if, after a permutation of rows, $\boldsymbol{S}$ and $\boldsymbol{R}$ have the same combinatorial structure. Note that a combinatorial invariant statistic is, in particular, symmetric in its vector arguments and affine invariant.

The example has shown that with four points in general position in $\mathbb{R}^2$ there exist two classes of combinatorial equivalence: Either all four are extreme points or one of them lies in the interior of the convex hull; see Figure 4.1.

As can be seen from simple examples, the zonoid depth is *not* combinatorial invariant; but several other depths have this property (Theorem 4.11).

It is most interesting whether and how the combinatorial equivalence of data is connected with the notion of the lift zonoid:

Theorem 4.10 (Combinatorial structure and lift zonotope) *Consider two data sets in* $\mathbb{R}^d$, $\{\boldsymbol{x}_1, \dots, \boldsymbol{x}_n\}$ *and* $\{\boldsymbol{y}_1, \dots, \boldsymbol{y}_n\}$, *both of them in general position. Then the following two restrictions are equivalent:*

(i) $\boldsymbol{S} = [\boldsymbol{x}_1, \ldots, \boldsymbol{x}_n]$ *and* $\boldsymbol{R} = [\boldsymbol{y}_1, \ldots, \boldsymbol{y}_n]$ *have the same combinatorial structure.*

(ii) *The lift zonotopes of the empirical distributions on* $\boldsymbol{S}$ *and* $\boldsymbol{R}$, $\widehat{Z}(\mu_{\boldsymbol{S}})$ *and* $\widehat{Z}(\mu_{\boldsymbol{R}})$, *are combinatorial equivalent, which means that their face lattices are isomorphic.*

Proof. See Björner et al. (1999), Proposition 2.2.2.

With the notion of Radon partitions we are able to consider another depth notion, which is due to Koshevoy (1999b). Assume that $\boldsymbol{y}, \boldsymbol{x}_1, \ldots, \boldsymbol{x}_n$ are in general position.

$$d_{dMaj}(\boldsymbol{y}|\boldsymbol{x}_1, \ldots, \boldsymbol{x}_n) = \frac{1}{\binom{n}{d+1}} \#\Big\{\{i_1, \ldots, i_{d+1}\} : \boldsymbol{y} \text{ is in the minor part of the Radon partition of } \{\boldsymbol{y}, \boldsymbol{x}_{i_1}, \ldots \boldsymbol{x}_{i_{d+1}}\}\Big\}.$$

Here $\boldsymbol{y}$ is said to be in the *minor part of the Radon partition* of the set $\{\boldsymbol{y}, \boldsymbol{x}_{i_1}, \ldots \boldsymbol{x}_{i_{d+1}}\}$ if the number of points that belong to the same index set as $\boldsymbol{y}$ is less or equal to $[(d+2)/2]$. As usual, $[\beta]$ denotes the largest integer $\leq \beta$.

Theorem 4.11 (Depth invariance in the point) *Let* $\boldsymbol{x}_1, \ldots, \boldsymbol{x}_n$ *be data in general position in* $\mathbb{R}^d$ *and let* $\boldsymbol{y}$ *and* $\boldsymbol{z} \in \mathbb{R}^d$. *Assume that* $\mathcal{C}([\boldsymbol{x}_1, \ldots, \boldsymbol{x}_n, \boldsymbol{y}]) = \mathcal{C}([\boldsymbol{x}_1, \ldots, \boldsymbol{x}_n, \boldsymbol{z}])$. *Then*

$$\text{depth}(\boldsymbol{y}|\boldsymbol{x}_1, \ldots, \boldsymbol{x}_n) = \text{depth}(\boldsymbol{z}|\boldsymbol{x}_1, \ldots, \boldsymbol{x}_n) \tag{4.27}$$

and

$$\text{depth}(\boldsymbol{x}_i|\boldsymbol{x}_1, \ldots, \boldsymbol{x}_n, \boldsymbol{y}) = \text{depth}(\boldsymbol{x}_i|\boldsymbol{x}_1, \ldots, \boldsymbol{x}_n, \boldsymbol{z}) \quad \textit{for all} \quad i = 1, \ldots, n, \tag{4.28}$$

where depth *is either* $d_H, d_{Sim}, d_{Maj}, d_{dMaj}$, *or* d_{CHP}.

Proof. Obvious for halfspace depth and simplicial depth. See Koshevoy (1999b) for the rest. Q.E.D.

Theorem 4.11 says that, if the data matrices $[\boldsymbol{x}_1, \ldots, \boldsymbol{x}_n, \boldsymbol{y}]$ and $[\boldsymbol{x}_1, \ldots, \boldsymbol{x}_n, \boldsymbol{z}]$ have the same combinatorial structure, then the depth of each $\boldsymbol{x}_i$ remains the same when $\boldsymbol{y}$ is exchanged for $\boldsymbol{z}$. This holds for the halfspace depth, the simplicial depth, the majority depth, the dual majority depth, and the convex-hull peeling depth. It is seen from simple examples that the zonoid depth, the Mahalanobis depth, the $\mathbb{L}_2$-depth, and the Oja depth do not have this property. From repeated application of Theorem 4.11 obtain:

Corollary 4.12 (Combinatorial invariance of depths) *Let* $\boldsymbol{x}_1, \ldots, \boldsymbol{x}_n$ *and* $\boldsymbol{u}_1, \ldots, \boldsymbol{u}_n$ *be data in* $\mathbb{R}^d$*, both in general position, and let* $\boldsymbol{y}$ *and* $\boldsymbol{z} \in \mathbb{R}^d$*. Assume that* $\mathcal{C}([\boldsymbol{x}_1, \ldots, \boldsymbol{x}_n]) = \mathcal{C}([\boldsymbol{u}_1, \ldots, \boldsymbol{u}_n])$ *and* $\mathcal{C}([\boldsymbol{x}_1, \ldots, \boldsymbol{x}_n, \boldsymbol{y}]) = \mathcal{C}([\boldsymbol{u}_1, \ldots, \boldsymbol{u}_n, \boldsymbol{z}])$*. Then*

$$\mathrm{depth}(\boldsymbol{y}|\boldsymbol{x}_1, \ldots, \boldsymbol{x}_n) = \mathrm{depth}(\boldsymbol{z}|\boldsymbol{u}_1, \ldots, \boldsymbol{u}_n)\,, \tag{4.29}$$

where depth *is either* $d_H, d_{Sim}, d_{Maj}, d_{dMaj}$*, or* d_{CHP}*.*

The corollary says that these five depths, halfspace, simplicial, majority, dual majority, and convex-hull peeling depth measure, are combinatorial invariant statistics. Combinatorial equivalent data are ranked by them in the same way. In other words, the depth of a point $\boldsymbol{y}$ with respect to an empirical distribution measures nothing than the combinatorial structure of the set consisting of the point $\boldsymbol{y}$ and the support of the distribution.

4.5 Computation of the zonoid depth

It is important to have special algorithms and computer codes to compute the depth of a given point in a data cloud. This section presents an algorithm of Dyckerhoff et al. (1996) that computes the zonoid depth by solving a linear program. If n data points in $\mathbb{R}^d$ are given, the linear program has $n+1$ variables and $n+d+1$ constraints. Although this linear program can be solved easily for moderate sizes of n and d, the problem becomes computationally intractable for large values of n or d if the standard simplex algorithm is used.

To overcome this problem, a more subtle approach is proposed in Dyckerhoff et al. (1996). In fact, the $n+1$ constraints have a very special structure and do not depend on the data. The algorithm exploits this structure by using a Dantzig-Wolfe decomposition. In the initialization step of the algorithm a basic feasible solution of the linear program is searched for. Then the iteration is started. In each step of the algorithm, n real numbers have to be sorted and a simplex tableau of size $(d+1) \times (d+1)$ has to be updated. Thus, for fixed d, each step of the algorithm has complexity $O(n \log n)$. The average overall complexity of the algorithm depends of course on the number of iterations needed. Simulations suggest that for fixed d the complexity lies below $O(n^2 \log n)$. Further the algorithm creates two sequences of upper and lower bounds for the depth which can be used as a stopping criterion for the algorithm.

We consider the $n \times d$ data matrix

$$R = \begin{pmatrix} r_1 \\ \vdots \\ r_n \end{pmatrix}$$

and denote $\boldsymbol{\lambda} = (\lambda_1, \ldots, \lambda_n)$, $\mathbf{1} = (1, \ldots, 1)$, $\mathbf{0} = (0, \ldots, 0) \in \mathbb{R}^n$.

The zonoid depth of a point y in $\mathbb{R}^d$ with respect to the empirical distribution on $r_1, \ldots, r_n$ can be computed from Formula (4.2) as follows:

$$\left.\begin{array}{rcl} \text{Minimize} & \gamma & \text{s.t.} \\ & & \\ \boldsymbol{\lambda} R & = & y \\ \boldsymbol{\lambda} \mathbf{1}' & = & 1 \\ \gamma \mathbf{1} - \boldsymbol{\lambda} & \geq & \mathbf{0} \\ \boldsymbol{\lambda} & \geq & \mathbf{0} \end{array}\right\} \quad \text{(LP)}$$

(LP) is a linear program in the real variables $\lambda_1, \ldots, \lambda_n$ and γ. If γ^* is an optimal solution, then

$$d_Z(y|r_1, \ldots, r_n) = \frac{1}{n\gamma^*}.$$

If (LP) has no feasible solution, it is clear that $y \notin conv(\{r_1, \ldots, r_n\})$. The algorithm exploits the special structure of the set of constraints by a Dantzig-Wolfe decomposition. (LP) can be written as

$$\left.\begin{array}{rcl} \text{Minimize} & \gamma & \text{s.t.} \\ & & \\ \boldsymbol{\lambda} R & = & y \\ (\boldsymbol{\lambda}, \gamma) & \in & W \end{array}\right\} \quad \text{(LP')}$$

where

$$W = \left\{ (\lambda_1, \ldots, \lambda_n, \gamma) \in \mathbb{R}^{n+1} : \sum_{i=1}^{n} \lambda_i = 1, 0 \leq \lambda_j \leq \gamma \leq 1 \text{ for all } j \right\}.$$

Because W is a polytope, any point in W is a convex combination of its extreme points. Fortunately the extreme points of W are explicitly known.

Proposition 4.13 (Feasible polytope) *The set V of extreme points of W is given by*

$$V = \left\{ \frac{1}{|J|}(\delta_J, 1)' : \emptyset \neq J \subset \{1, \ldots, n\} \right\},$$

where

$$\boldsymbol{\delta}_J = (\delta_J(1), \delta_J(2), \ldots, \delta_J(n)), \qquad \delta_J(k) = \begin{cases} 1 & \text{if } k \in J, \\ 0 & \text{if } k \notin J. \end{cases}$$

The proof is left to the reader. By Proposition 4.13, (LP′) can be decomposed as follows. The *master problem*, with variables β_J, $\emptyset \neq J \subset \{1, \ldots, n\}$, is given by:

$$\left.\begin{array}{lrcl} \text{Minimize} & \sum_J \frac{1}{|J|}\beta_J & & \\ \text{subject to} & \sum_J \frac{1}{|J|}\beta_J \boldsymbol{\delta}_J \boldsymbol{R} & = & \boldsymbol{y} \\ & \sum_J \beta_J & = & 1 \\ & \beta_J & \geq & 0 \text{ for all } J \end{array}\right\} \quad \text{(MP)}$$

In every simplex step of (MP) a new pivot column is selected by solving a subproblem (SP):

$$\left.\max_J \frac{1}{|J|}(\boldsymbol{\delta}_J \boldsymbol{R}\boldsymbol{w}' - 1) + \alpha \quad \right\} \quad \text{(SP)}$$

Here $\boldsymbol{w} \in \mathbb{R}^d$ and $\alpha \in \mathbb{R}$ are the simplex multipliers of the master problem.

If the maximum objective of the subproblem is greater than zero and maximized at $J = J^*$, the new pivot column for the master problem is calculated as

$$\boldsymbol{B}^{-1}\left(\frac{1}{|J^*|}\boldsymbol{\delta}_{J^*}\boldsymbol{R}, 1\right)',$$

where $\boldsymbol{B}^{-1}$ is the basis inverse of the master problem. Then the pivot row for the simplex step is determined by the usual minimal ratio choice and the tableau is updated. This process is continued until the maximum objective of the subproblem equals zero. The algorithm is stopped if the current solution of the master problem is optimal.

Additionally, the algorithm generates an increasing sequence of lower bounds for the data depth and a (not necessarily decreasing) sequence of upper bounds.

The algorithm is rather fast. Calculating the zonoid depth of a point on a standard PC, e.g., in dimension $d = 5$ takes less than half a second if $n = 1000$ and less than two seconds if $n = 2000$.

4.6 Notes

Sections 4.1 and 4.2 are largely based on Koshevoy and Mosler (1997b). Section 4.4 is extracted from Dyckerhoff et al. (1996).

The zonoid depth and many other depths vanish outside the convex hull of the distribution's support. But certain applications, e.g. two-sample tests, suggest the use of a depth which is nonconstant outside the convex hull of the data. Such an extended depth should be a depth function in the sense of Section 5.1 and be continuous on the point and on the distribution. Various such extensions are possible for different applications. The principal difference between these extensions is whether they are positive everywhere or vanish outside some superset of the convex hull. E.g., a convex combination of the zonoid depth with some other depth can be used (Hoberg, 2002); a convex combination of the zonoid depth with the Mahalanobis depth is positive everywhere.

We have seen that the Oja median gives rise to a notion of depth, the Oja depth. Vardi and Zhang (2000) provide a general approach to define a depth based on a given notion of median and present further examples.

There are many geometric relations between the Oja depth and the halfspace depth on one side and the lift zonoid on the other side, in particular (Koshevoy, 200xa):

1. The Oja depth of x with respect to μ equals the length of the vector $(1, x)$ multiplied by the d-dimensional volume of a proper projection of the lift zonoid of μ. It is the projection on the hyperplane orthogonal to $(1, x)$. In other words, the Oja depth equals the mixed volume of the segment $[0, (1, x)]$ and the lift zonoid. From this and a variant of Alexandrov's uniqueness theorem for mixed volumes a uniqueness property of the Oja depth is derived (Koshevoy, 200xa): The Oja depth determines the underlying distribution uniquely among all distributions having compact support of full dimension.

2. Consider again, for some L-continuous μ, the projection of the lift zonoid of μ on the hyperplane orthogonal to $(1, x)$. The halfspace depth of x equals the height of the inverse image of the boundary of this projection. This relation yields a homeomorphism between halfspace depth contours and zonoid depth contours for all values of the halfspace depth that are smaller than the Tukey median. (The zonoid depth of x in μ equals the height of the intersection of the straight line through $(1, x)$ with the lift zonoid.)

It is obvious in dimension one that there exists a transformation of the real line that relates the halfspace depth with the zonoid depth: The zonoid depth with respect to a given distribution is the halfspace depth with respect to the transformed distribution. It has been shown (Koshevoy and Mosler, 1997b) that also in higher dimensions the zonoid depth equals twice Tukey's data depth of a properly transformed distribution. However, no closed expression of this transformation is known in dimension two and more.

A further characterization of the halfspace depth is given in Carrizosa (1996). The asymptotic distribution of the halfspace depth is derived in Nolan (1992) for the bivariate case and in Massé (1999) for the general case.

The combinatorial structure of a finite number of points can be characterized in several alternative ways. It is also encoded by the oriented matroid of the points; see e.g. Björner et al. (1999). There is a close relation between the oriented matroid and the set of all minimal Radon partitions: The pair $(K, \mathcal{C}(\boldsymbol{S}))$ is an affine oriented matroid. Two data matrices $\boldsymbol{S} = [\boldsymbol{x}_1, \dots, \boldsymbol{x}_n]$ and $\boldsymbol{R} = [\boldsymbol{u}_1, \dots, \boldsymbol{u}_n]$ have the same oriented matroid if and only if they are combinatorial equivalent.

5

Inference based on data depth

by Rainer Dyckerhoff

Tests which are based on the rank statistic play an important role in univariate nonparametric statistics. The rank statistic of a sample assigns to each sample variable its position in the order statistic. If one tries to generalize rank tests to the multivariate case, one is faced with the problem of defining a multivariate order statistic. Because of the absence of a natural linear order on $\mathbb{R}^d$ it is not clear how to define a multivariate order or rank statistic in a meaningful way.

One approach to overcome this problem is the notion of data depth. Data depth is a concept which measures the 'depth' of a point in a given data cloud. Using this concept, a multivariate order statistic can be defined by ordering the data points according to their depth in a suitable data set. If the rank statistic is defined with respect to this ordering, a point which is far from the center has a low rank whereas points near the center have high ranks.

In this chapter two multivariate rank tests that use a data depth are described. In both tests the null hypothesis is homogeneity, that is, that two samples come from the same (multivariate) distribution. The first test uses the sum of the ranks of the first sample as a test statistic. Here the ranks are assigned according to the depth in the *pooled* sample. This is a situation analogous to the usual univariate Wilcoxon test. The distribution of the test under the null hypothesis is the same as the null distribution of the univariate Wilcoxon test. It will be seen that this test has high power on scale alternatives whereas it is not able to detect differences in location between the two distributions. Although the null distribution of the test does not depend on the depth which is used, it influences the power of the test. This influence is investigated in simulation studies.

The second test we discuss was proposed by Liu (1992) and Liu and Singh (1993). In this test ranks are assigned to points of the *second sample* according to their depth in the *first sample*. Liu and Singh (1993) have shown that the distribution of the test statistic under the null hypothesis is asymptotic normal for certain depths. This test detects differences in location between the two samples as well as a greater scale of the first sample compared to the second sample.

In Section 5.1 the abstract concept of data depth is shortly discussed. The tests mentioned above are presented in detail in Sections 5.2 and 5.3. Also some basic properties of the tests are developed. In Section 5.4 some classical tests which are natural competitors to the depth tests are surveyed. A new rank test which is based on distances between observations is proposed in Section 5.6. Finally the simulation studies are discussed in Section 5.7.

5.1 General notion of data depth

Data depth is a concept that measures how 'deep' a point $\boldsymbol{y}$ is located in a given data cloud $\boldsymbol{x}_1, \ldots, \boldsymbol{x}_n \in \mathbb{R}^d$ or a given distribution P. Therefore, data depth can be used to order multivariate data. Different notions of data depth have been proposed in the literature; see the preceding Chapter 4. Let us define a data depth as a function that satisfies the following four conditions.

Affine invariance. For a regular $d \times d$-matrix $\boldsymbol{A}$ and $\boldsymbol{b} \in \mathbb{R}^d$ it holds

$$d(\boldsymbol{y}\boldsymbol{A} + \boldsymbol{b}|\boldsymbol{x}_1\boldsymbol{A} + \boldsymbol{b}, \ldots, \boldsymbol{x}_n\boldsymbol{A} + \boldsymbol{b}) = d(\boldsymbol{y}|\boldsymbol{x}_1, \ldots, \boldsymbol{x}_n).$$

Vanishing at infinity. For every sequence $(\boldsymbol{y}_k)_{k \in \mathbb{N}}$ with $\lim_{k \to \infty} ||\boldsymbol{y}_k|| = \infty$ it holds $\lim_{k \to \infty} d(\boldsymbol{y}_k|\boldsymbol{x}_1, \ldots, \boldsymbol{x}_n) = 0$.

Monotone on rays. For each $\boldsymbol{y}$ of maximal depth and each $\boldsymbol{r} \in S^{d-1}$, the function $\mathbb{R}_+ \to \mathbb{R}_+$, $\lambda \mapsto d(\boldsymbol{y} + \lambda\boldsymbol{r}|\boldsymbol{x}_1, \ldots, \boldsymbol{x}_n)$, is monotone decreasing.

Upper semicontinuity. For each $\alpha > 0$ the set

$$D_\alpha(\boldsymbol{x}_1, \ldots, \boldsymbol{x}_n) := \{\boldsymbol{y} \in \mathbb{R}^d \mid d(\boldsymbol{y}|\boldsymbol{x}_1, \ldots, \boldsymbol{x}_n) \geq \alpha\}$$

is closed.

Definition 5.1 *Every function that satisfies the above four conditions is called a* data depth.

In the simulation studies (see Section 5.6 below) we used the Mahalanobis depth (Mahalanobis, 1936), the halfspace depth (Tukey, 1975) and the zonoid depth (Koshevoy and Mosler, 1997b). These three depths all satisfy the above conditions, whereas other proposals such as the simplicial depth of Liu (1990) or the majority depth of Singh (1991) do not. The Mahalanobis depth, the halfspace depth and the zonoid depth have already been defined in Sections 4.1 and 4.3 above, so we do not repeat their definitions here.

The level sets $D_\alpha(\boldsymbol{x}_1, \dots, \boldsymbol{x}_n)$ defined above are called *trimmed regions*. From these conditions several important properties follow. In particular, the trimmed regions of any given depth are affine equivariant, compact and starshaped with respect to every deepest point. Further, for centrally symmetric distributions the center of symmetry has maximal depth. For elliptical distributions[1] the trimmed regions are ellipsoids parallel to the density contours.

Every depth defines a dispersion order $\preceq^{depth}$ between random vectors $\boldsymbol{X}$ and $\boldsymbol{Y}$ via

$$\boldsymbol{X} \preceq^{depth} \boldsymbol{Y} \quad \Longleftrightarrow \quad D_\alpha(\boldsymbol{X}) \subset D_\alpha(\boldsymbol{Y}) \quad \text{for every } \alpha.$$

We will call this order *depth order*.

The depth order is reflexive and transitive. The depth order $\preceq^M$ that results from the Mahalanobis depth can be easily characterized through the moments of the random vectors:

$$\boldsymbol{X} \preceq^M \boldsymbol{Y} \quad \Longleftrightarrow \quad \boldsymbol{\mu}_{\boldsymbol{X}} = \boldsymbol{\mu}_{\boldsymbol{Y}},\ \boldsymbol{\Sigma}_{\boldsymbol{Y}} - \boldsymbol{\Sigma}_{\boldsymbol{X}} \text{ is positive semidefinite.}$$

The depth order $\preceq^Z$ that results from the zonoid depth is the lift zonoid order, which is discussed in Section 1.6 and in Chapter 8. The order $\preceq^H$ which is defined by the halfspace depth cannot be characterized in such a nice way. However, the halfspace depth ordering is closely connected to the well-known peakedness ordering, see (Dyckerhoff, 2002, Section 3.2).

For elliptical distributions of the same type and with the same center $\boldsymbol{a}$, all depth orders coincide. If $\boldsymbol{X} \sim \mathcal{E}(\boldsymbol{a}, \boldsymbol{\Sigma}_{\boldsymbol{X}}, \psi)$ and $\boldsymbol{Y} \sim \mathcal{E}(\boldsymbol{a}, \boldsymbol{\Sigma}_{\boldsymbol{Y}}, \psi)$, then for every depth holds that

$$\boldsymbol{X} \preceq^{depth} \boldsymbol{Y} \quad \Longleftrightarrow \quad \boldsymbol{\Sigma}_{\boldsymbol{Y}} - \boldsymbol{\Sigma}_{\boldsymbol{X}} \text{ is positive semidefinite.}$$

[1] For the definition of elliptical distributions, see Section 2.3.5.

5.2 Two-sample depth test for scale

Let $\boldsymbol{X}_1, \ldots, \boldsymbol{X}_m \overset{iid}{\sim} F$ and $\boldsymbol{Y}_1, \ldots, \boldsymbol{Y}_n \overset{iid}{\sim} G$ be two independent samples from multivariate distributions F and G. Further, let the random variables $\boldsymbol{X}_1, \ldots, \boldsymbol{X}_m, \boldsymbol{Y}_1, \ldots, \boldsymbol{Y}_n$ be independent. We want to test the null hypothesis $H_0 : F = G$. In what follows denote the size of the pooled sample by N, i.e., $N = m + n$.

The *depth test* which we propose here is in some sense an analogue of the usual univariate Wilcoxon rank sum test. In fact, we replace the linear order of the real numbers by the order induced by the depth in the pooled sample. That means we order the data points with respect to their depths in the pooled sample.

Let 'd' be a data depth. For $\boldsymbol{z} \in \mathbb{R}^d$ we define

$$J(\boldsymbol{z}) = d(\boldsymbol{z} | \boldsymbol{X}_1, \ldots, \boldsymbol{X}_m, \boldsymbol{Y}_1, \ldots, \boldsymbol{Y}_n) .$$

Thus, $J(\boldsymbol{z})$ is the depth of $\boldsymbol{z}$ in the pooled sample $\boldsymbol{X}_1, \ldots, \boldsymbol{X}_m, \boldsymbol{Y}_1, \ldots, \boldsymbol{Y}_n$. Now, define the rank $R(\boldsymbol{X}_i)$ of an observation as the usual univariate rank of $J(\boldsymbol{X}_i)$ with respect to the set $\{J(\boldsymbol{X}_1), \ldots, J(\boldsymbol{X}_m), J(\boldsymbol{Y}_1), \ldots, J(\boldsymbol{Y}_n)\}$. The rank $R(\boldsymbol{Y}_i)$ is analogously defined. If the depth is not continuous, it may happen with positive probability that some of the values $\{J(\boldsymbol{X}_1), \ldots, J(\boldsymbol{X}_m), J(\boldsymbol{Y}_1), \ldots, J(\boldsymbol{Y}_n)\}$ are tied. To resolve these ties we use a random tie breaking scheme. If k observations have the same depth and the ranks $i, i+1, \ldots, i+k-1$ have to be assigned to these observations, each permutation of these ranks will be assigned to these observations with equal probability.

The test statistic is defined as the sum of the ranks of the first sample,

$$T = \sum_{i=1}^{m} R(\boldsymbol{X}_i) .$$

The following theorem gives the distribution of T under the null hypothesis.

Theorem 5.1 *Assume that the random tie breaking scheme is used. Then, under $H_0 : F = G$, the distribution of T is the same as the distribution of the univariate Wilcoxon test, i.e., the distribution of the sum of m numbers randomly drawn from the integers $\{1, \ldots, m+n\}$ without replacement.*

Proof. Let

$$\boldsymbol{Z}_i = \begin{cases} \boldsymbol{X}_i & \text{if } i \leq m , \\ \boldsymbol{Y}_{i-m} & \text{if } i > m . \end{cases}$$

Further, let $\boldsymbol{S} = (S_1, \ldots, S_{m+n})$ be a random vector which assumes all permutations of the integers $1, 2, \ldots, m+n$ with equal probability $1/(m+n)!$. We define an order on the pairs $(\boldsymbol{Z}_i, S_i)$ by

$$(\boldsymbol{Z}_i, S_i) \preceq (\boldsymbol{Z}_j, S_j) \iff J(\boldsymbol{Z}_i) < J(\boldsymbol{Z}_j) \text{ or } (J(\boldsymbol{Z}_i) = J(\boldsymbol{Z}_j) \text{ and } S_i \leq S_j).$$

For $i \neq j$ we have $(\boldsymbol{Z}_i, S_i) \neq (\boldsymbol{Z}_j, S_j)$ since $S_i \neq S_j$. Thus, there are no ties. If we define the rank of $\boldsymbol{X}_i$ as the rank of $(\boldsymbol{Z}_i, S_i)$ in all $(\boldsymbol{Z}_j, S_j)$, $j = 1, \ldots, m+n$, this assignment of ranks is equivalent to the random tie breaking scheme.

Under H_0 the pairs $(\boldsymbol{Z}_i, S_i)$, $i = 1, \ldots, m+n$, are identically distributed and exchangeable. Thus, each rank constellation has the same probability $1/(m+n)!$. Therefore, T has the same distribution as $\sum_{i=1}^{m} T_i$, where $T_1, \ldots, T_m$ denote a sample without replacement from the integers $1, \ldots, m+n$. But this is exactly the distribution of the univariate Wilcoxon statistic under the null hypothesis. Q.E.D.

If the two distributions differ only by a location parameter, the center of *all* data points lies somewhere between the centers of the two distributions. Therefore, the ranks of the first sample have approximately the same distribution as the ranks of the second sample. Thus, this test is not able to detect differences in location between the two distributions.

We now assume that both distributions have the same center $\boldsymbol{\mu}$ but that the first distribution is more dispersed about the center than the second. In this case points of the first sample tend to be farther away from the center than points of the second sample. Thus, one should expect that in this case the value of the test statistic is small. Analogously, the test statistic should be large when the first distribution is less dispersed about the common center than the second one.

Thus, the null hypothesis is to be rejected, when the value of the test statistic is either very small or very large. According to the above Theorem 5.1 critical values for this test can be found in the usual Wilcoxon test tables or, for large sample sizes, by the respective normal approximation.

An important property of a test is *affine invariance*. This means that the test statistic is invariant against affine transformation of the sample variables. More formally, let $\boldsymbol{A}$ be a regular $d \times d$ matrix and $\boldsymbol{b} \in \mathbb{R}^d$. Then it should hold that

$$\begin{aligned} &T(\boldsymbol{X}_1, \ldots, \boldsymbol{X}_m, \boldsymbol{Y}_1, \ldots, \boldsymbol{Y}_n) \\ &\qquad = T(\boldsymbol{X}_1\boldsymbol{A} + \boldsymbol{b}, \ldots, \boldsymbol{X}_m\boldsymbol{A} + \boldsymbol{b}, \boldsymbol{Y}_1\boldsymbol{A} + \boldsymbol{b}, \ldots, \boldsymbol{Y}_n\boldsymbol{A} + \boldsymbol{b}). \end{aligned}$$

The above described test possesses this property.

Theorem 5.2 *The test statistic T is affine invariant.*

Proof. Since we consider only depths that are affine invariant, it holds that

$$d(\boldsymbol{X}_i\boldsymbol{A}+\boldsymbol{b}|\boldsymbol{X}_1\boldsymbol{A}+\boldsymbol{b},\dots,\boldsymbol{X}_m\boldsymbol{A}+\boldsymbol{b},\boldsymbol{Y}_1\boldsymbol{A}+\boldsymbol{b},\dots,\boldsymbol{Y}_n\boldsymbol{A}+\boldsymbol{b}) \\ = d(\boldsymbol{X}_i|\boldsymbol{X}_1,\dots,\boldsymbol{X}_m,\boldsymbol{Y}_1,\dots,\boldsymbol{Y}_n)$$

Thus, the original and the transformed sample variables get the same ranks, which proofs the assertion. Q.E.D.

Because of the affine invariance one can always restrict to the case that both distributions have their centers in the origin. If F and G are elliptical distributions, that differ only in their scale parameters $\boldsymbol{\Sigma_X}$ and $\boldsymbol{\Sigma_Y}$, in investigating the power, one can restrict oneself always to the case that $\boldsymbol{\Sigma_X}=\boldsymbol{I}_d$ and $\boldsymbol{\Sigma_Y}=\boldsymbol{D}$ is a diagonal matrix. This is made precise in the following theorem.

Theorem 5.3 *Let $\boldsymbol{X}\sim\mathcal{E}(\boldsymbol{a},\boldsymbol{\Sigma_X},\psi)$, $\boldsymbol{Y}\sim\mathcal{E}(\boldsymbol{a},\boldsymbol{\Sigma_Y},\psi)$, $\boldsymbol{\Sigma_X},\boldsymbol{\Sigma_Y}$ positive definite. Further, let $\lambda_1\geq\dots\geq\lambda_d$ be the eigenvalues of $\boldsymbol{\Sigma_Y}\boldsymbol{\Sigma_X}^{-1}$. Then the distribution of T depends only on ψ and $\lambda_1,\dots,\lambda_d$.*

Proof. It holds that $\boldsymbol{\Sigma_X}=\boldsymbol{Q}_1\boldsymbol{\Lambda}\boldsymbol{Q}_1'$, where $\boldsymbol{Q}_1$ is orthogonal and $\boldsymbol{\Lambda}$ is the diagonal matrix of eigenvalues of $\boldsymbol{\Sigma_X}$. Further, the diagonal elements of $\boldsymbol{\Lambda}$ are strictly positive. Let $\boldsymbol{\Sigma_X}^{-1/2}=\boldsymbol{Q}_1\boldsymbol{\Lambda}^{-1/2}\boldsymbol{Q}_1{}'$. Then

$$(\boldsymbol{X}-\boldsymbol{a})\boldsymbol{\Sigma_X}^{-1/2}\sim\mathcal{E}(\boldsymbol{0},\boldsymbol{I}_d,\psi),$$
$$(\boldsymbol{Y}-\boldsymbol{a})\boldsymbol{\Sigma_X}^{-1/2}\sim\mathcal{E}(\boldsymbol{0},\boldsymbol{\Sigma_X}^{-1/2}\boldsymbol{\Sigma_Y}\boldsymbol{\Sigma_X}^{-1/2},\psi).$$

But $\boldsymbol{\Sigma_X}^{-1/2}\boldsymbol{\Sigma_Y}\boldsymbol{\Sigma_X}^{-1/2}$ is positive definite, too. Thus, $\boldsymbol{\Sigma_X}^{-1/2}\boldsymbol{\Sigma_Y}\boldsymbol{\Sigma_X}^{-1/2}=\boldsymbol{Q}_2\boldsymbol{D}\boldsymbol{Q}_2'$ where $\boldsymbol{Q}_2$ is orthogonal and $\boldsymbol{D}$ is the diagonal matrix of eigenvalues of $\boldsymbol{\Sigma_X}^{-1/2}\boldsymbol{\Sigma_Y}\boldsymbol{\Sigma_X}^{-1/2}$. Now, let $\boldsymbol{A}=\boldsymbol{\Sigma_X}^{-1/2}\boldsymbol{Q}_2$. Then

$$(\boldsymbol{X}-\boldsymbol{a})\boldsymbol{A}\sim\mathcal{E}(\boldsymbol{0},\boldsymbol{I}_d,\psi)\quad\text{and}\quad(\boldsymbol{Y}-\boldsymbol{a})\boldsymbol{A}\sim\mathcal{E}(\boldsymbol{0},\boldsymbol{D},\psi).$$

Because of affine invariance it follows that

$$T((\boldsymbol{X}_1-\boldsymbol{a})\boldsymbol{A},\dots,(\boldsymbol{X}_m-\boldsymbol{a})\boldsymbol{A},(\boldsymbol{Y}_1-\boldsymbol{a})\boldsymbol{A},\dots,(\boldsymbol{Y}_n-\boldsymbol{a})\boldsymbol{A}) \\ =_{st} T(\boldsymbol{X}_1,\dots,\boldsymbol{X}_m,\boldsymbol{Y}_1,\dots,\boldsymbol{Y}_n)$$

Therefore the distribution of $T(\boldsymbol{X}_1,\dots,\boldsymbol{X}_m,\boldsymbol{Y}_1,\dots,\boldsymbol{Y}_n)$ depends only on ψ and $\boldsymbol{D}$. It remains to show that the entries of $\boldsymbol{D}$ are the eigenvalues of $\boldsymbol{\Sigma_Y}\boldsymbol{\Sigma_X}^{-1}$. If $\boldsymbol{x}$ is an eigenvector of $\boldsymbol{\Sigma_X}^{-1/2}\boldsymbol{\Sigma_Y}\boldsymbol{\Sigma_X}^{-1/2}$ to the eigenvalue λ, then

$$\boldsymbol{x}\boldsymbol{\Sigma_X}^{-1/2}\boldsymbol{\Sigma_Y}\boldsymbol{\Sigma_X}^{-1/2}=\lambda\boldsymbol{x}\quad\Longleftrightarrow\quad\boldsymbol{x}\boldsymbol{\Sigma_X}^{-1/2}\boldsymbol{\Sigma_Y}\boldsymbol{\Sigma_X}^{-1}=\lambda\boldsymbol{x}\boldsymbol{\Sigma_X}^{-1/2}.$$

The second equality states that $\boldsymbol{y} = \boldsymbol{x}\boldsymbol{\Sigma}_{\boldsymbol{X}}^{-1/2}$ is an eigenvector of $\boldsymbol{\Sigma}_{\boldsymbol{Y}}\boldsymbol{\Sigma}_{\boldsymbol{X}}^{-1}$ to the eigenvalue λ. Thus, $\boldsymbol{\Sigma}_{\boldsymbol{X}}^{-1/2}\boldsymbol{\Sigma}_{\boldsymbol{Y}}\boldsymbol{\Sigma}_{\boldsymbol{X}}^{-1/2}$ and $\boldsymbol{\Sigma}_{\boldsymbol{Y}}\boldsymbol{\Sigma}_{\boldsymbol{X}}^{-1}$ have the same set of eigenvalues. Q.E.D.

To make the test practically applicable, one has to ensure that the depth can be efficiently calculated. Possible candidates for the depth are therefore the Mahalanobis depth, whose calculation is very simple, as well as the halfspace depth and the zonoid depth, for which good algorithms exist; see Rousseeuw and Ruts (1996) and Rousseeuw and Struyf (1998) for the halfspace depth and Dyckerhoff et al. (1996) for the zonoid depth.

In computing the test statistic, N values of the depth have to be evaluated. Thus, the time complexity is of order $O(N)C(N)$, where $C(N)$ denotes the complexity of computing the depth of a single point with respect to a data cloud of size N. However, all depths are computed with respect to the same data cloud, namely the pooled sample. For some depths this can result in a considerable simplification of the computations. E.g., the Mahalanobis depth of each point in a data cloud with respect to that data cloud can be computed with a time complexity of $O(N)$.

Besides the computability of the depth, an important aspect in choosing a specific depth for this test is the power of the resulting test. The simulation studies which will be presented in Section 5.6 show that the specific depth used in the test has a great influence on the power of the test.

5.3 Two-sample rank test for location and scale

In this section a depth-based test is described which goes back to Liu (1992) and Liu and Singh (1993). This is a rank test for the two-sample location-scale problem. In the depth test for the two-sample scale problem, ranks were assigned to the observations according to their *depth in the pooled sample.* In the test discussed here ranks are assigned to the Y-sample according to their *depth in the X-sample.* For $\boldsymbol{z} \in \mathbb{R}^d$ we define

$$J(\boldsymbol{z}) = d(\boldsymbol{z} \mid \boldsymbol{X}_1, \dots, \boldsymbol{X}_m, \boldsymbol{z}),$$

i.e., $J(\boldsymbol{z})$ is the depth of $\boldsymbol{z}$ in the union of the Y-sample and the point $\boldsymbol{z}$ itself. Now, define $R(\boldsymbol{Y}_j)$ as the rank of $J(\boldsymbol{Y}_j)$ with respect to the set $\{J(\boldsymbol{X}_1), \dots, J(\boldsymbol{X}_m), J(\boldsymbol{Y}_j)\}$. As in the case of the depth test for scale, some of the values $J(\boldsymbol{X}_1), \dots, J(\boldsymbol{X}_m), J(\boldsymbol{Y}_j)$ may be tied. Liu (1992) computes the rank as

$$R(\boldsymbol{Y}_j) = \#\{\boldsymbol{X}_i \mid J(\boldsymbol{X}_i) \leq J(\boldsymbol{Y}_j)\},$$

which means that $R(\boldsymbol{Y}_j)$ always assumes the highest possible rank if some of the values $J(\boldsymbol{X}_i)$ are tied with $J(\boldsymbol{Y}_j)$. However, we suggest to use a random tie breaking scheme as discussed in the section on the depth test for scale. The test statistic of the Liu test is defined by

$$T = \frac{1}{n} \sum_{j=1}^{n} \frac{R(\boldsymbol{Y}_i) - 1}{m}.$$

The term $(R(\boldsymbol{Y}_j) - 1)/m$ can be interpreted as the fraction of $\boldsymbol{X}_i$s that have a depth of at most $J(\boldsymbol{Y}_j)$ in the data cloud consisting of all $\boldsymbol{X}_i$s and $\boldsymbol{Y}_j$. Under H_0 and if the random tie breaking scheme is used each $R(\boldsymbol{Y}_j)$ is uniformly distributed on the integers $1, \ldots, m+1$. Therefore, $(R(\boldsymbol{Y}_j) - 1)/m$ is uniformly distributed on $\{0, \frac{1}{m}, \ldots, \frac{m-1}{m}, 1\}$. This implies that under H_0 the expectation of T equals $1/2$. Note that this condition is only approximately true if the tie breaking scheme in Liu (1992) is used.

On the other hand if there is a difference in location between the two samples then the $\boldsymbol{Y}_j$s will be further away from the center of the X-sample and thus have low ranks. The same holds when the Y-sample is more scattered than the X-sample. Thus, low values of T suggest that there is a shift in location or an increase in scale from $\boldsymbol{X}$ to $\boldsymbol{Y}$. One would reject the null $H_0 : F = G$ if the test statistic has low values.

If the $R(\boldsymbol{Y}_j)$s were independent then the exact as well as the asymptotic distribution could be easily derived. However, it is clear that the $R(\boldsymbol{Y}_j)$s are dependent which makes it difficult to find the null distribution of T in general. Liu and Singh (1993) were able to derive the limiting distribution of T in special cases. In fact, they were able to prove that under some weak assumptions the limiting distribution of the standardized test statistic is standard normal if one of the following two conditions is satisfied:

1. The depth used in the test is the Mahalanobis depth.

2. F and G are univariate distributions and the depth used in the test is the halfspace depth, the simplicial depth or the majority depth.

Liu and Singh (1993) conjecture that in general the limit distribution is standard normal. We have conducted several simulation studies on this; their results support the conjecture.

As in the scale test, we conclude that the Liu test is affine invariant. This is obvious from the fact that a depth function is affine invariant.

To compute the test statistic a large number of depths has to be computed. More precisely we have to compute all the depths

$$\begin{aligned} &d(\boldsymbol{Y}_j \,|\, \boldsymbol{X}_1, \ldots, \boldsymbol{X}_m, \boldsymbol{Y}_j), \quad j = 1, \ldots, n, \\ &d(\boldsymbol{X}_i \,|\, \boldsymbol{X}_1, \ldots, \boldsymbol{X}_m, \boldsymbol{Y}_j), \quad i = 1, \ldots, m, \; j = 1, \ldots, n. \end{aligned}$$

Thus, $(m+1)n$ depth values have to be computed. The complexity of the computation of the test statistic depends, of course, on the depth which is used. If m and n go to infinity in such a way that m/n has a limit $\lambda \in (0, \infty)$, the complexity of computing the test statistic is of order $O(N^2)C(N)$, where $C(N)$ denotes the complexity of computing the depth. Contrary to the scale test, the depths have to be computed with respect to different data clouds. This makes the computations more complicated than in the scale test.

5.4 Classical two-sample tests

Before we discuss the simulations and their results in the next section we will give a short overview on some tests which are natural competitors to the depth tests discussed in the previous two sections. These tests were used in the simulations to show how the depth tests perform in comparison with existing tests.

5.4.1 Box's M test

Under normality, i.e., if we assume that $\boldsymbol{X}_1, \ldots, \boldsymbol{X}_m \overset{iid}{\sim} N(\boldsymbol{\mu}_1, \boldsymbol{\Sigma}_1)$ and $\boldsymbol{Y}_1, \ldots, \boldsymbol{Y}_n \overset{iid}{\sim} N(\boldsymbol{\mu}_2, \boldsymbol{\Sigma}_2)$, the hypothesis $H_0 : \boldsymbol{\Sigma}_1 = \boldsymbol{\Sigma}_2$ can be tested using a likelihood ratio test. The likelihood ratio test statistic is given by

$$T_{\boldsymbol{X},\boldsymbol{Y}} = m \log \frac{\det(\boldsymbol{S}_{\boldsymbol{X},\boldsymbol{Y}})}{\det(\boldsymbol{S}_{\boldsymbol{X}})} + n \log \frac{\det(\boldsymbol{S}_{\boldsymbol{X},\boldsymbol{Y}})}{\det(\boldsymbol{S}_{\boldsymbol{Y}})},$$

where $\boldsymbol{S}_{\boldsymbol{X}}$ and $\boldsymbol{S}_{\boldsymbol{Y}}$ denote the sample covariance matrices of the $\boldsymbol{X}_i$ and the $\boldsymbol{Y}_i$ respectively, and $\boldsymbol{S}_{\boldsymbol{X},\boldsymbol{Y}} = \frac{1}{N}(m\boldsymbol{S}_{\boldsymbol{X}} + n\boldsymbol{S}_{\boldsymbol{Y}})$. Under H_0 this statistic has an asymptotic χ^2-distribution with $\frac{1}{2}d(d+1)$ degrees of freedom. Instead of $T_{\boldsymbol{X},\boldsymbol{Y}}$, Box (1949) proposed to use the test statistic

$$M_{\boldsymbol{X},\boldsymbol{Y}} = \gamma \left((m-1) \log \frac{\det(\boldsymbol{S}^*_{\boldsymbol{X},\boldsymbol{Y}})}{\det(\boldsymbol{S}^*_{\boldsymbol{X}})} + (n-1) \log \frac{\det(\boldsymbol{S}^*_{\boldsymbol{X},\boldsymbol{Y}})}{\det(\boldsymbol{S}^*_{\boldsymbol{Y}})} \right),$$

where

$$\gamma = 1 - \frac{2d^2 + 3d - 1}{6(d+1)} \left(\frac{1}{m-1} + \frac{1}{n-1} - \frac{1}{N-2} \right)$$

and $\boldsymbol{S}^*_{\boldsymbol{X}}$, $\boldsymbol{S}^*_{\boldsymbol{Y}}$ and $\boldsymbol{S}^*_{\boldsymbol{X},\boldsymbol{Y}}$ are the unbiased estimators

$$\boldsymbol{S}^*_{\boldsymbol{X}} = \frac{m}{m-1} \boldsymbol{S}_{\boldsymbol{X}}\,, \quad \boldsymbol{S}^*_{\boldsymbol{Y}} = \frac{n}{n-1} \boldsymbol{S}_{\boldsymbol{Y}}\,, \quad \boldsymbol{S}^*_{\boldsymbol{X},\boldsymbol{Y}} = \frac{N}{N-2} \boldsymbol{S}_{\boldsymbol{X},\boldsymbol{Y}}\,.$$

The test statistic $M_{\boldsymbol{X},\boldsymbol{Y}}$ has an asymptotic χ^2-distribution with $\frac{1}{2}d(d+1)$ degrees of freedom, too. To be fair, it should be noted that Box's M test is consistent on a much broader alternative than our rank test. In fact, the alternative of Box's M test is $H_1 : \boldsymbol{\Sigma}_1 \neq \boldsymbol{\Sigma}_2$, whereas the appropriate alternative of the scale test would be $H_1 : \boldsymbol{\Sigma}_1 - \boldsymbol{\Sigma}_2$ is semidefinite.

Note that Box's M test is affine invariant. If all the data points are transformed according to $\boldsymbol{x} \mapsto \boldsymbol{x}\boldsymbol{A} + \boldsymbol{b}$, where $\boldsymbol{A}$ is a regular matrix, then $\det(\boldsymbol{S}_{\boldsymbol{X}\boldsymbol{A}+\boldsymbol{b}}) = \det(\boldsymbol{A}')\det(\boldsymbol{S}_{\boldsymbol{X}})\det(\boldsymbol{A}) = \det^2(\boldsymbol{A})\det(\boldsymbol{S}_{\boldsymbol{X}})$. In the same way it follows that $\det(\boldsymbol{S}_{\boldsymbol{Y}\boldsymbol{A}+\boldsymbol{b}}) = \det^2(\boldsymbol{A})\det(\boldsymbol{S}_{\boldsymbol{Y}})$ and $\det(\boldsymbol{S}_{\boldsymbol{X}\boldsymbol{A}+\boldsymbol{b},\boldsymbol{Y}\boldsymbol{A}+\boldsymbol{b}}) = \det^2(\boldsymbol{A})\det(\boldsymbol{S}_{\boldsymbol{X},\boldsymbol{Y}})$. Therefore

$$T_{\boldsymbol{X}\boldsymbol{A}+\boldsymbol{b},\boldsymbol{Y}\boldsymbol{A}+\boldsymbol{b}} = T_{\boldsymbol{X},\boldsymbol{Y}}\,, \qquad M_{\boldsymbol{X}\boldsymbol{A}+\boldsymbol{b},\boldsymbol{Y}\boldsymbol{A}+\boldsymbol{b}} = M_{\boldsymbol{X},\boldsymbol{Y}}\,.$$

Box's M is easy to compute. The computation of the covariance matrices and their determinants can be done in $O(N)$. Thus, the test statistic can be computed with a complexity of $O(N)$.

5.4.2 Friedman-Rafsky test

The *Friedman-Rafsky test*, also called *minimal spanning tree test* or shortly MSTT, was proposed by Friedman and Rafsky (1979) and can be seen as a generalization of the univariate run test of Wald and Wolfowitz (1940). The run test starts with sorting the pooled sample in ascending order. Each observation is then replaced by an 'x' or a 'y' depending on the sample to which the observation belongs. The test statistic is the number of *runs* in this sequence. A run is a consecutive sequence of identical labels which is preceded or followed by a different label or by no label.

Let us shortly review some notions of graph theory. A *graph* consists of two sets, a set of *nodes* and a set of node pairs called *edges*. Two nodes a and b in a graph are said to be *connected* if there is a sequence $z_0, \ldots, z_k$ of nodes such that $a = z_0$, $b = z_k$ and each pair (z_{i-1}, z_i), $i = 1, \ldots, k$, is an edge. A graph is said to be connected if each pair of nodes is connected. A graph contains a *cycle* if there is a sequence $z_0, \ldots, z_k = z_0$ of nodes such that every pair (z_{i-1}, z_i), $i = 1, \ldots, k$, is an edge. A connected graph that contains no cycles is called a *tree.* A *subgraph* of a given graph is a graph whose node and edge sets are subsets of the given graph. Two subgraphs of a given graph are *disjoint* if they have no node in common. A *spanning tree* of a given graph is a tree that has a node set equal to the node set of the given graph and an edge set that is a subset of the edges of the given graph. A subgraph of a tree is a tree and is called *subtree.* A weighted graph is a graph where each edge is assigned a real number. A *minimal spanning tree* of a weighted graph is a spanning tree for which the sum of the weights of the edges is a minimum.

Given two samples $\boldsymbol{x}_1, \ldots, \boldsymbol{x}_m$ and $\boldsymbol{y}_1, \ldots, \boldsymbol{y}_n$ in $\mathbb{R}^d$ consider the graph whose nodes are the points of the pooled sample and whose edge set is the set of *all* pairs of data points in the pooled sample. Each edge is given a weight which equals the Euclidean distance between the nodes (or any other sensible distance). The minimal spanning tree of this graph is called the minimal spanning tree of the pooled sample.

The minimal spanning tree test proceeds as follows. In the first step the minimal spanning tree of the pooled sample is constructed. In the second step all edges which consist of points originating from different samples are removed. The test statistic R is the number of disjoint subtrees that result.

If we apply this procedure to two univariate samples, the test statistic R is exactly the number of runs. Thus, the minimal spanning tree test can be seen as a generalization of the run test.

The expectation of R under H_0 is given by

$$E(R) = \frac{2mn}{N} + 1 .$$

The variance of R under H_0 depends on the number of edge pairs in the minimal spanning tree that share a common node. We denote this number by C. The conditional variance of R under H_0 given C is given by

$$V(R \,|\, C) = \frac{2mn}{N(N-1)} \left\{ \frac{2mn - N}{N} + \frac{C - N + 2}{(N-2)(N-3)} \left[N(N-1) - 4mn + 2 \right] \right\} .$$

In the univariate case holds always $C = N - 2$, but in general C depends on the form of the underlying distribution. Therefore, unlike in the univariate case (the run test), the distribution of R under the null $H_0 : F = G$ is not distribution free. However, we can condition on the observed pooled sample and apply a permutation test. Since the minimal spanning tree depends only on the pooled sample, C is fixed given the pooled sample. Thus, $V(R\,|\,C)$ is the variance of R under the permutation distribution. Friedman and Rafsky (1979) showed that the asymptotic permutation distribution of

$$W = \frac{R - E(R)}{\sqrt{V(R\,|\,C)}}$$

is standard normal. Thus, the null hypotesis $H_0 : F = G$ will be rejected if W is smaller than $-u_{1-\alpha}$ where u_α is the α-quantile of the standard normal distribution. The minimal spanning tree test is constructed to have good power on location as well as on scale alternatives.

It should be noted that the minimal spanning tree test is not affine invariant. Since the Euclidean distance changes under an affine-linear transformation of the data, the minimal spanning tree and thus the test statistic will also change. However, the minimal spanning tree test is invariant with respect to affine orthogonal transformations and transformations of general scale, $\boldsymbol{x} \mapsto \beta\boldsymbol{x}$, $\beta > 0$.

The complexity of computing the test statistic depends mainly on the complexity of computing the minimal spanning tree. Whereas the classical algorithms require a computation time of $O(N^2)$, more recent algorithms can compute the minimal spanning tree in $O(N \log N)$ time. Shamos and Hoey (1975) have presented an algorithm which computes the minimal spanning tree of a data set in the plane in $O(N \log N)$. For dimension $d \geq 3$, Bentley and Friedman (1975) and Rohlf (1977) have developed algorithms with computation time $O(N \log N)$ on the average.

5.4.3 Hotelling's T^2 test

If we assume that both the X- and the Y-sample stem from two multivariate normal distribution, with identical covariance matrices, then the hypothesis $H_0 : \boldsymbol{\mu_X} = \boldsymbol{\mu_Y}$ can be tested using a likelihood ratio test. The likelihood ratio test statistic is given by

$$T^2_{\boldsymbol{X}.\boldsymbol{Y}} = \frac{mn}{N}(\overline{\boldsymbol{x}} - \overline{\boldsymbol{y}})(\boldsymbol{S}^*_{\boldsymbol{X}.\boldsymbol{Y}})^{-1}(\overline{\boldsymbol{x}} - \overline{\boldsymbol{y}})' \,.$$

where $\boldsymbol{S}^*_{\boldsymbol{X}.\boldsymbol{Y}}$ is the unbiased estimator $\boldsymbol{S}^*_{\boldsymbol{X}.\boldsymbol{Y}} = \frac{1}{N-2}(m\boldsymbol{S_X} + n\boldsymbol{S_Y})$. Here $\boldsymbol{S_X}$ and $\boldsymbol{S_Y}$ denote the sample covariance matrices of the $\boldsymbol{X}_i$ and the $\boldsymbol{Y}_i$ respectively. The statistic $T^2_{\boldsymbol{X},\boldsymbol{Y}}$ is known as Hotelling's T^2.

Under H_0 the quantity $\frac{N-d-1}{(N-2)d}T^2$ has an F-distribution with d and $N-d-1$ degrees of freedom. Therefore the null hypothesis $H_0 : \boldsymbol{\mu_X} = \boldsymbol{\mu_Y}$ is rejected if

$$T^2_{\boldsymbol{X},\boldsymbol{Y}} > \frac{(N-2)d}{N-d-1} F_{d,N-d-1}(1-\alpha)\,,$$

where $F_{d,N-d-1}(1-\alpha)$ denotes the $(1-\alpha)$-quantile of the F-distribution with d and $N-d-1$ degrees of freedom.

Hotelling's T^2 test is affine invariant. This is obvious from the equations

$$\overline{\boldsymbol{x}\boldsymbol{A}+\boldsymbol{b}} = \overline{\boldsymbol{x}}\boldsymbol{A}+\boldsymbol{b} \quad \text{and} \quad \boldsymbol{S}_{\boldsymbol{X}\boldsymbol{A}+\boldsymbol{b}} = \boldsymbol{A}'\boldsymbol{S}_{\boldsymbol{X}}\boldsymbol{A}\,.$$

Therefore

$$\begin{aligned} T^2_{\boldsymbol{X}\boldsymbol{A}+\boldsymbol{b},\boldsymbol{Y}\boldsymbol{A}+\boldsymbol{b}} &= \frac{mn}{N}\left[(\overline{\boldsymbol{x}}-\overline{\boldsymbol{y}})\boldsymbol{A}\right](\boldsymbol{A}'\boldsymbol{S}^*_{\boldsymbol{X},\boldsymbol{Y}}\boldsymbol{A})^{-1}\left[(\overline{\boldsymbol{x}}-\overline{\boldsymbol{y}})\boldsymbol{A}\right]' \\ &= \frac{mn}{N}(\overline{\boldsymbol{x}}-\overline{\boldsymbol{y}})\boldsymbol{A}\boldsymbol{A}^{-1}(\boldsymbol{S}^*_{\boldsymbol{X},\boldsymbol{Y}})^{-1}\boldsymbol{A}'^{-1}\boldsymbol{A}'(\overline{\boldsymbol{x}}-\overline{\boldsymbol{y}})' \\ &= T^2_{\boldsymbol{X},\boldsymbol{Y}}\,. \end{aligned}$$

Hotelling's T^2 is easy to compute since it involves only the calculation of the mean vectors and the covariance matrices, each of which can be done in $O(N)$ time.

5.4.4 Puri-Sen test

In this section we discuss a test which was proposed by Puri and Sen (1971). Like the tests based on depths this test is a rank test. But instead of the scalar ranks in the depth tests, the Puri-Sen test uses vector ranks, i.e., each observation is assigned a vector of ranks instead of one single rank in the depth tests. The Puri-Sen test is a test for the multisample problem. In what follows we give the formulation of the Puri-Sen test for the special case of two samples.

Let X_{ij} be the j-th component of $\boldsymbol{X}_i$, $i = 1, \ldots, m, j = 1, \ldots, d$, and accordingly for the Y-sample. We assign ranks for each component separately. By R^X_{ij} denote the rank of X_{ij} in $X_{1j}, \ldots, X_{mj}, Y_{1j}, \ldots, Y_{nj}$. R^Y_{ij} is to be understood in the same way. Thus, each observation $\boldsymbol{X}_i$ or $\boldsymbol{Y}_i$ is assigned a vector of ranks $\boldsymbol{R}^X_i = (R^X_{i1}, \ldots, R^X_{id})$ or $\boldsymbol{R}^Y_i = (R^Y_{i1}, \ldots, R^Y_{id})$, respectively.

The rank matrix $\boldsymbol{R}$ is then given by

$$\boldsymbol{R} = \begin{pmatrix} R_{11}^X & \cdots & R_{1d}^X \\ \vdots & & \vdots \\ R_{m1}^X & \cdots & R_{md}^X \\ R_{11}^Y & \cdots & R_{1d}^Y \\ \vdots & & \vdots \\ R_{n1}^Y & \cdots & R_{nd}^Y \end{pmatrix}.$$

Each column of the rank matrix is a permutation of the integers $1, \ldots, n$. By this, there are $(N!)^d$ possible realizations of the rank matrix. Since, in general, the d components of $\boldsymbol{X}_i$ are dependent, the ranks of $\boldsymbol{X}_i$ will also be dependent. Therefore, even under $H_0 : F = G$, the distribution of $\boldsymbol{R}$ will depend on the true underlying distribution F. Thus, $\boldsymbol{R}$ is not distribution free. As in the minimal spanning tree test, we can still apply a permutation test by conditioning on the observed pooled sample. Under the permutation distribution, all $N!$ permutations of the rows of $\boldsymbol{R}$ have equal probability.

Puri and Sen (1971) define the following quantities. The *mean rank vector of the X-sample* is defined by

$$\overline{\boldsymbol{R}}^X = \frac{1}{m} \sum_{i=1}^{m} \boldsymbol{R}_i^X$$

and the mean rank vector of the Y-sample is defined analogously. The *total mean rank vector* is given by

$$\overline{\boldsymbol{R}} = \frac{1}{N} \left[\sum_{i=1}^{m} \boldsymbol{R}_i^X + \sum_{i=1}^{n} \boldsymbol{R}_i^Y \right] = \frac{m}{N} \overline{\boldsymbol{R}}^X + \frac{n}{N} \overline{\boldsymbol{R}}^Y.$$

Since every column of the rank matrix is a permutation of the integers $1, \ldots, N$ the sum of all ranks in a row is $\frac{1}{2}N(N+1)$ and $\overline{R}_j = \frac{1}{2}(N+1)$ for each $j = 1, \ldots, d$.

Further, the empirical covariance matrix of the rank vectors is given by $\boldsymbol{V} = (v_{ij})_{i,j=1,\ldots,d}$, where

$$v_{ij} = \frac{1}{N} \left[\sum_{k=1}^{m} R_{ki}^X R_{kj}^X + \sum_{k=1}^{n} R_{ki}^Y R_{kj}^Y \right] - \left(\frac{N+1}{2} \right)^2.$$

Under H_0, the expectation and the covariance of the mean rank vector $\overline{\boldsymbol{R}}^X$ under the permutation distribution are given by

$$E[\overline{\boldsymbol{R}}^X] = \overline{\boldsymbol{R}} \quad \text{and} \quad \operatorname{Cov}(\overline{\boldsymbol{R}}^X) = \frac{1}{N-1} \frac{n}{m} \boldsymbol{V}.$$

Since under H_0 the mean rank vectors should be close to the total mean rank vector, the test statistic is based on the difference between the mean rank vector of the X-sample and the total mean rank vector. The test statistic is defined as the quadratic form

$$Q = (\overline{\boldsymbol{R}}^X - \overline{\boldsymbol{R}})\mathrm{Cov}(\overline{\boldsymbol{R}}^X)^{-1}(\overline{\boldsymbol{R}}^X - \overline{\boldsymbol{R}})' .$$

This finally yields

$$Q = (N-1)\frac{m}{n}(\overline{\boldsymbol{R}}^X - \overline{\boldsymbol{R}})\boldsymbol{V}^{-1}(\overline{\boldsymbol{R}}^X - \overline{\boldsymbol{R}})' .$$

Puri and Sen (1971) have shown that, under the null, the asymptotic permutation distribution of Q is a χ^2-distribution with d degrees of freedom. Thus, the Puri-Sen test rejects the null if Q is greater than the $(1-\alpha)$-quantile of a χ^2-distribution with d degrees of freedom.

We have tacitly assumed that $\boldsymbol{V}$ is regular. However, it may happen that $\boldsymbol{V}$ is indeed singular. In that case one can work with the highest order nonsingular minor of $\boldsymbol{V}$ and, by this, gets a test statistic similar to Q.

The test as described above is able to detect changes in location between the two samples. If the univariate ranks are assigned in a different way, e.g., as in the Siegel-Tukey test, the test is also capable of detecting changes in scale between the two distributions.

It is obvious from the construction of the test that the Puri-Sen test is neither affine invariant nor orthogonal invariant. However, it is invariant against monotone transformations of the coordinate axes.

To calculate the Puri-Sen test statistic the univariate ranks have to be calculated. This requires the sorting of each component of the pooled sample. Since sorting N items has a complexity of $O(N \log N)$, the computation of the rank matrix has a time complexity of $O(N \log N)$. The computation of $\overline{\boldsymbol{R}}^X$ and $\boldsymbol{V}$ can be done in $O(N)$ time and the computation of the quadratic form is of order $O(1)$. Therefore the overall time complexity is of order $O(N \log N)$.

5.5 A new Wilcoxon distance test

In this section we present another test for the two-sample location problem. To the best of our knowledge this test has not appeared in the literature before. The proposed test is a rank test, too. However, the ranks are not assigned to the observations themselves but to the distances between pairs of observations.

We call the distances between two observations in the X-sample or in the Y-sample *intra-sample distances* and the distances between $\boldsymbol{X}$- and $\boldsymbol{Y}$-observations *inter-sample distances*. That is, the intra-sample distances are

$$\|\boldsymbol{x}_i - \boldsymbol{x}_j\|,\ i,j = 1,\ldots,m, i < j, \quad \text{and} \quad \|\boldsymbol{y}_i - \boldsymbol{y}_j\|,\ i,j = 1,\ldots,n, i < j,$$

and the inter-sample distances are

$$\|\boldsymbol{x}_i - \boldsymbol{y}_j\|,\ i = 1,\ldots,m, j = 1,\ldots,n.$$

We sort all the inter- and intra-sample distances in ascending order and assign ranks to the distances. The test statistic is simply the sum of the ranks of the inter-sample distances,

$$T = \sum_{i=1}^{m}\sum_{j=1}^{n} R(\|\boldsymbol{x}_i - \boldsymbol{y}_j\|),$$

where $R(\|\boldsymbol{x}_i - \boldsymbol{y}_j\|)$ denotes the rank of $\|\boldsymbol{x}_i - \boldsymbol{y}_j\|$ in the sequence of all distances. If both samples stem from identical distributions, then the inter- and intra-sample distances are identically distributed. On the other hand, if both samples have different locations then, on the average, the inter-sample distances should be larger than the intra-sample distances. Thus, we would reject the null, if the sum of the ranks of the inter-sample distances is large.

Under H_0 the $\binom{N}{2}$ distances are identically distributed. However, they are not independent. Thus, the null distribution of T is not the distribution of the usual Wilcoxon rank sum statistic. As in the case of the minimal spanning tree test or the Puri-Sen test, we can still perform a permutation test. Let $\boldsymbol{Z}_i$, $i = 1,\ldots,N$, denote the pooled sample, i.e.,

$$\boldsymbol{Z}_i = \begin{cases} \boldsymbol{X}_i & \text{if } i \le m, \\ \boldsymbol{Y}_{i-m} & \text{if } i > m. \end{cases}$$

Further, for every two-element subset $I = \{i,j\}$ of $\{1,\ldots,N\}$ let R_I be the rank of $\|\boldsymbol{Z}_i - \boldsymbol{Z}_j\|$. By a straightforward but tedious calculation the first two moments under the permutation distribution are obtained:

$$E(T) = \frac{mn}{2}\left[\binom{m+n}{2} + 1\right]$$

and

$$E(T^2) = \frac{mn}{\binom{m+n}{2}}\left[S_1 + S_2 + \frac{4(m-1)(n-1)}{(m+n-2)(m+n-3)}S_3\right],$$

where

$$S_1 = \sum_I R_I^2 = \frac{1}{6}\binom{m+n}{2}\left[\binom{m+n}{2}+1\right]\left[2\binom{m+n}{2}+1\right],$$

$$S_2 = \sum_{I,J:|I\cap J|=1} R_I R_J ,$$

$$S_3 = \sum_{I,J:|I\cap J|=0} R_I R_J .$$

Thus, computing the variance under the permutation distribution requires to consider all pairs R_I, R_J of ranks. Since there are approximately $N^4/8$ of these pairs, computing the variance has a complexity of $O(N^4)$. In our opinion it seems more appropriate to simulate the critical values of the test statistic by resampling, as described in the following.

There are $\binom{m+n}{m}$ possibilities to split the pooled sample in an X- and a Y-sample. We choose a large number of these possible splittings and compute the value of the test statistic for each such splitting. This yields an estimate for the true permutation distribution from which we can compute critical values.

Since the test is based on Euclidean distances between observations, it is clear that the Wilcoxon distance test is not invariant against affine linear transformations. However, it is still invariant against affine orthogonal transformations.

The computation of the $\binom{N}{2}$ distances can be done in $O(N^2)$ time. The following sorting of the distances has a complexity of $O(N^2 \log N)$ and the computation of the rank sum has again complexity $O(N^2)$. Thus, the overall complexity of computing the test statistic is of order $O(N^2 \log N)$.

5.6 Power comparison

For our power studies we assume that both samples are distributed either according to a multivariate normal distribution or according to a multivariate Cauchy distribution.

For the multivariate normal distribution we use the notation $N(\boldsymbol{\mu}, \boldsymbol{\Sigma})$, where $\boldsymbol{\mu}$ is the expectation vector and $\boldsymbol{\Sigma}$ ist the covariance matrix. The generation of multivariate normal distributed pseudo-random vectors can be easily accomplished. Assume that $U_1, \ldots, U_d$ are independent and identically (univariate) standard normal distributed random variables. Then $(U_1, \ldots, U_d)$ is

d-variate standard normal distributed, $N(\mathbf{0}, \boldsymbol{I}_d)$.. Further, if $\boldsymbol{\Sigma} = \boldsymbol{A}'\boldsymbol{A}$ is the Cholesky decomposition of $\boldsymbol{\Sigma}$, then $\boldsymbol{X} = \boldsymbol{\mu} + (U_1, \ldots, U_d)\boldsymbol{A}$ is a multivariate normal distributed random vector, $\boldsymbol{X} \sim N(\boldsymbol{\mu}, \boldsymbol{\Sigma})$.

A d-variate random vector is said to be multivariate standard Cauchy distributed, shortly $\boldsymbol{X} \sim C(\mathbf{0}, \boldsymbol{I}_d)$, if it has a density

$$f(x_1, \ldots, x_d) = \frac{\Gamma(\frac{d+1}{2})}{\pi^{(d+1)/2}} \left[1 + \sum_{i=1}^{d} x_i^2\right]^{-\frac{d+1}{2}} .$$

Let $\boldsymbol{\mu} \in \mathbb{R}^d$ and $\boldsymbol{\Sigma}$ be a symmetric positive definite $d \times d$-matrix. Then a d-variate random vector $\boldsymbol{X}$ is said to have a multivariate Cauchy distribution with parameters $\boldsymbol{\mu}$ and $\boldsymbol{\Sigma}$, shortly $\boldsymbol{X} \sim C(\boldsymbol{\mu}, \boldsymbol{\Sigma})$, if it is distributed as $\boldsymbol{YA} + \boldsymbol{\mu}$, where $\boldsymbol{Y} \sim C(\mathbf{0}, \boldsymbol{I}_d)$ and $\boldsymbol{A}$ is a matrix such that $\boldsymbol{A}'\boldsymbol{A} = \boldsymbol{\Sigma}$. For the multivariate Cauchy distribution no moments exist. The multivariate Cauchy distribution has the following stochastic representation. Let $\boldsymbol{Z} \sim N(\mathbf{0}, \boldsymbol{I}_d)$ and $S \sim N(0, 1)$, and let $\boldsymbol{Z}$ and S be independent. Then $\boldsymbol{Z}/S \sim C(\mathbf{0}, \boldsymbol{I}_d)$. Further, if $\boldsymbol{\Sigma} = \boldsymbol{A}'\boldsymbol{A}$ is the Cholesky decomposition of $\boldsymbol{\Sigma}$, then $\boldsymbol{\mu} + \frac{1}{S}\boldsymbol{ZA}$ has a multivariate Cauchy distribution, $\boldsymbol{X} \sim C(\boldsymbol{\mu}, \boldsymbol{\Sigma})$. Using this representation, it is easy to generate multivariate Cauchy distributed pseudo-random vectors from random numbers.

In the simulations of the power for the scale test we used the Mahalanobis depth, the halfspace depth and the zonoid depth. Both samples were drawn either from a d-variate normal distribution or from a d-variate Cauchy distribution, $d = 2, 3$. For the samples we used equal sample sizes, namely, $m = n = 10, 25, 50$. As discussed earlier, in the simulations of the power, one can restrict oneself to the case that the scale matrix of the first sample is the identity matrix and the scale matrix of the second sample is a diagonal matrix. According to this, we replaced one or more diagonal elements of the scale matrix by a parameter σ. This σ was varied in the range from $\sigma = 0.3$ to $\sigma = 3$. The figures show the power functions $G(\sigma)$.

For the normal distribution there was no great difference between the three depth tests. Completely different was the situation for the Cauchy distribution, where one can see a clear ranking of the three tests. In all these situations the halfspace depth test had the highest power, the zonoid depth test performed slightly worse, whereas the Mahalanobis depth test was clearly inferior.

Further, we compared the power functions of these three depth tests with the power function of Box's M test, the Puri-Sen test (with Siegel-Tukey ranks) and the Friedman-Rafsky test. The simulations showed that the rank tests based on depth performed in the case of normal distributed samples nearly as good as Box's parametric test. The two classic nonparametric tests (Puri-Sen and Friedman-Rafsky) were clearly inferior, although the power of the Puri-Sen test approached that of the depth tests when the sample size was increased. However, it became apparent that Box's M test is very sensitive to deviations from the normality assumption. In the case of samples from Cauchy distributions, the size of Box's M test was always above 80% for a significance level of $\alpha = 5\%$. This, of course, forbids the use of Box's M test when the assumption of normality of the two samples cannot be warranted. In the case of Cauchy distributions, the depth tests and the Puri-Sen test had nearly identical power functions, while the Friedman-Rafsky test was clearly inferior.

Our simulations show that the use of nonparametric tests is preferable in these situations. This holds the more since the presented rank tests – besides other nonparametric tests – constitute a good alternative.

For simulating the power of the location test, we used the same setup as for the scale tests. Both samples were drawn from d-variate normal or from d-variate Cauchy distributions, $d = 2, 3$. We used the same sample sizes as before, namely, $m = n = 10, 25, 50$. The scale matrices of the samples were always chosen as the identity matrices. In the first sample $\boldsymbol{\mu_X}$ was equal to $\mathbf{0}$, while in the second sample $\boldsymbol{\mu_Y}$ was equal to $(\Delta, \mathbf{0})$. Δ was varied in the range from $\Delta = 0$ to $\Delta = 3$. The figures show the power functions $G(\Delta)$.

We first studied the influence of the depth that was used on the power of the Liu test. For the depth we used the Mahalanobis depth, the halfspace depth and the zonoid depth. Since our results were similar for dimensions 2 and 3, we just show the diagrams for $d = 2$. In general, there was no great difference between the three depth tests. For the normal distribution with sample sizes $m = n = 10$, the Mahalanobis depth performed best, although for larger sample sizes the gap between the Mahalanobis depth and the other two depths closed. For the case of the Cauchy distribution with small sample sizes, all three depths performed nearly identical. Nevertheless, for $m = n = 50$ the halfspace depth gave slightly better results than the other two depths.

We compared the Liu test also with Hotelling's T^2 test, the Puri-Sen test, the Friedman-Rafsky test and the Wilcoxon distance test. For samples from normal distributions, Hotelling's T^2, the Wilcoxon distance and the Puri-Sen test had nearly identical power functions. The Friedman-Rafsky and the Liu test performed clearly worse than the other three tests. However, for all three sample sizes the Friedman-Rafsky test was better than the Liu test, altough the power function of the Liu test approached that of the Friedman-Rafsky test when samples sizes were increasing. The situation was different in the case of Cauchy distributions. There, Hotelling's T^2 test was very conservative and had a size of under 2% when the nominal significance level was 5%. For all sample sizes the Wilcoxon distance test had clearly the best and the Liu test had the worst power. For sample sizes $m = n = 10$ the Friedman-Rafsky test was better than the Puri-Sen test, whereas for sample sizes 25 and 50 the converse was true.

To conclude, we saw that the depth tests for location were in all circumstances inferior to the other nonparametric test. This contrasts with the situation for the scale tests, where the depth tests were clearly competitive. Interesting was the good performance of the Wilcoxon distance test. Since this test is not affine invariant, its power has still to be checked for a wider spectrum of scale matrices and distributions than we did in the present simulation study.

The following figures exhibit simulated power functions of the tests discussed in the previous sections. The first six figures focus on the two-sample scale problem. Figures 5.1 to 5.4 show the influence of the depth which is used on the power of the depth test for scale. In Figures 5.5 and 5.6 the power of the depth test for scale is compared with several other tests for the two-sample scale problem. The last four figures show power functions for location alternatives. Power functions for the Liu test when used with different depths are given in Figures 5.7 and 5.8. Finally, a comparison of the Liu test with other tests for the two-sample location problem is shown in Figures 5.9 and 5.10.

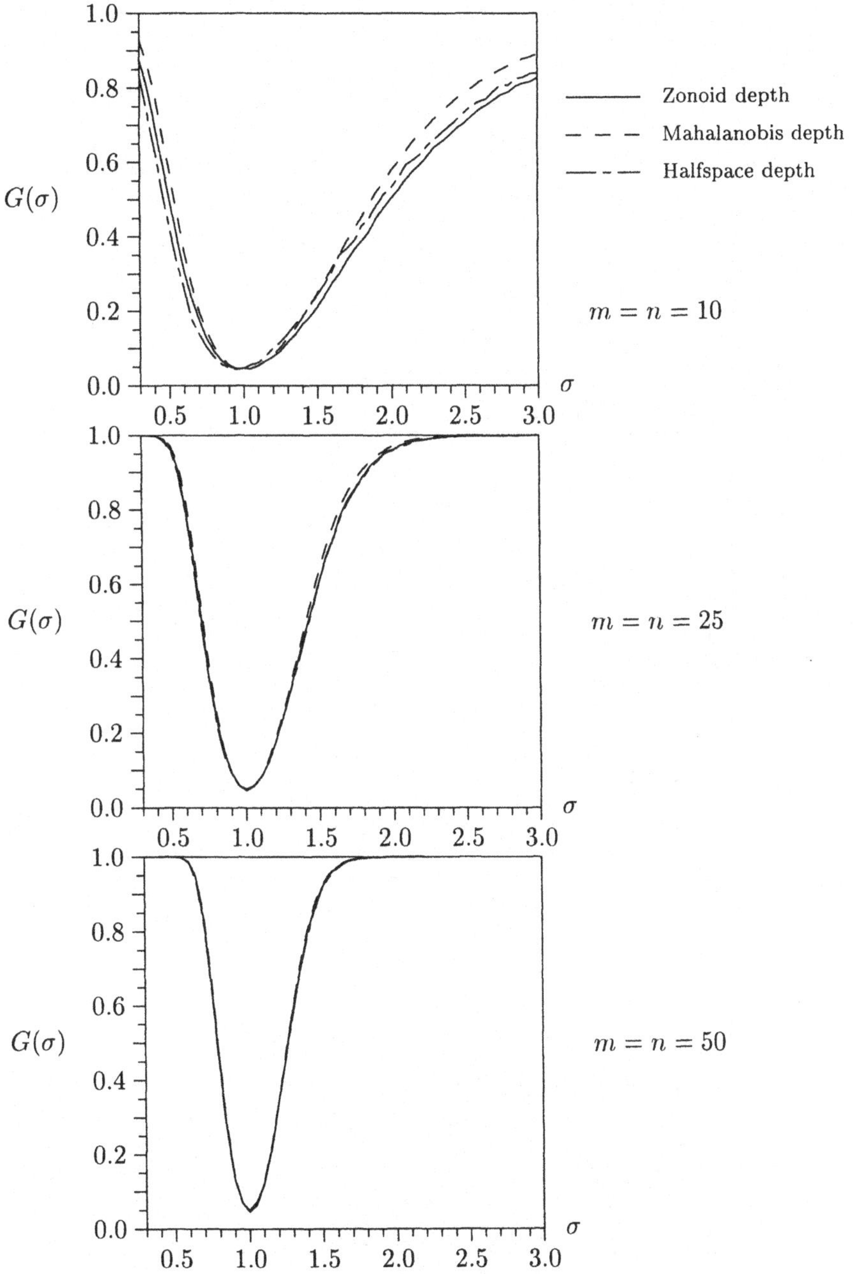

FIGURE 5.1: Power $G(\sigma)$ of scale tests based on different depths; bivariate normal distribution, $\mu_1 = \mu_2 = (0, 0)$, $\Sigma_1 = I_2$, $\Sigma_2 = \sigma^2 I_2$.

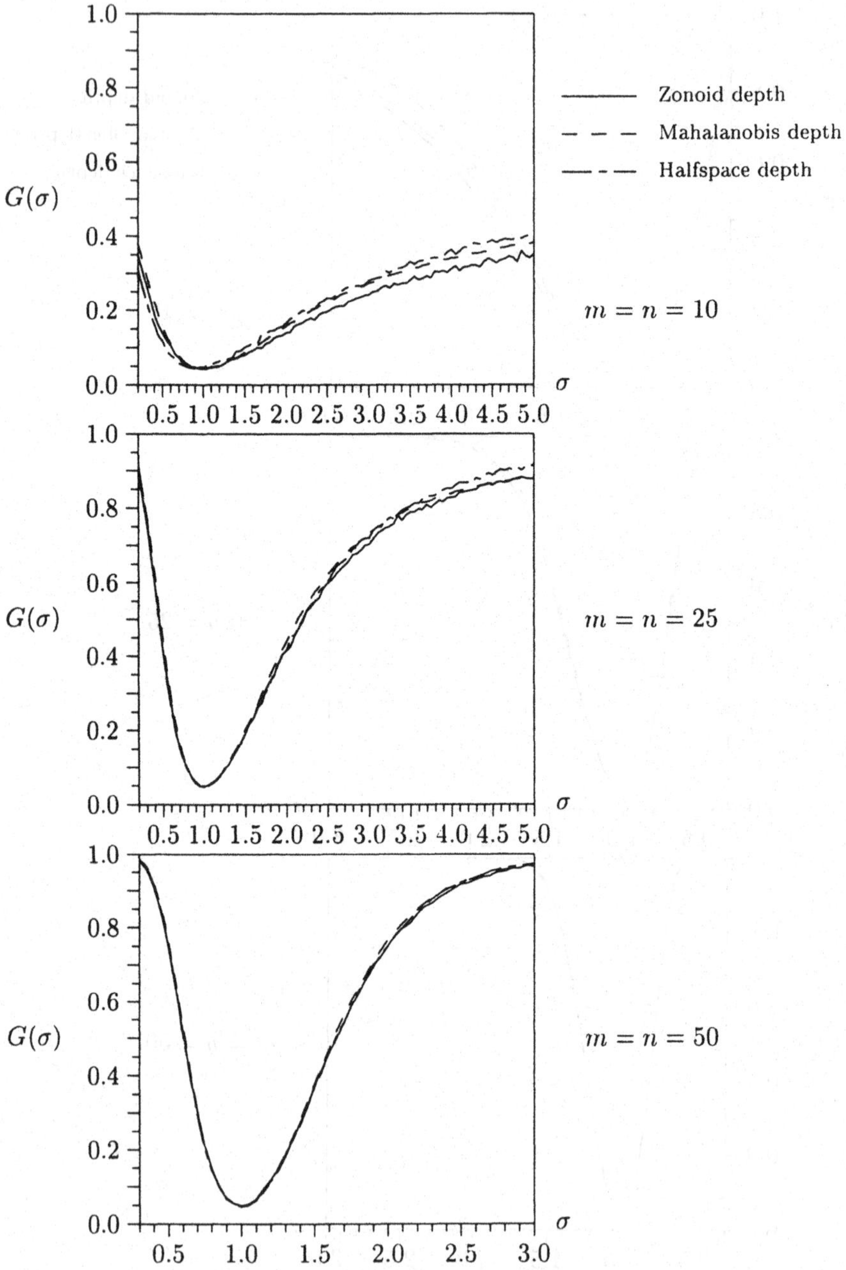

FIGURE 5.2: Power $G(\sigma)$ of scale tests based on different depths; bivariate normal distribution.
$\mu_1 = \mu_2 = (0,0)$, $\Sigma_1 = I_2$, $\Sigma_2 = \begin{pmatrix} \sigma^2 & 0 \\ 0 & 1 \end{pmatrix}$.

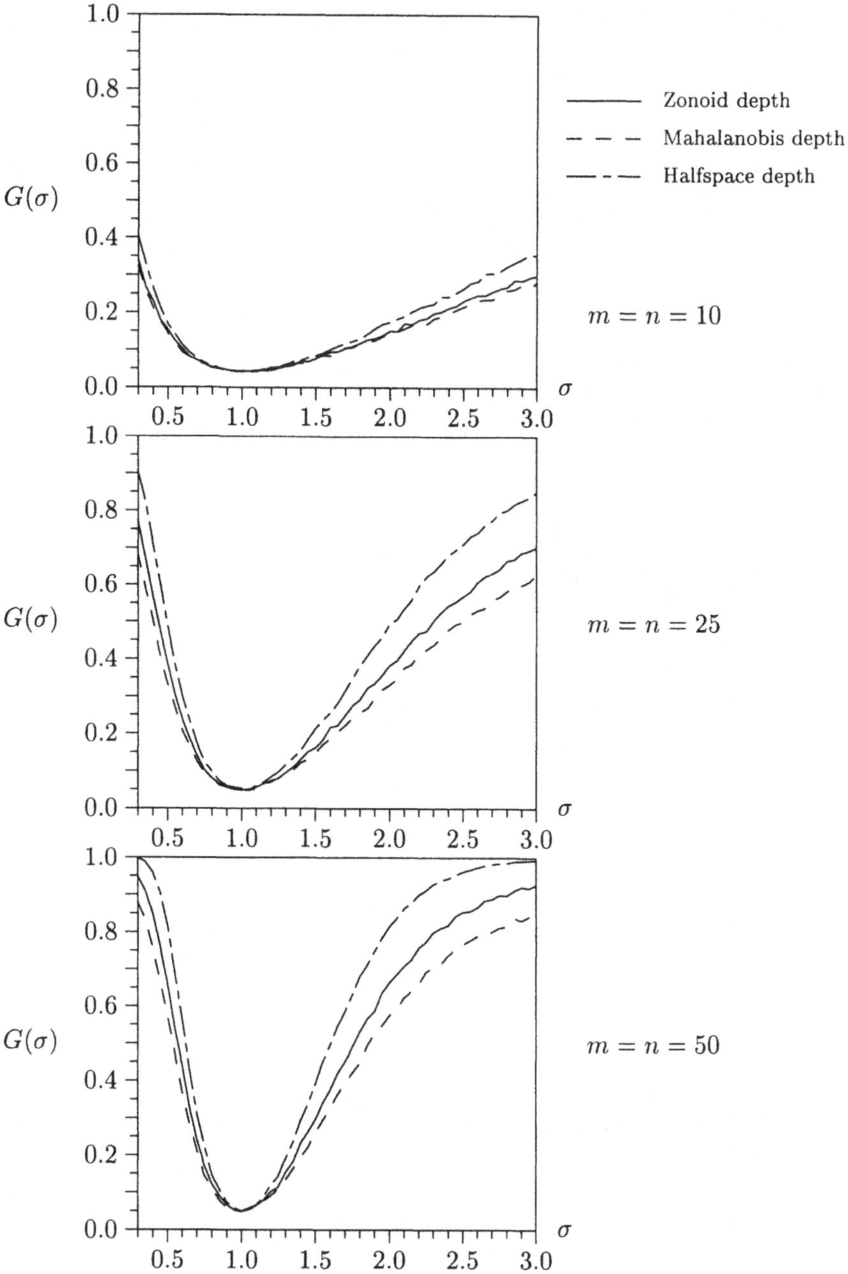

FIGURE 5.3: Power $G(\sigma)$ of scale tests based on different depths; bivariate Cauchy distribution, $\mu_1 = \mu_2 = (0,0)$, $\Sigma_1 = I_2$, $\Sigma_2 = \sigma^2 I_2$.

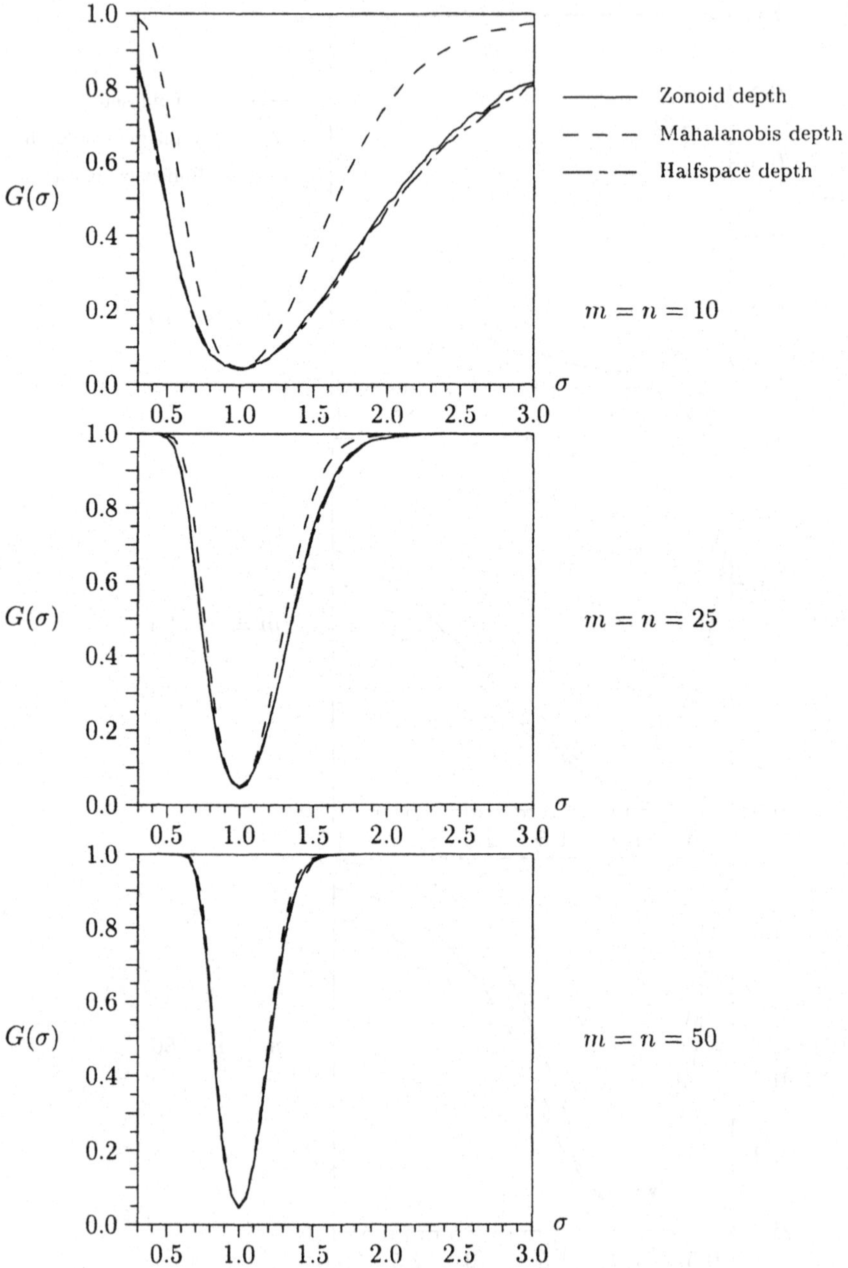

FIGURE 5.4: Power $G(\sigma)$ of scale tests based on different depths; trivariate normal distribution, $\mu_1 = \mu_2 = (0,0,0)$, $\Sigma_1 = I_3$, $\Sigma_2 = \sigma^2 I_3$.

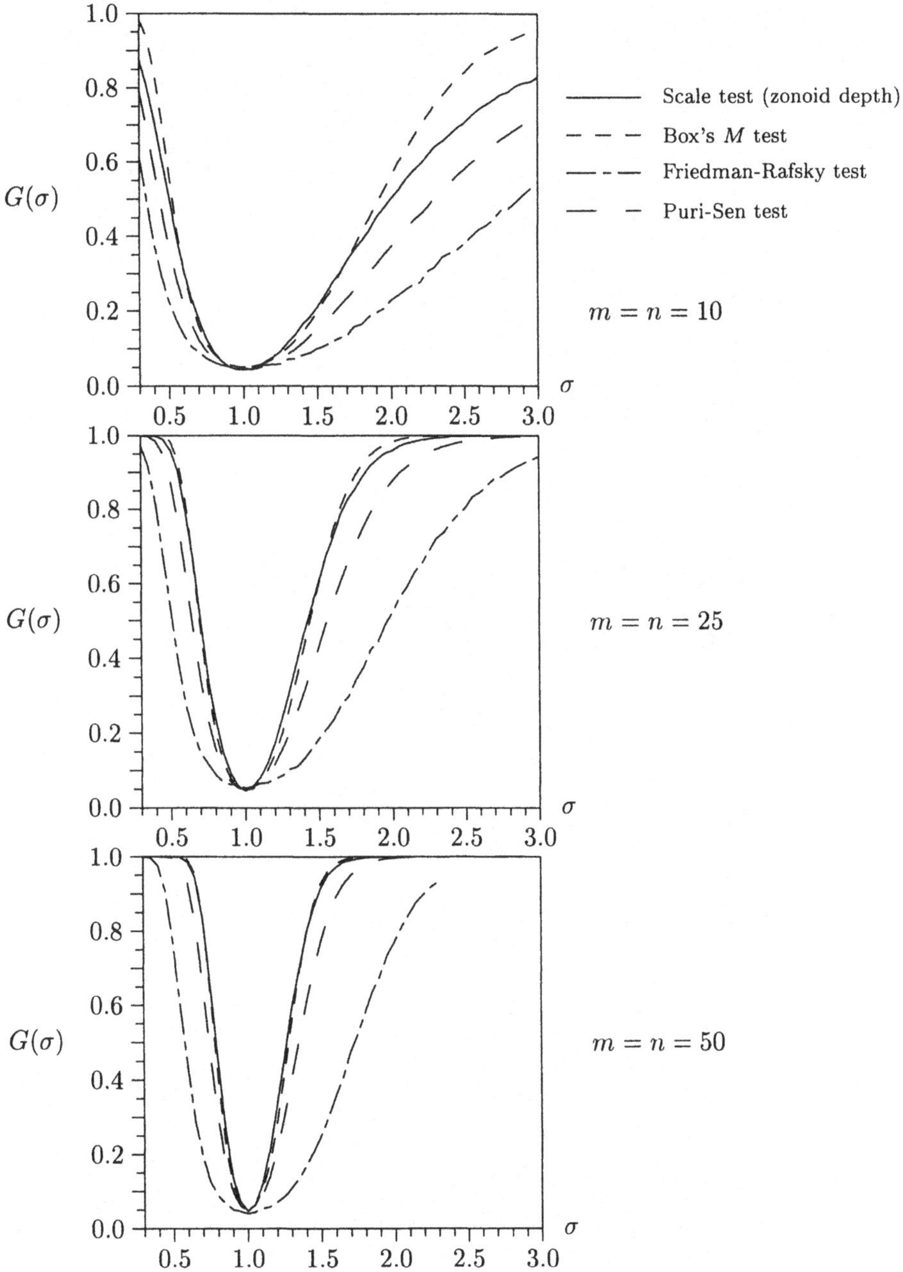

FIGURE 5.5: Power $G(\sigma)$ of scale tests;
bivariate normal distribution,
$\mu_1 = \mu_2 = (0,0)$, $\Sigma_1 = I_2$, $\Sigma_2 = \sigma^2 I_2$.

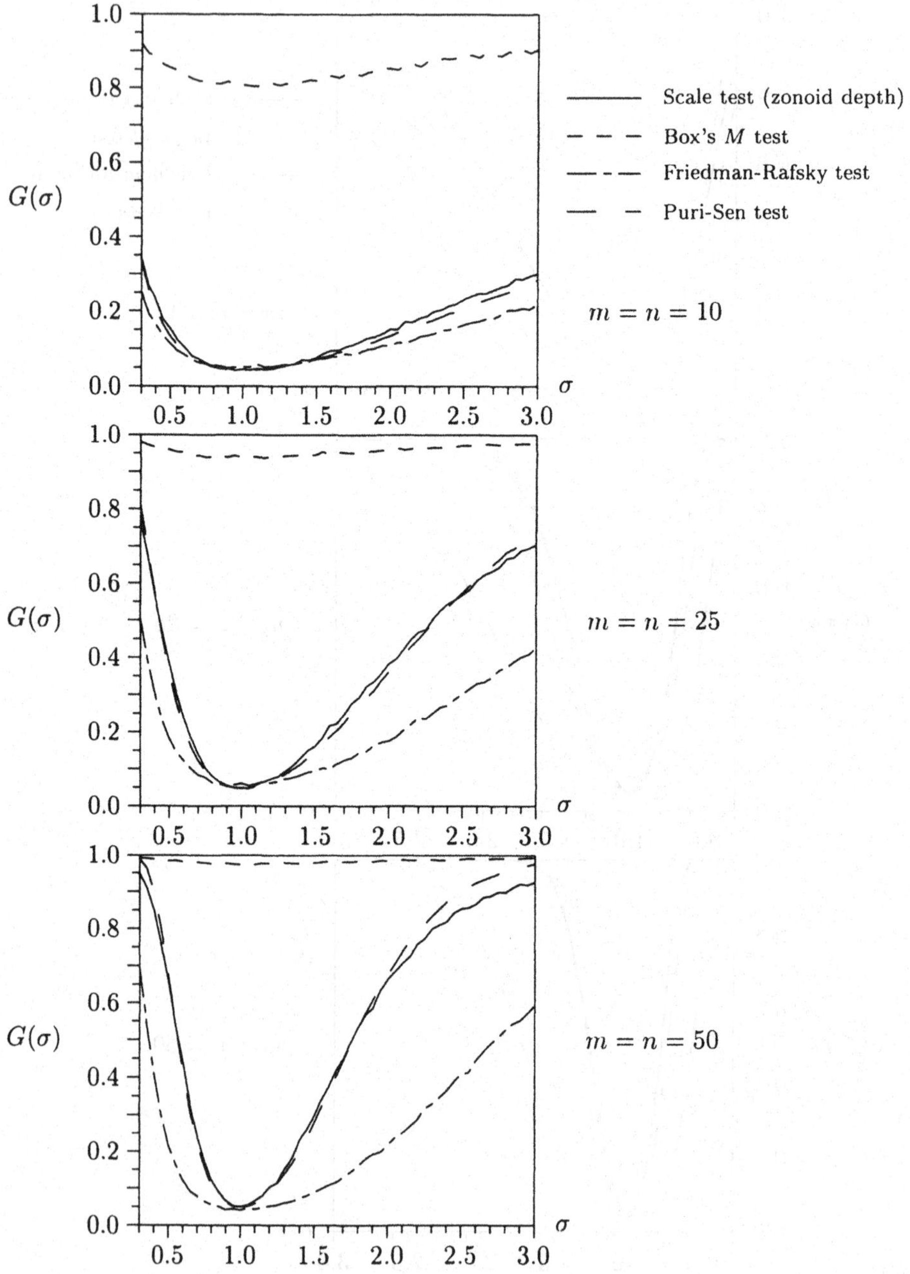

FIGURE 5.6: Power $G(\sigma)$ of scale tests;
bivariate Cauchy distribution,
$\mu_1 = \mu_2 = (0, 0)$, $\Sigma_1 = I_2$, $\Sigma_2 = \sigma^2 I_2$.

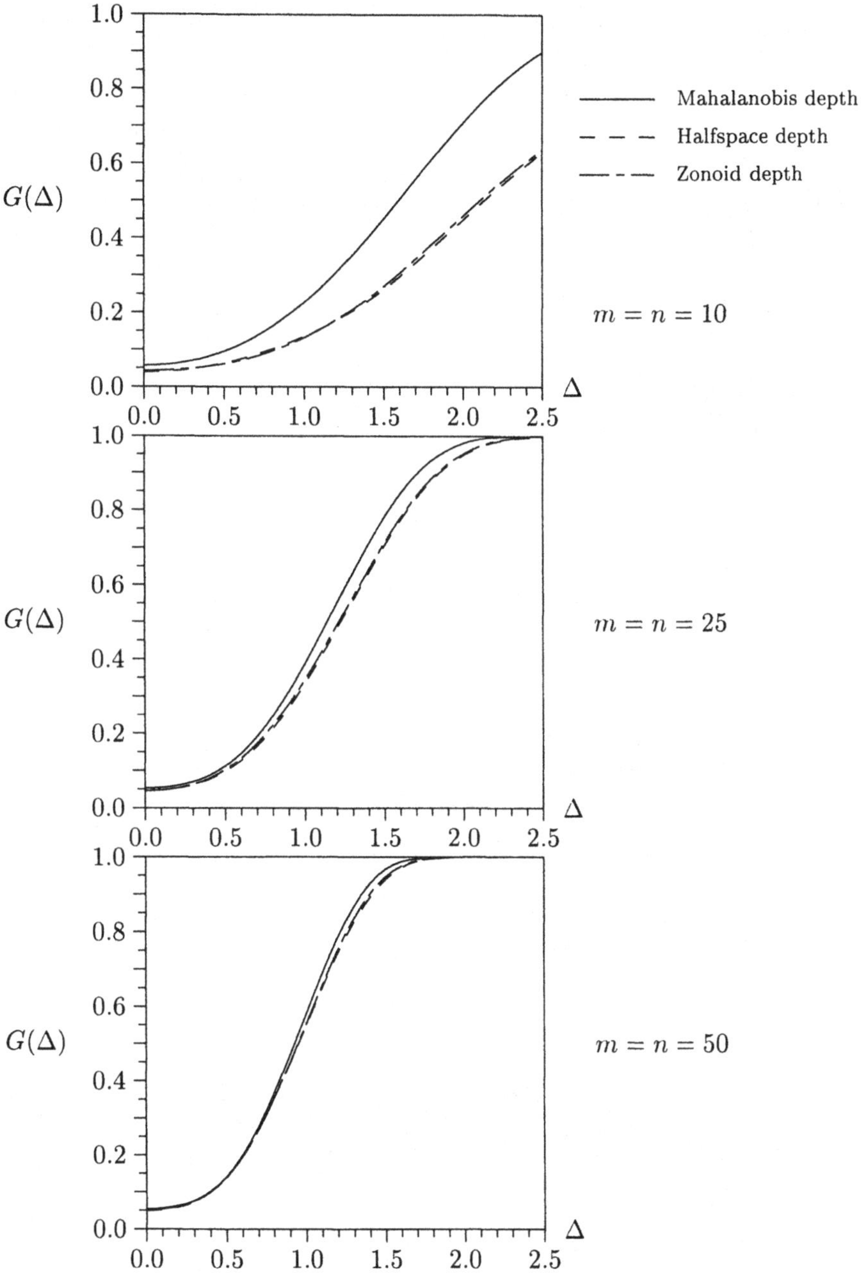

FIGURE 5.7: Power $G(\Delta)$ of Liu tests based on different depths; bivariate normal distribution, $\mu_1 = (0,0)$, $\mu_2 = (\Delta, 0)$, $\Sigma_1 = \Sigma_2 = I_2$.

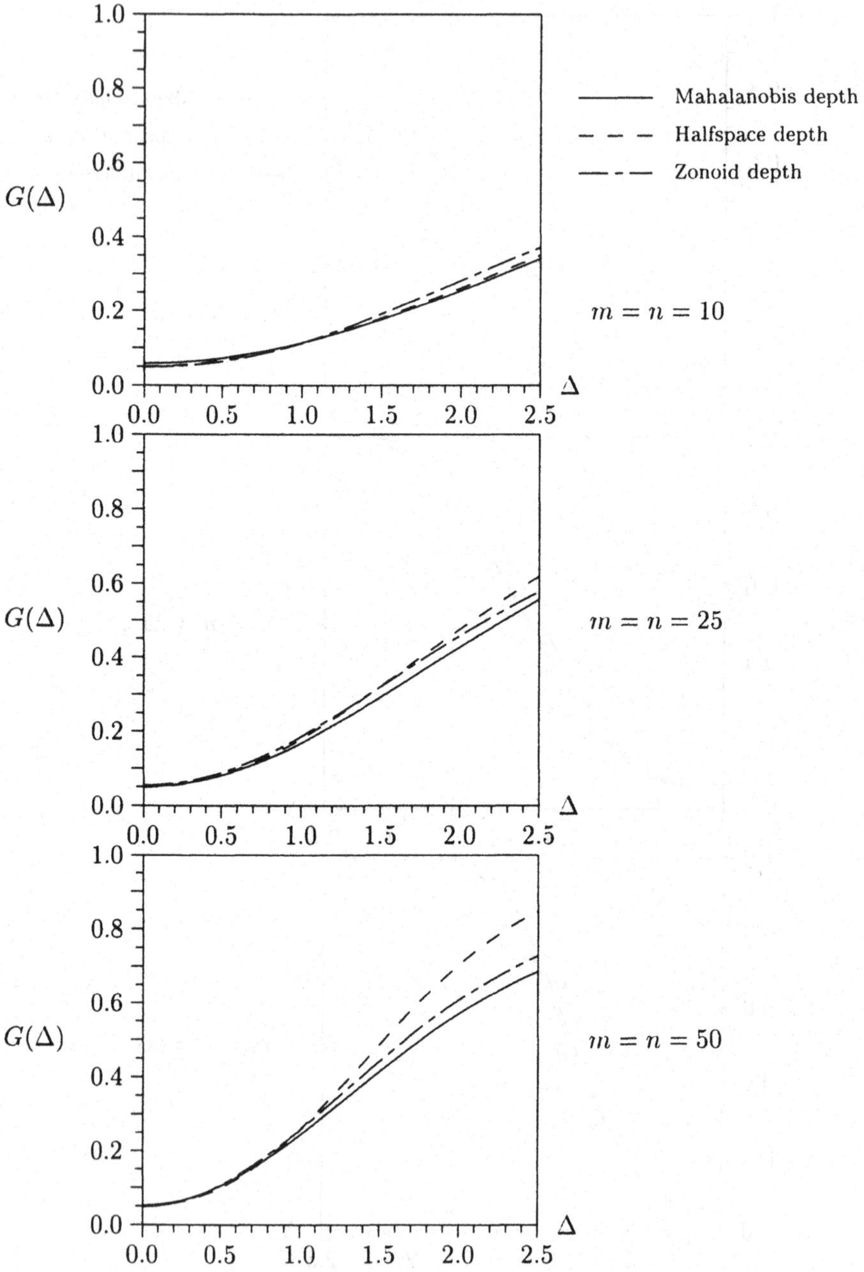

FIGURE 5.8: Power $G(\Delta)$ of Liu tests based on different depths; bivariate Cauchy distribution, $\mu_1 = (0,0)$, $\mu_2 = (\Delta, 0)$, $\Sigma_1 = \Sigma_2 = I_2$.

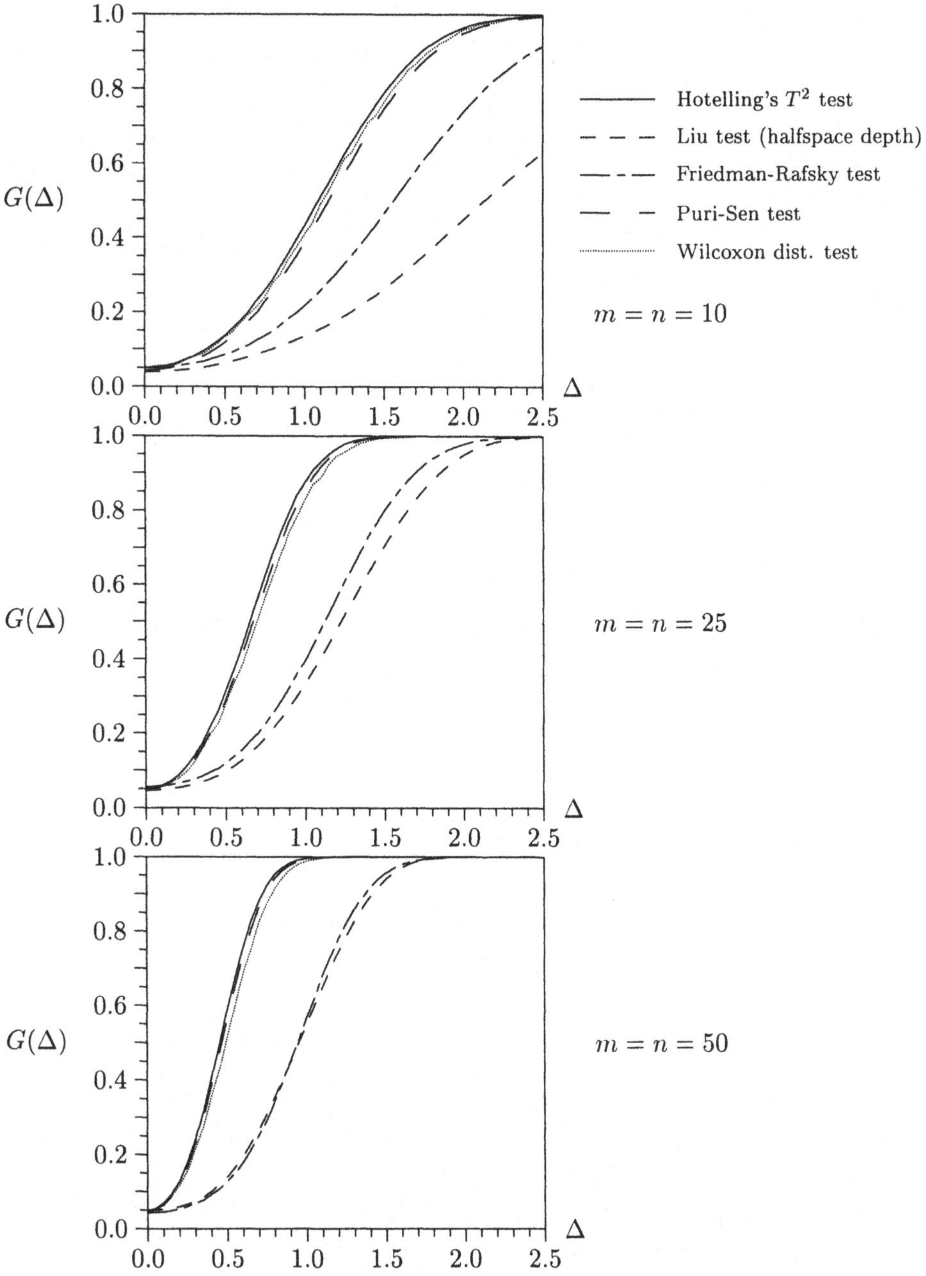

FIGURE 5.9: Power $G(\Delta)$ of location tests; bivariate normal distribution, $\mu_1 = (0, 0)$, $\mu_2 = (\Delta, 0)$, $\Sigma_1 = \Sigma_2 = I_2$.

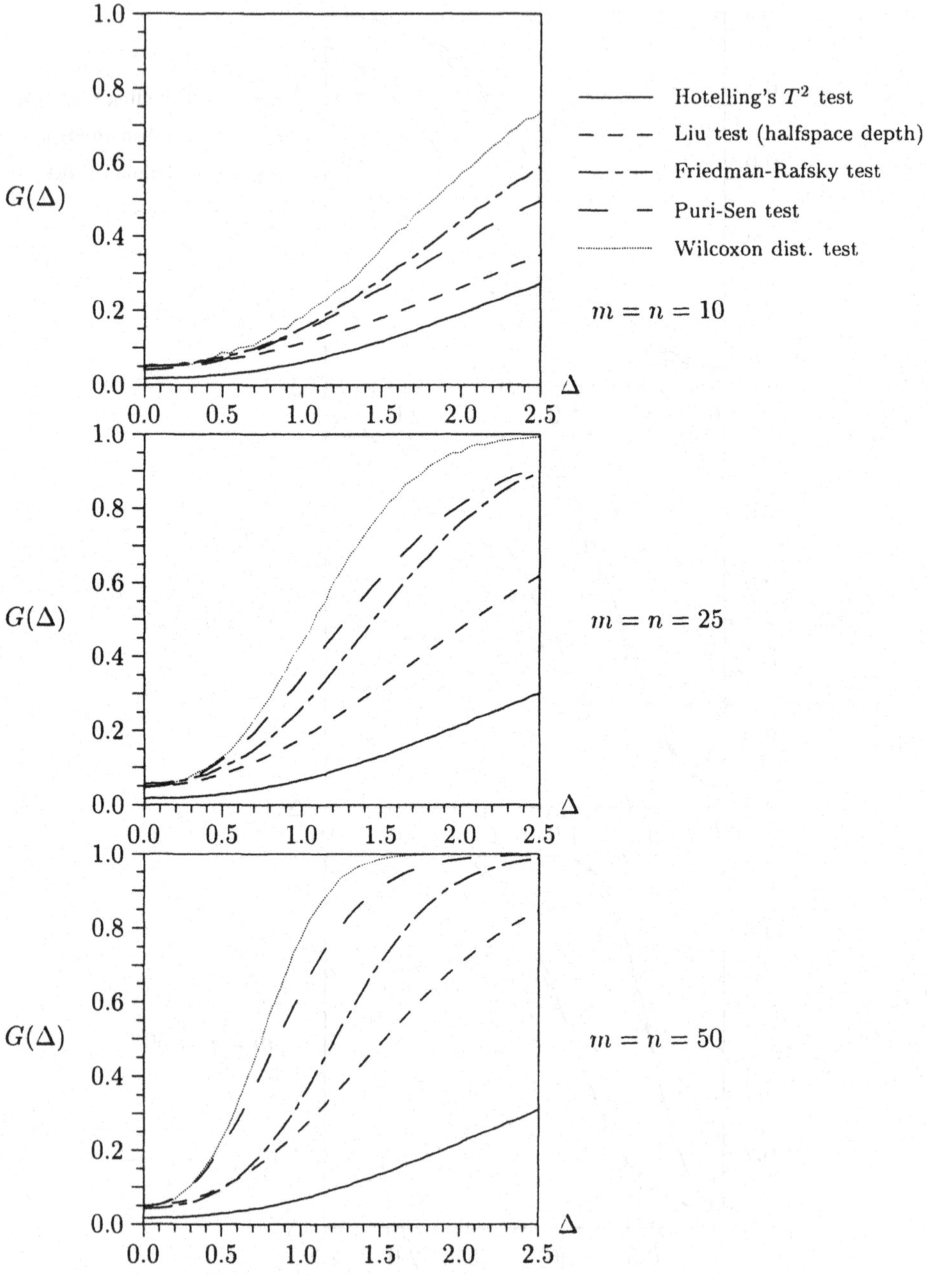

FIGURE 5.10: Power $G(\Delta)$ of location tests; bivariate Cauchy distribution, $\mu_1 = (0,0)$, $\mu_2 = (\Delta, 0)$, $\Sigma_1 = \Sigma_2 = I_2$.

5.7 Notes

Different sets of postulates on a data depth have been given in Liu (1990) and Zuo and Serfling (2000a). The main difference to our Definition 5.1 is that Zuo and Serfling (2000a) do not require that a depth be upper semicontinuous. The properties of a data depth in the sense of Definition 5.1 are further studied in Dyckerhoff (2002).

Another class of multivariate nonparametric tests is constructed using multivariate extensions of sign, rank and signed-rank which are based on the Oja median, see Brown and Hettmansperger (1987) and Hettmansperger et al. (1992). In contrast to the ranks considered here, their multivariate ranks are, however, vector valued. Based on these concepts generalizations of univariate sign and rank tests have been developed, see, e.g., Brown and Hettmansperger (1987), Brown et al. (1992b), Hettmansperger et al. (1994), Hettmansperger and Oja (1994), Möttönen and Oja (1995), Hettmansperger et al. (1997), Hettmansperger et al. (1998). The efficiency of these tests is investigated in, e.g, Möttönen et al. (1997) and Möttönen et al. (1998).

Further applications of data depth in multivariate statistics include the development of multivariate control charts (Liu, 1995), construction of confidence regions (Yeh and Singh, 1997), determination of limiting P-values in testing hypotheses (Liu and Singh, 1997) and cluster analysis (Hoberg, 2000).

6

Depth of hyperplanes

So far, notions of data depths have been discussed that measure how deep a single point is located in a distribution. In many applications, particularly two-sample problems, a statistic is needed that measures how deep a set of points is located in a distribution. For this, the notion of data depth of a set or, more precisely, a data matrix has to be developed. Like in the case of point data depth, many notions are possible and, depending on the application, useful.

In this chapter the the depth of a hyperplane in a sample is introduced, which has applications in two-sample tests for homogeneity. To compare two samples, the mean hyperplane depth (MHD) of one with respect to the other can be calculated. Several descriptive statistics based on the MHD are proposed. We explore the properties of the depth of a hyperplane and the mean hyperplane depth, in particular, their combinatorial invariance and their relation to majority depth. The MHD statistics are also used in significance testing. We discuss such tests and provide a small power study in elliptical distributions.

Given a hyperplane and a finite number of points in $\mathbb{R}^d$, the *depth of the hyperplane* is defined as the minimal portion of points which are in one of the two closed halfspaces supported by the hyperplane. This is called the sample version of the depth; the notion extends to a population version. The depth of a hyperplane is easily calculated. It is invariant under affine transformations of $\mathbb{R}^d$ and gives rise to sample functions that are useful in multivariate data analysis. Two samples can be compared by the *mean hyperplane depth* (MHD) of one with respect to the other data. For this, one considers the depth, with respect to the second sample, of all hyperplanes generated by points of the first sample and takes the mean of these depths. We use the MHD as a basic statistic to measure how close the data of the first sample are located to the center of the second sample. Particularly the two samples can be the same; in general they have an arbitrary intersection.

The MHD is related to the majority depth of points. We define the *extended majority depth* of a point in a sample with respect to a second sample and show that the MHD equals one minus the mean extended majority depth of points. If the two samples are the same, the extended majority depth reduces to the usual majority depth. In this way the MHD is an estimator of the expected majority depth of a sample. The majority depth is combinatorial invariant, it depends only on the combinatorial structure of data; see Section 4.4 above. Here a notion is introduced of combinatorial invariance in pairs of data matrices. The extended majority depth and the MHD are shown to be *combinatorial invariant in pairs.*

The MHD of a sample with respect to itself has a particular meaning. It takes its maximum if all points in the sample are extreme points, and it becomes smaller the more points are located in inner 'chambers', that is, polytopes spanned by other sample points as vertices. Thus, the MHD measures an aspect of the data which is called their *combinatorial dispersion.* In order to analyse whether two samples are homogeneous or not, several statistics are proposed that rest on the MHD. One of them detects differences in the dispersion, the other one shifts in the location of the data.

Beyond their descriptive use, the MHD statistics are employed in statistical inference. We consider them in a probabilistic setting and calculate their distributions as well as the power of the resulting tests of significance for several families of elliptical distributions.

In Section 6.1 the depth of a hyperplane and the MHD of a sample are defined. Section 6.2 exhibits the relation between the MHD and the majority depth and investigates properties of both depth notions. Section 6.3 introduces the notion of combinatorial invariance in pairs and Section 6.4 discusses the combinatorial dispersion of a sample. Several MHD statistics are proposed in 6.5, and their distribution and the power of corresponding significance tests are given in Section 6.6.

6.1 Depth of a hyperplane and MHD of a sample

In the sequel the d-variate data $\boldsymbol{s}_1, \ldots, \boldsymbol{s}_k$ is always considered as a data matrix $\boldsymbol{S} = [\boldsymbol{s}_1, \ldots, \boldsymbol{s}_k]$ having rows $\boldsymbol{s}_1, \ldots, \boldsymbol{s}_k$ in $\mathbb{R}^d$. $|\boldsymbol{S}|$ is the number of points (= rows) of $\boldsymbol{S}$ and, for any subset M of $\mathbb{R}^d$, $|\boldsymbol{S} \cap M|$ is the number of rows of $\boldsymbol{S}$ that lie in M.

Definition 6.1 (Depth of a hyperplane) *For a hyperplane H in $\mathbb{R}^d$ and a data matrix $\boldsymbol{S}$ define the* depth of H in $\boldsymbol{S}$ *by*

$$d_{Hyp}(H|\boldsymbol{S}) = \frac{1}{|\boldsymbol{S}|} \min\{|\boldsymbol{S} \cap H_+|, |\boldsymbol{S} \cap H_-|\}, \tag{6.1}$$

where H_+ and H_- are the two closed halfspaces bounded by H.

In words, the depth of a hyperplane in the data $\boldsymbol{S}$ amounts to the smaller portion of points in $\boldsymbol{S}$ that lie on one side of the hyperplane.

The depth of a hyperplane extends the usual univariate mid-rank to the multivariate case: In dimension one a hyperplane is a point, $z \in \mathbb{R}$, and $d_{Hyp}(z|\boldsymbol{S})$ is the smaller portion of the data $\boldsymbol{S}$ lying above or below z. The maximum depth, equal to $1/2$, is attained at the median of $\boldsymbol{S}$ if $|\boldsymbol{S}|$ is even. On rays from the median the depth decreases; it vanishes outside the convex hull of the rows of $\boldsymbol{S}$. If $d = 1$, $\boldsymbol{S} = [s_1, \ldots, s_k]$, and $z = s_j$ for some j, the depth $d_{Hyp}(z|\boldsymbol{S})$ equals the *mid-rank* of z in $s_1, \ldots, s_k$.

For a hyperplane H and a probability measure μ on $\mathbb{R}^d$ we similarly define the *depth of H in μ*,

$$d_{Hyp}(H|\mu) = \min\{\mu(H_+), \mu(H_-)\}. \tag{6.2}$$

The two definitions are related by

$$d_{Hyp}(H|\boldsymbol{S}) = d_{Hyp}(H|\mu_{\boldsymbol{S}}),$$

where $\mu_{\boldsymbol{S}}$ denotes the empirical probability measure on the rows of $\boldsymbol{S}$. Definition (6.1) is called the *sample version*, and definition (6.2) the *population version* of the depth of a hyperplane.

It follows immediately from the definitions that the depth of a hyperplane is affine invariant, i.e. if both H and $\boldsymbol{S}$ (or μ) are subject to the same affine transformation the value of the depth does not change. Moreover, as we will see below, the depth of a hyperplane is invariant to much more general transformations which leave the combinatorial structure of the data unchanged.

An oriented hyperplane in $\mathbb{R}^d$ is described by a vector $\boldsymbol{p} \in S^{d-1}$, its direction, and a number $\alpha \in \mathbb{R}$, its distance from the origin. The set of hyperplanes can be seen as the set of oriented hyperplanes modulo the orientation and therefore be parameterized by elements of the factor space $(S^{d-1} \times \mathbb{R})/\{1, -1\}$.

The numerical calculation of the hyperplane depth is easy. If H is the affine hull of the points $\boldsymbol{a}_1, \ldots \boldsymbol{a}_d$, $H = \text{aff}(\boldsymbol{a}_1, \ldots \boldsymbol{a}_d)$, then

$$d_{Hyp}(\text{aff}(\boldsymbol{a}_1, \ldots \boldsymbol{a}_d)|\boldsymbol{S}) = \frac{1}{|\boldsymbol{S}|} \min(N_+, N_-), \tag{6.3}$$

$$\begin{aligned} \text{where} \quad N_+ &= \left|\left\{\boldsymbol{s} \in \boldsymbol{S} : \det(M(\boldsymbol{s}, \boldsymbol{a}_1, \ldots, \boldsymbol{a}_d)) \geq 0\right\}\right|, \\ N_- &= \left|\left\{\boldsymbol{s} \in \boldsymbol{S} : \det(M(\boldsymbol{s}, \boldsymbol{a}_1, \ldots, \boldsymbol{a}_d)) \leq 0\right\}\right|, \\ M(\boldsymbol{s}, \boldsymbol{a}_1, \ldots, \boldsymbol{a}_d) &= \begin{bmatrix} 1 & \boldsymbol{s} \\ 1 & \boldsymbol{a}_1 \\ \vdots & \vdots \\ 1 & \boldsymbol{a}_d \end{bmatrix}. \end{aligned}$$

Now we are able to define my principal statistic, the *mean hyperplane depth* of some data $\boldsymbol{r}_1, \ldots, \boldsymbol{r}_n$ with respect to some other data $\boldsymbol{s}_1, \ldots, \boldsymbol{s}_k$ in $\mathbb{R}^d$.

Definition 6.2 (Mean hyperplane depth) *Consider two data matrices $\boldsymbol{R} = [\boldsymbol{r}_1, \ldots, \boldsymbol{r}_n] \in \mathbb{R}^{n \times d}$ and $\boldsymbol{S} = [\boldsymbol{s}_1, \ldots, \boldsymbol{s}_k] \in \mathbb{R}^{k \times d}$ and let $\mathcal{G}(\boldsymbol{R})$ denote the set of all hyperplanes which are affine hulls of rows of $\boldsymbol{R}$.*

$$g(\boldsymbol{R}|\boldsymbol{S}) = \frac{1}{\binom{n}{d}} \sum_{H \in \mathcal{G}(\boldsymbol{R})} d_{Hyp}(H|\mu_S). \tag{6.4}$$

is called the mean hyperplane depth *(shortly,* MHD*) of $\boldsymbol{R}$ with respect to $\boldsymbol{S}$.*

Since the depth of a hyperplane is affine invariant, the MHD has the same property.

Calculation of the MHD is easy for moderately sized d. For larger d, the MHD can be approximately calculated by sampling from $\mathcal{G}(\boldsymbol{R})$.

6.2 Properties of MHD and majority depth

Liu and Singh (1993) have proposed a quality index which is based on majority depth or some other notion of data depth. They have advocated its use for detecting increasing variance and/or location shift; see the two-sample tests discussed in Section 5.3 above.

The MHD is related to majority depth (Liu and Singh, 1993) in the following way. For a given point $\boldsymbol{y}$ and two data matrices $\boldsymbol{R} = [\boldsymbol{r}_1, \ldots, \boldsymbol{r}_n]$ and $\boldsymbol{S} = [\boldsymbol{s}_1, \ldots, \boldsymbol{s}_k]$ in $\mathbb{R}^d$, define the *extended majority depth of $\boldsymbol{y}$ in $\boldsymbol{S}$ with respect to $\boldsymbol{R}$*,

$$d_{eMaj}(\boldsymbol{y}; \boldsymbol{R}|\boldsymbol{S}) = \frac{1}{\binom{n}{d}} \left|\{H \in \mathcal{G}(\boldsymbol{R}) : \boldsymbol{y} \in \mathrm{Maj}(H; \boldsymbol{S})\}\right|. \tag{6.5}$$

The *major side* $\mathrm{Maj}(H; \boldsymbol{S})$ *of H in $\boldsymbol{S}$* is that halfspace, H_+ or H_-, which contains more rows of $\boldsymbol{S}$; if both contain the same number of rows of $\boldsymbol{S}$, the major side is $\mathbb{R}^d$. Note that, if $\boldsymbol{R} = \boldsymbol{S}$,

$$d_{Maj}(\boldsymbol{y}|\boldsymbol{R}) = d_{eMaj}(\boldsymbol{y}; \boldsymbol{R}|\boldsymbol{R})$$

is the usual majority depth of $\boldsymbol{y}$ in $\boldsymbol{R}$. The population version, for two probability measures μ and ν on $\mathbb{R}^d$, is defined by

$$d_{eMaj}(\boldsymbol{y}; \mu|\nu) = P(\boldsymbol{y} \in \mathrm{Maj}(\mathrm{aff}(\boldsymbol{X}_1, \ldots, \boldsymbol{X}_d); \nu)), \tag{6.6}$$

where

$$\mathrm{Maj}(H; \nu) = \begin{cases} H_+ & \text{if} \quad \nu(H_+) > \nu(H_-), \\ H_- & \text{if} \quad \nu(H_+) < \nu(H_-), \\ \mathbb{R}^d & \text{if} \quad \nu(H_+) = \nu(H_-), \end{cases} \tag{6.7}$$

and $\boldsymbol{X}_1, \ldots, \boldsymbol{X}_d$ are i.i.d. random vectors distributed as μ.

Note that $\mathrm{Maj}(\mathrm{aff}(\boldsymbol{X}_1, \ldots, \boldsymbol{X}_d); \nu)$ is a random halfspace. Definitions (6.6) and (6.5) are connected by $d_{eMaj}(\boldsymbol{y}; \boldsymbol{R}|\boldsymbol{S}) = d_{eMaj}(\boldsymbol{y}; \mu_{\boldsymbol{R}}|\mu_{\boldsymbol{S}})$. The population version of majority depth is

$$d_{Maj}(\boldsymbol{y}|\mu) = d_{eMaj}(\boldsymbol{y}; \mu|\mu). \tag{6.8}$$

Proposition 6.1 (Mean extended majority depth) *Let $[\boldsymbol{r}_1, \ldots, \boldsymbol{r}_n, \boldsymbol{s}_1, \ldots, \boldsymbol{s}_k]$ be in general position*[1]*. Then*

$$1 - g(\boldsymbol{R}|\boldsymbol{S}) = \frac{1}{k} \sum_{j=1}^{k} d_{eMaj}(\boldsymbol{s}_j; \boldsymbol{R}|\boldsymbol{S}). \tag{6.9}$$

The proposition reveals the close relationship between the mean hyperplane depth and the extended majority depth: If for each $\boldsymbol{s} \in \boldsymbol{S}$ the extended

[1] A subset of $\mathbb{R}^d$ is *in general position* if no more than d of its points lie on the same hyperplane.

majority depth in $\boldsymbol{S}$ is calculated with respect to the data $\boldsymbol{R}$ and the mean of these depth values is taken, the mean equals one minus the MHD of $\boldsymbol{R}$ in $\boldsymbol{S}$.

Proof. Since all points are in general position, $\mathcal{G}(\boldsymbol{R})$ possesses exactly $\binom{n}{d}$ elements and, for any $H \in \mathcal{G}(\boldsymbol{R})$ and $\boldsymbol{s}_j \in \boldsymbol{S}$, there holds $\boldsymbol{s}_j \notin H$. Therefore

$$d_{eMaj}(\boldsymbol{s}_j; \boldsymbol{R}|\boldsymbol{S}) = \frac{1}{\binom{n}{d}} \sum_{H \in \mathcal{G}(\boldsymbol{R})} 1(\boldsymbol{s}_j|\mathrm{Maj}(H; \boldsymbol{S})),$$

where $1(\boldsymbol{s}_j|\mathrm{Maj}(H;\boldsymbol{S})) = 1$ or 0 indicates whether $\boldsymbol{s}_j$ is an element of the set $\mathrm{Maj}(H;\boldsymbol{S})$ or not. Further,

$$\begin{aligned} d_{Hyp}(H|\boldsymbol{S}) &= \frac{1}{k}\sum_{j=1}^{k}[1 - 1(\boldsymbol{s}_j|\mathrm{Maj}(H;\boldsymbol{S}))] \\ &= 1 - \frac{1}{k}\sum_{j=1}^{k} 1(\boldsymbol{s}_j|\mathrm{Maj}(H;\boldsymbol{S})). \end{aligned}$$

Inserting this into (6.4) yields

$$\begin{aligned} g(\boldsymbol{R}|\boldsymbol{S}) &= \frac{1}{\binom{n}{d}} \sum_{H \in \mathcal{G}(\boldsymbol{R})} \left[1 - \frac{1}{k}\sum_{j=1}^{k} 1(\boldsymbol{s}_j|\mathrm{Maj}(H;\boldsymbol{S}))\right] \\ &= 1 - \frac{1}{k}\sum_{j=1}^{k} \frac{1}{\binom{n}{d}} \sum_{H \in \mathcal{G}(\boldsymbol{R})} 1(\boldsymbol{s}_j|\mathrm{Maj}(H;\boldsymbol{S})) \\ &= 1 - \frac{1}{k}\sum_{j=1}^{k} d_{eMaj}(\boldsymbol{s}_j; \boldsymbol{R}|\boldsymbol{S}). \end{aligned}$$

Q.E.D.

Liu and Singh (1993) have shown some properties of the majority depth, mostly restricted to elliptical distributions. In this section results are presented about majority depth and extended majority depth for general distributions.

Proposition 6.2 (Bounds and attained values) *There holds:*

(i) $0 < d_{Maj}(\boldsymbol{y}|\boldsymbol{R}) \le 1$, *and the upper bound is sharp,*

(ii) $0 \le d_{eMaj}(\boldsymbol{y}; \boldsymbol{R}|\boldsymbol{S}) \le 1$, *and the bounds are sharp.*

(iii) *The function* $\hat{d} : \boldsymbol{y} \mapsto \binom{n}{d} d_{eMaj}(\boldsymbol{y}; \boldsymbol{R}|\boldsymbol{S}), \boldsymbol{y} \in \mathbb{R}^d$, *attains every integer value between its minimum and its maximum.*

Proof. (i) and (ii): $d_{Maj}(\mathbf{0}|\boldsymbol{R}) = 1$ if the rows of $\boldsymbol{R}$ are rotation symmetric on the sphere S^{d-1}. Simple examples show that the extended majority depth can be 0. The rest is obvious.
(iii): Consider the set of hyperplanes $\mathcal{G}(\boldsymbol{R})$. Connected components of the complement $\mathbb{R}^d \setminus \cup_{H\in\mathcal{G}(R)} H$ are called *chambers.* In each chamber the function $\hat{d}$ is constant. If $\boldsymbol{y}$ crosses exactly one of the hyperplanes, $\hat{d}(\boldsymbol{y})$ changes by either $+1$ or -1 or remains constant. We conclude that $\hat{d}$ attains every integer value between its minimum and its maximum. Q.E.D.

Next, let us discuss continuity properties of the extended majority depth. It is obvious from the definition (6.6) that, given the data $\boldsymbol{R}$, $d_{eMaj}(\boldsymbol{y};\mu_{\boldsymbol{R}}|\nu)$ is continuous with respect to ν under weak convergence.

For continuity with respect to μ, we have to investigate the dependence of $P(\boldsymbol{y} \in \text{Maj}(\text{aff}(\boldsymbol{X}_1,\ldots,\boldsymbol{X}_d);\nu))$ on the random hyperplane $H_\mu = \text{aff}(\boldsymbol{X}_1,\ldots,\boldsymbol{X}_d)$, where $\boldsymbol{X}_1,\ldots,\boldsymbol{X}_d$ are i.i.d. random vectors having distribution μ. Given $\mu,\nu \in \mathcal{P}_1$, let

$$\gamma_\mu(\boldsymbol{S}|\nu) = P(\boldsymbol{S} \cap \text{Maj}(H_\mu;\nu) \neq \emptyset), \quad \boldsymbol{S} \subset \mathbb{R}^d.$$

The function $\boldsymbol{S} \mapsto \gamma_\mu(\boldsymbol{S}|\nu)$ is a Choquet capacity and the extended majority depth equals its values at single rows, $d_{eMaj}(\boldsymbol{y};\mu|\nu) = \gamma_\mu([\boldsymbol{y}]|\nu)$. We introduce the *hyperplane convergence* of measures as follows: A sequence of random closed sets converges if the corresponding capacities converge at any closed set in $\mathbb{R}^d$ (Matheron, 1975). Define $\mu^n \to_{Hyp} \mu$ if the corresponding random hyperplanes H_{μ^n} converge to H_μ, where H_μ is the affine hull of an i.i.d. sample $\boldsymbol{X}_1,\ldots,\boldsymbol{X}_d$ from μ and H_{μ^n} is the affine hull of an i.i.d. sample $\boldsymbol{X}_1^{(n)},\ldots,\boldsymbol{X}_d^{(n)}$ from μ^n. Because of the semi-continuity of the capacity function conclude:

Proposition 6.3 (Continuity) *For any $\boldsymbol{y}$ and ν, the extended majority depth, $d_{eMaj}(\boldsymbol{y};\mu|\nu)$, is continuous on μ under hyperplane convergence of μ.*

6.3 Combinatorial invariance

Next, let us investigate the invariance of the extended majority depth and of the MHD if the data are changed to combinatorial equivalent ones. We define a combinatorial structure of data matrices in pairs and demonstrate that the extended majority depth and, therefore, also the MHD depend only on this structure.

Recall from Chapter 4 that two data matrices are combinatorial equivalent, that is, have the same combinatorial structure, if their minimal Radon partions, modulo permutations of the data points, coincide. A statistic is combinatorial invariant if it is constant on combinatorial equivalent data.

To state the invariance of MHD we need a notion of combinatorial equivalence of data matrices in pairs. If $\boldsymbol{R}$ is an $n \times d$ matrix and $\boldsymbol{U}$ is an $m \times d$ matrix, notate the pooled matrix by

$$[\boldsymbol{R}, \boldsymbol{U}] = \begin{bmatrix} \boldsymbol{R} \\ \boldsymbol{U} \end{bmatrix}.$$

Definition 6.3 (Combinatorial equivalence and invariance in pairs) (i) *Let* $\boldsymbol{R}$ *and* $\boldsymbol{R}^*$ *be* $n \times d$ *matrices and* $\boldsymbol{U}$ *and* $\boldsymbol{U}^*$ *be* $m \times d$ *matrices. They are said to be* combinatorial equivalent in pairs *if the rows of the pooled matrices* $[\boldsymbol{R}, \boldsymbol{U}]$ *and* $[\boldsymbol{R}^*, \boldsymbol{U}^*]$ *are in general position and there exists a permutation* σ *of* $\{1, \ldots, n\}$ *and a permutation* ρ *of* $\{1, \ldots, m\}$ *such that* $\mathcal{C}(\boldsymbol{R}) = \mathcal{C}(\boldsymbol{R}^*_\sigma)$ *and* $\mathcal{C}([\boldsymbol{R}, \boldsymbol{U}]) = \mathcal{C}([\boldsymbol{R}^*_\sigma, \boldsymbol{U}^*_\rho])$ *holds.*
(ii) *A statistic depending on* $\boldsymbol{R}$ *and* $\boldsymbol{U}$ *is* combinatorial invariant in pairs *if it attains the same value for any two* $\boldsymbol{R}, \boldsymbol{U}$ *and* $\boldsymbol{R}^*, \boldsymbol{U}^*$ *that are combinatorial equivalent in pairs.*

If $\boldsymbol{R}, \boldsymbol{U}$ and $\boldsymbol{R}^*, \boldsymbol{U}^*$ are combinatorial equivalent in pairs, obviously also $\mathcal{C}(\boldsymbol{U}) = \mathcal{C}(\boldsymbol{U}^*_\rho)$ is true. For $m = 0$, the notion reduces to combinatorial invariance as defined in Chapter 4. Now we will see that the extended majority depth as well as the MHD satisfy combinatorial invariance in pairs.

Proposition 6.4 (Extended majority depth) *The extended majority depth,* $d_{eMaj}(\boldsymbol{r}_i; \boldsymbol{R}; [\boldsymbol{R}, \boldsymbol{U}])$ *depending on* $\boldsymbol{R}$ *and* $\boldsymbol{U}$, *is combinatorial invariant in pairs, for* $i = 1, \ldots, n$.

Proof. The proposition follows essentially from the fact that an affine oriented matroid is uniquely characterized by its maximal covectors (see Björner et al. (1999)). To be more explicit, let the rows of $[\boldsymbol{R}, \boldsymbol{U}]$ be in general position and $\boldsymbol{R}, \boldsymbol{U}$ and $\boldsymbol{R}^*, \boldsymbol{U}^*$ be combinatorial equivalent in pairs. Then there exist permutations σ and ρ with $\mathcal{C}([\boldsymbol{R}, \boldsymbol{U}]) = \mathcal{C}([\boldsymbol{R}^*_\sigma, \boldsymbol{U}^*_\rho])$. By $h(\boldsymbol{r}_i) = \boldsymbol{x}^*_{\sigma(i)}$ and $h(\boldsymbol{u}_j) = \boldsymbol{u}^*_{\rho(j)}$, the permutations induce a one-to-one mapping h of the rows of $[\boldsymbol{R}, \boldsymbol{U}]$ to the rows of $[\boldsymbol{R}^*, \boldsymbol{U}^*]$. If $\boldsymbol{S}$ is the support of a minimal Radon partition of $[\boldsymbol{R}, \boldsymbol{U}]$, then $h(\boldsymbol{S})$ is the support of the same minimal Radon partition of $[\boldsymbol{R}^*, \boldsymbol{U}^*]$. Recall that $\boldsymbol{S}$ and $h(\boldsymbol{S})$ have exactly $d+2$ elements. Consider a hyperplane $H \in \mathcal{G}(\boldsymbol{R})$, $H = \text{aff}(\boldsymbol{r}_{i_1}, \ldots, \boldsymbol{r}_{i_d})$ and let $H^* = \text{aff}(h(\boldsymbol{r}_{i_1}), \ldots, h(\boldsymbol{r}_{i_d}))$. From the unique characterization of an affine oriented matroid by its maximal covectors follows that a point $\boldsymbol{r}_i$ (or $\boldsymbol{s}_j$) is on the major side of H if and only if $h(\boldsymbol{r}_i)$ (respectively $h(\boldsymbol{s}_j)$) is on the major side of H^*. We conclude the proposition. Q.E.D.

Corollary 6.5 (Majority depth, MHD)
(i) *The majority depth $d_{Maj}(\boldsymbol{r}_i|\boldsymbol{R})$ depending on $\boldsymbol{R}$ is combinatorial invariant for any $i = 1, \dots, n$.*

(ii) *The mean hyperplane depth $g(\boldsymbol{R}|[\boldsymbol{R}, \boldsymbol{U}])$ depending on $\boldsymbol{R}$ and $\boldsymbol{U}$ is combinatorial invariant in pairs.*

6.4 Measuring combinatorial dispersion

The MHD of a sample with respect to itself can be seen as a measure of *combinatorial dispersion.* The MHD is minimal if every point of the sample is an extreme point of the convex hull. (E.g., in dimension $d = 2$ the minimum MHD equals $2 + (n-3)/4$ if the number n of rows in $\boldsymbol{S}$ is odd, and $2 + (n-2)^2/(4(n-1))$ if n is even.) It takes its maximum if only $d+1$ data points are in the convex hull and in every successive convex hull peeling exactly $d+1$ extreme points are removed. In this way the MHD describes how far the data points are away from being extremal and to which degree they are included in bounded polyedric chambers that have other data points as vertices.

Definition 6.4 (MHD-dispersion) *The MHD of the data $\boldsymbol{R}$ with respect to themselves, $g(\boldsymbol{R}|\boldsymbol{R})$, is called the MHD-dispersion of $\boldsymbol{R}$.*

Proposition 6.1 specializes to the following relation between the mean of usual majority depths and the MHD-dispersion of the data:

Corollary 6.6 (Mean of majority depth)

$$1 - g(\boldsymbol{R}|\boldsymbol{R}) = \frac{1}{n}\sum_{i=1}^{n} d_{Maj}(\boldsymbol{r}_i|\boldsymbol{R}). \tag{6.10}$$

In other words, the MHD-dispersion of a sample equals one minus the mean majority depth of its points.

Observe that, instead of the majority depth, any other combinatorial invariant depth notion can be employed in this way to measure the combinatorial dispersion of the data.

6.5 MHD statistics

Given two data matrices, $\boldsymbol{R} = [\boldsymbol{r}_1, \ldots, \boldsymbol{r}_n]$ and $\boldsymbol{U} = [\boldsymbol{u}_1, \ldots, \boldsymbol{u}_m]$, let us consider the MHD of $\boldsymbol{R}$ in the pooled matrix $[\boldsymbol{R}, \boldsymbol{U}]$.

In dimension $d = 1$ the statistic

$$M_1 = g(\boldsymbol{R}|[\boldsymbol{R}, \boldsymbol{U}])$$

is the mean mid-rank of the $\boldsymbol{r}$'s in the pooled sample. If both data clouds have the same location, as measured by the median, a large value of M_1 means that the $\boldsymbol{u}$'s are more dispersed than the $\boldsymbol{r}$'s while a small value means the opposite. If the data are arbitrarily located, again, a large value says that the $\boldsymbol{u}$'s are more dispersed than the $\boldsymbol{r}$'s. A small value says that either the $\boldsymbol{r}$'s are more dispersed than the $\boldsymbol{u}$'s (in this case, $g(\boldsymbol{R}|[\boldsymbol{R}, \boldsymbol{U}])$ is large) or that the two data clouds differ in location (in this case, $g(\boldsymbol{R}|[\boldsymbol{R}, \boldsymbol{U}])$ is small). This carries over to dimensions greater than one.

To compare two samples $\boldsymbol{R}$ and $\boldsymbol{U}$ in $\mathbb{R}^d$ we investigate a symmetric version of M_1,

$$M_2 = g(\boldsymbol{R}|[\boldsymbol{R}, \boldsymbol{U}]) - g(\boldsymbol{U}|[\boldsymbol{R}, \boldsymbol{U}]) . \tag{6.11}$$

Since the MHD is affine invariant, so are M_1 and M_2. As has been shown above, extreme values of M_1 indicate that the two samples are not homogeneous but differ in scale and/or location. If M_2 is large, this suggests that the $\boldsymbol{u}$'s are more dispersed than the $\boldsymbol{r}$'s and location is about the same. If M_2 is small the opposite applies, with $\boldsymbol{r}$ and $\boldsymbol{u}$ interchanged.

As a shift in location is generally not detected by M_2, let us introduce another MHD statistic,

$$M_3 = \frac{g(\boldsymbol{R}|[\boldsymbol{R}, \boldsymbol{U}])}{g(\boldsymbol{R}|\boldsymbol{R})} .$$

A small value of M_3 indicates a location shift while a large value indicates that the $\boldsymbol{u}$'s are less dispersed than the $\boldsymbol{r}$'s. Significance tests for homogeneity of two samples are investigated in the next section.

6.6 Significance tests and their power

Beyond their descriptive use, MHD statistics can be employed for inference in a probabilistic setting. The statistic $g(\boldsymbol{R}|\boldsymbol{R})$ is a linear mid-rank statistic, where the majority depth serves as a mid-rank. When $\boldsymbol{X}$ is a random vector distributed as μ, $g(\boldsymbol{R}|\boldsymbol{R})$ is an estimator of the expectation $E[d_{Maj}(\boldsymbol{X}|\mu)]$. In particular, if μ is an elliptical distribution[2] this estimator is consistent; this follows from the uniform convergence of the sample majority depth to its population version (Liu and Singh, 1993) for an elliptical distribution.

Suppose that $\boldsymbol{r}_1, \ldots, \boldsymbol{r}_n$ and $\boldsymbol{u}_1, \ldots, \boldsymbol{u}_m$ are independent samples, $\boldsymbol{r}_i$ distributed as μ, $\boldsymbol{u}_j$ distributed as ν. We want to test whether the two distributions are the same, $H_0 : \mu = \nu$. The following proposition is obvious.

Proposition 6.7 (Homogeneity) *Let $m = n$. Then, under $H_0 : \mu = \nu$,*

$$g(\boldsymbol{R}|[\boldsymbol{R}, \boldsymbol{U}]) \stackrel{d}{=} g(\boldsymbol{U}|[\boldsymbol{R}, \boldsymbol{U}]).$$

Here $\stackrel{d}{=}$ denotes equality in distribution. It follows that the distribution of M_2 is symmetric around 0 and, in particular, $\mathrm{E}[M_2] = 0$ if the expectation exists.

In the sequel assume that μ and ν belong to an elliptical family, that is, are elliptical distributions with the same characteristic generator, ϕ. If $\boldsymbol{X}$ has an elliptical distribution, $\mathcal{E}(\boldsymbol{a}, \boldsymbol{B}\boldsymbol{B}', \phi)$, there exists a random vector $\boldsymbol{Z}$ distributed as $\mathcal{E}(\boldsymbol{0}, \boldsymbol{I}_d, \phi)$ such that $\boldsymbol{X} = \boldsymbol{a} + \boldsymbol{Z}\boldsymbol{B}$ where $\boldsymbol{B}\boldsymbol{B}' = \Sigma$ and $\boldsymbol{I}_d$ is the identity matrix. Special elliptical distribution families are the normal family, the Cauchy family and the family of distributions that are uniform on ellipsoids. From the affine invariance of the MHD it follows that, under the null hypothesis $\mu = \nu$, the distribution of each MHD statistic does not depend on $\boldsymbol{a}$ and Σ.

Figure 6.1 exhibits the probability density of M_2 for the normal, the Cauchy and the ellipsoid-uniform families, when $d = 2$ and $m = n = 20$. It is seen that the densities differ among the three families. In fact, the MHD statistics are not distribution free but depend on ϕ. Critical values can be computed by Monte-Carlo simulation for every given generator ϕ and significance level α.

A power study has been performed of the M_2-test in three elliptical families (normal, Cauchy and ellipsoid-uniform) under different dispersion alternatives. Due to the affine invariance of the MHD statistics, the study restricts on alternatives (μ, ν) where Σ_μ is the identity matrix and Σ_ν is diagonal. Some power results are shown in Figures 6.2 and 6.3. The tests are one-sided with level $\alpha = .05$. We present the power function for the homoscedastic scale alternative (Figure 6.2) and a heteroscedastic one (Figure 6.3). For other heteroscedastic scale alternatives the power looks very similar.

[2]For the definition of elliptical distributions and for references, see Section 3.4 above.

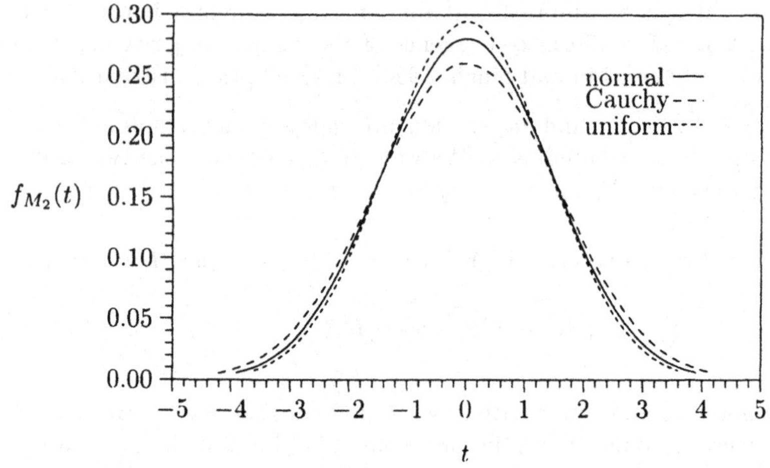

FIGURE 6.1: Density f_{M_2} of M_2 for three elliptical families.

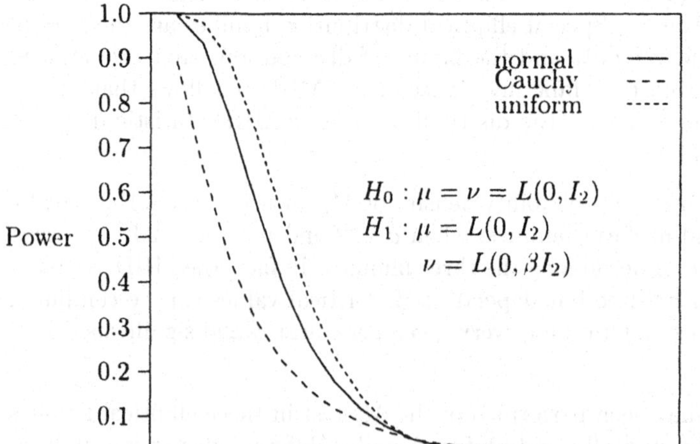

FIGURE 6.2: Power of the M_2-test for homoscedastic scale alternatives depending on β.

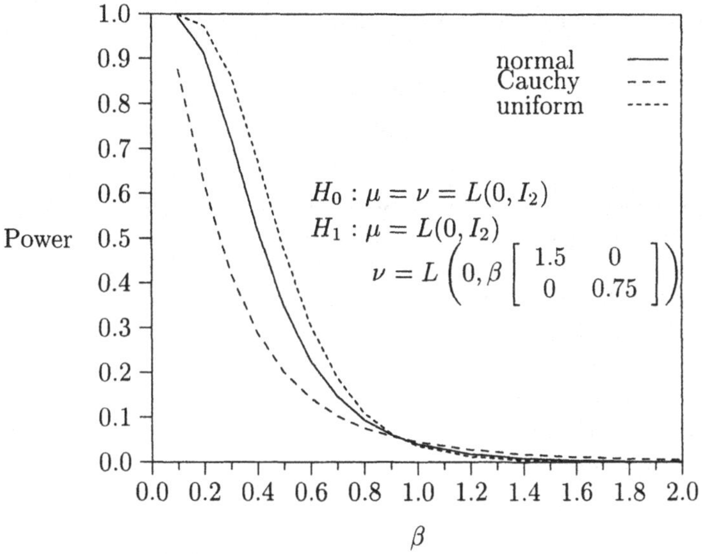

FIGURE 6.3: Power of the M_2-test for heteroscedastic scale alternatives depending on β.

In addition, under normality the one-sided M_2-test has been compared with a standard parametric test for scale, the Box test. It comes out that for uncorrelated homoscedastic normal alternatives the Box test obtains much better power. For heteroscedastic normal alternatives the Box test is still better but sometimes not much. For nonnormal alternatives the M_2-test can be very competitive. From the power study we conclude that the M_2 statistic is able to detect heteroscedastic scale and correlation alternatives in normal and particularly in non-normal elliptical families.

Concerning location alternatives we obtain the following, negative result for our statistic M_2.

Proposition 6.8 (Elliptical shift) *Let $m = n$. If μ and ν are elliptical and μ is a shift of ν, then*

$$g(\boldsymbol{R}|[\boldsymbol{R}, \boldsymbol{U}]) \stackrel{d}{=} g(\boldsymbol{U}|[\boldsymbol{R}, \boldsymbol{U}])$$

Proof. Without loss of generality, let μ and ν be spherical. Then the proposition follows from a simple symmetry argument. Q.E.D.

The proposition says that, at least in an elliptical family, a location shift cannot be discovered by M_2.

To detect a difference in location, let us use another MHD statistic,

$$M_3 = \frac{g(\boldsymbol{R}|[\boldsymbol{R}, \boldsymbol{U}])}{g(\boldsymbol{R}|\boldsymbol{R})}.$$

A small value of M_3 indicates a location shift while a large value indicates that the $\boldsymbol{u}$'s are less dispersed than the $\boldsymbol{r}$'s. Location alternatives are tested by one-sided use of M_3. The power of the M_3-test is exhibited in Figure 6.4. It is rather good for the elliptical Cauchy family and it increases with the heaviness of the distribution tails.

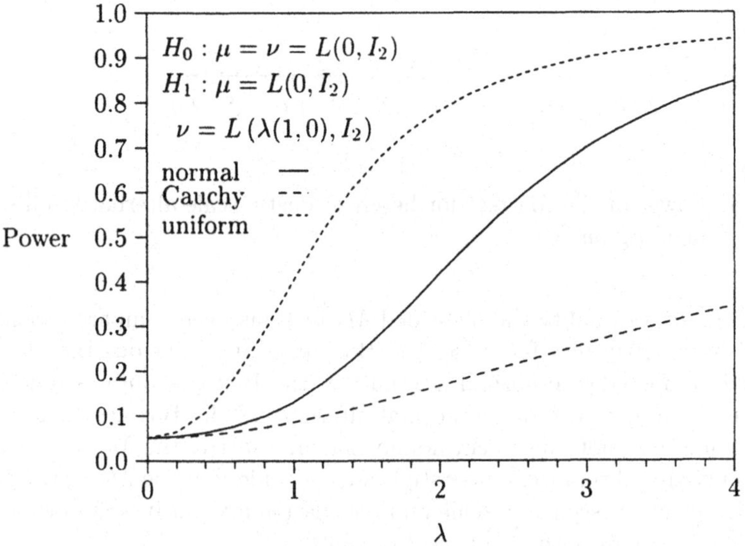

FIGURE 6.4: Power of the M_3-test for location.

The MHD statistics are a kind of multivariate rank statistics. Like other statistics based on multivariate ranks, they are not distribution free. Distribution free tests can be constructed by considering the rank of the depth instead of the depth itself.

Finally, note that an advantage of the hyperplane depth is its naturalness and easy computability, and an disadvantage is the complicated distribution of the random halfspaces generated by the data. This hints at finding random sets that have a 'simpler' distribution and applying a similar depth with respect to them.

6.7 Notes

This chapter is an extended and revised version of Koshevoy and Mosler (1999a). The notion of combinatorial dispersion is new.

As another robust measure of dispersion, Koshevoy et al. (2002) propose a scatter matrix estimate that is based on the zonotope. They call it the *zonoid covariance matrix*, as it is the usual covariance matrix of the centers of the zonotope's facets, and derive its influence function and its asymptotic distribution.

7

Volume statistics

The volume of the lift zonoid is a parameter that reflects the dispersion of a distribution. The volume is zero if the distribution concentrates at one point, it increases if the distribution becomes more diverse. For an empirical distribution the lift zonoid volume is easily calculated; it serves as a statistic to describe the dispersion of the data.

In dimension $d = 1$ the lift zonoid is the region between the *generalized Lorenz curve* and a curve symmetric to it, the *dual generalized Lorenz curve.* Its volume is known as the *Gini mean difference.* Because of this, we can say that the usual Gini mean difference equals the volume of the lift zonoid and is consistent with the lift zonoid order; see Section 7.1. This observation hints at a multivariate generalization of the Gini mean difference, which will be considered in the sequel.

The multivariate Gini mean difference defined below (in Section 7.4) is based on 'expanded' volumes of lift zonoids and named the *volume-Gini mean difference.* It is characterized as the average volume of the lift zonoids corresponding to all $2^d - 1$ marginal distributions. The volume-Gini mean difference is shown to inherit the properties of the univariate notion plus a *ceteris paribus* property. In particular, this statistic adopts its minimum value ($= 0$) if and only if the distribution is concentrated at one point. The usual univariate *Gini index* equals the Gini mean difference of the relative data, which are the original data 'scaled' by their mean. A multivariate *volume-Gini index* will be defined as the scaled version of the mean difference, and similar properties will be derived.

Firstly, in Section 7.2 the volume of the lift zonoid is introduced as a statistic that measures the dispersion of a theoretical or empirical distribution. The lift zonoid volume statistic is used in Section 7.3 to establish bounds for the expected volume of a random convex hull, that is, the convex hull of independent and identically distributed random points. Lower and upper bounds are based on the volume of the lift zonoid of the distribution from which the points are sampled. A related statistic is the volume of an α-trimmed region; it measures the dispersion of a (depending on α) central part of the distribution. The lift zonoid volume is a weighted mean of the volumes of all zonoid trimmed regions.

Among other applications, the lift zonoid volume and the volumes of zonoid trimmed regions serve as inhomogeneity statistics in cluster analysis (Section 7.5). The basic problem of cluster analysis is to divide a finite set S of points into k sets ('clusters') whose convex hulls are disjoint. Many clustering procedures can be regarded as strategies for minimizing a certain function that measures the inhomogeneity of clusters. The *hypervolume approach*, due to Hardy and Rasson (1982) and Rasson (1996), considers the convex hulls of the clusters and minimizes the sum of their volumes. Instead Hoberg (2000) replaces the convex hull by the lift zonoid of each cluster and minimizes the sum of lift zonoid volumes. The procedure is called *zonoid clustering*.

To measure the dependency that is contained in a random vector $\boldsymbol{X}$, volumes of trimmed regions can be employed, too; see Section 7.6. Consider two random vectors $\boldsymbol{X}$ and $\boldsymbol{Y}$ that have the same univariate marginal distributions. A d-variate random vector $\boldsymbol{X}$ is called more dependent than another random vector $\boldsymbol{Y}$ that has the same univariate marginals if the volume of each central region is smaller with $\boldsymbol{X}$ than with $\boldsymbol{Y}$. This dependence order coincides with the determinant ordering of correlation matrices if X and Y have arbitrary distributions and the central regions are Mahalanobis. It coincides with a similar determinant ordering if X and Y belong to an elliptical family of distributions and the central regions are arbitrary. If the regions are zonoid regions, the dependence order implies the ordering of lift zonoid volumes. Alternatively, the dependence order is applied to the copulae of the given distributions. Further, generalized correlation indices are presented which increase with the dependence order.

7.1 Univariate Gini index

This section shortly surveys the Gini mean difference and the Gini index of a univariate distribution. Let $F : \mathbb{R} \to [0, 1]$ be a given probability distribution function on $\mathbb{R}$ that has finite expectation $\epsilon(F) = \int_{-\infty}^{\infty} x dF(x) \neq 0$.

Definition 7.1 (Gini mean difference, Gini index)

$$M(F) = \frac{1}{2} \int_{\mathbb{R}} \int_{\mathbb{R}} |x - y| dF(x) dF(y) \tag{7.1}$$

is the Gini mean difference *of* F.

$$R(F) = \frac{M(F)}{|\epsilon(F)|}$$

is the Gini index *of* F.

According to the definition, $M(F)$ is half of the mean Euclidean distance between two independent random variables which both are distributed as F. $R(F)$ is the mean Euclidean distance divided by twice the absolute value of the expectation. Let $Q(s) = \min\{x : F(x) \geq s\}, s \in]0,1]$, denote the quantile function of F, and

$$GL_F(t) = \int_0^t Q(s) ds \,, \quad t \in [0,1] \,. \tag{7.2}$$

GL_F is the *generalized Lorenz function* and its graph is the *generalized Lorenz curve* of F. The graph of its symmetric counterpart,

$$\overline{GL}_F(t) = \int_{1-t}^1 Q(s) ds \,, \quad 0 \leq t \leq 1 \,,$$

is the *dual generalized Lorenz curve.* Observe that $GL_F(1) = \epsilon(F)$ and $\overline{GL}_F(t) = \epsilon(F) - GL_F(1-t)$ for $0 \leq t \leq 1$. The convex set bordered by the generalized Lorenz curve and its dual is the lift zonoid.

The *Lorenz curve* of F is the graph of

$$L_F(t) = \begin{cases} \frac{1}{\epsilon(F)} GL_F(t) & \text{if } \epsilon(F) > 0 \,, \\ 1 - \frac{1}{\epsilon(F)} GL_F(1-t) & \text{if } \epsilon(F) < 0 \,, \end{cases} \tag{7.3}$$

while the *dual Lorenz curve* is the graph of $\overline{L}_F(t) = 1 - L_F(1-t), t \in [0,1]$. The area between the Lorenz curve and its dual has been named the *Lorenz zonoid* of F (Koshevoy and Mosler, 1996). Note that the Lorenz zonoid of F is the lift zonoid of the scaled distribution, *viz.* the distribution $\widetilde{F}$ of a random variable $\widetilde{X} = X/\epsilon(F)$ where X has distribution F. Figure 7.1 exhibits the lift zonoids and Lorenz zonoids of two empirical distributions having supports $\{1,2,3\}$ and $\{1,-0.5,-2\}$, respectively.

The Lorenz zonoid and the lift zonoid give rise to two stochastic orders which are defined for dimension $d \geq 1$. A distribution F is larger than another

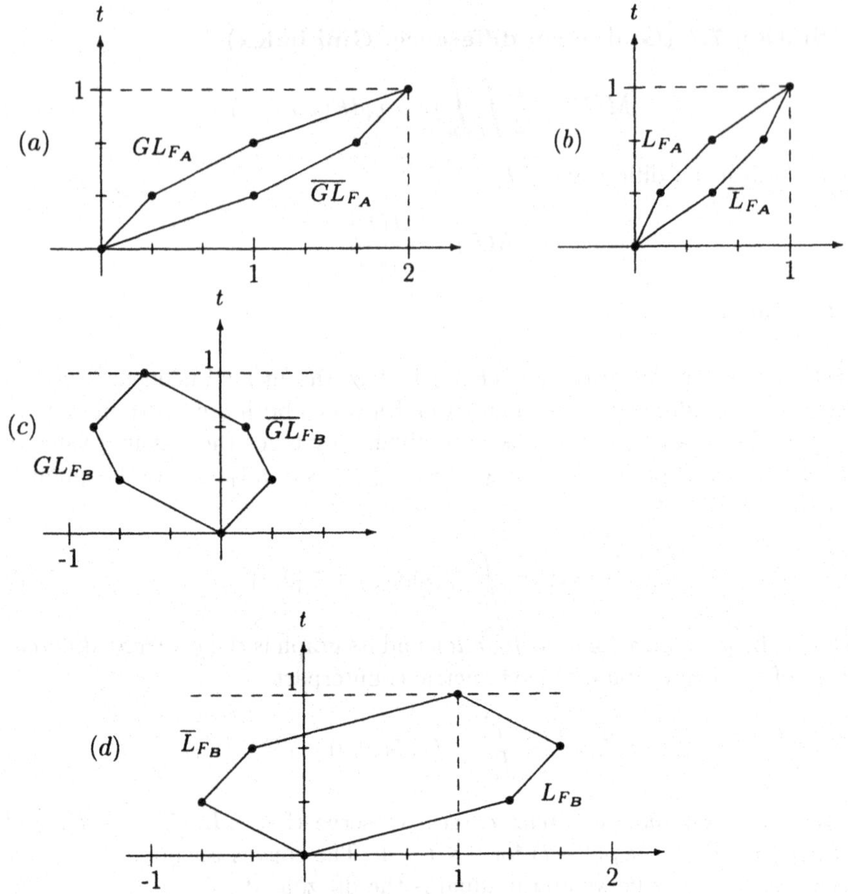

FIGURE 7.1: Lift zonoids, (a) and (c), and Lorenz zonoids, (b) and (d), for $F_{\boldsymbol{A}}$ and $F_{\boldsymbol{B}}$, respectively, where $\boldsymbol{A} = [1, 2, 3]$ and $\boldsymbol{B} = [1, -0.5, -2]$.

distribution G in the *Lorenz order* if the Lorenz zonoid of F includes that of G as a subset. F is larger than G in the *lift zonoid order* if the lift zonoid of F includes that of G. By this, the lift zonoid order can also be named the *generalized Lorenz order*. For details on this order, see Chapter 8.

The following well known proposition establishes a relation between the Gini index and the Lorenz curve.

Proposition 7.1 (Area) *Let $F(0) = 0$. Then*

(i) *$M(F)$ equals the area between the generalized Lorenz curve and its dual,*

(ii) *$R(F)$ equals the area between the Lorenz curve and its dual.*

As will be demonstrated in Section 7.4 (Remark 1), the assumption $F(0) = 0$ can be dropped, and Proposition 7.1 holds for any F that has finite, respectively, finite and nonzero expectation.

The special case of an empirical distribution is particularly important. Consider the empirical distribution, $F_{\boldsymbol{A}}$, on $a_1, \ldots, a_n$ and the sample mean $\overline{a} = n^{-1}(a_1 + \ldots + a_n)$. The Lorenz curve of $F_{\boldsymbol{A}}$ is obtained by linear interpolation of the points

$$\frac{1}{n}\left(k, \frac{1}{\overline{a}}\sum_{i=1}^{k} a_{(i)}\right), \quad k = 1, \ldots, n,$$

in two-space.

$$M(a_1, \ldots, a_n) = M(F_{\boldsymbol{A}}) = \frac{1}{2n^2}\sum_{j=1}^{n}\sum_{i=1}^{n} |a_i - a_j| \tag{7.4}$$

is the *Gini mean difference* of the sample $a_1, \ldots, a_n$, and

$$R(a_1, \ldots, a_n) = R(F_{\boldsymbol{A}}) = \frac{1}{\overline{a}} M(a_1, \ldots, a_n) \tag{7.5}$$

is the *Gini index*, provided $\overline{a}$ is not zero. The Gini index of the sample $a_1, \ldots, a_n$ equals the Gini mean difference of the 'scaled' sample $a_1/\overline{a}, \ldots, a_n/\overline{a}$,

$$R(a_1, \ldots, a_n) = \frac{1}{2n^2}\sum_{j=1}^{n}\sum_{i=1}^{n} \left|\frac{a_i}{\overline{a}} - \frac{a_j}{\overline{a}}\right|. \tag{7.6}$$

We state several important properties of the Gini index and the Gini mean difference when they refer to empirical distributions of *nonnegative data*. Parts of them hold also for more general probability distributions and are proved later in the general multivariate case.

Proposition 7.2 (Properties of R, univariate)

(i) *Let* $(a_1, \dots, a_n) \in \mathbb{R}^d_+$ *with* $\sum_i a_i > 0$. *Then*

$$\begin{aligned} 0 &= R(\bar{a}, \dots, \bar{a}) \le R(a_1, \dots, a_n) \\ &\le R\left(0, \dots, 0, \sum_{i=1}^n a_i\right) = 1 - \frac{1}{n} < 1, \\ R(\beta a_1, \dots, \beta a_n) &= R(a_1, \dots, a_n) \quad \text{for } \beta > 0, \\ R(a_1 + \lambda, \dots, a_n + \lambda) &= \frac{\bar{a}}{\bar{a} + \lambda} R(a_1, \dots, a_n) \quad \text{for } \lambda > 0, \end{aligned} \tag{7.7}$$

(ii) *R is strictly increasing with the Lorenz order,*

(iii) *R is a continuous function* $\mathbb{R}^d_+ \to \mathbb{R}$.

Proposition 7.3 (Properties of M, univariate)

(i) *Let* $(a_1, \dots, a_n) \in \mathbb{R}^d_+$ *with* $\sum_i a_i > 0$. *Then*

$$\begin{aligned} 0 &= M(\bar{a}, \dots, \bar{a}) \le M(a_1, \dots, a_n) \\ &\le M\left(0, \dots, 0, \sum_{i=1}^n a_i\right) = \bar{a}(1 - \frac{1}{n}) < \bar{a}, \\ M(\beta a_1, \dots, \beta a_n) &= \beta M(a_1, \dots, a_n) \quad \text{for every } \beta > 0, \\ M(a_1 + \lambda, \dots, a_n + \lambda) &= M(a_1, \dots, a_n) \quad \text{for every } \lambda \in \mathbb{R}, \end{aligned}$$

(ii) *M is strictly increasing with the generalized Lorenz order,*

(iii) *M is a continuous function* $\mathbb{R}^d_+ \to \mathbb{R}$.

These and other properties have been investigated by many authors. For surveys and references, see Nygård and Sandström (1981), Giorgi (1990), and Giorgi (1992).

7.2 Lift zonoid volume

Let μ be a d-variate probability measure, $\mu \in \mathcal{P}_1$, and F its distribution function. By slight abuse of notation we say that $F \in \mathcal{P}_1$. From Theorem

2.37 in Section 2.4.3 follows that the volume of the lift zonoid is equal to

$$\mathrm{vol}_{d+1}\widehat{Z}(F) = \frac{1}{(d+1)!}\int_{\mathbb{R}^d}\cdots\int_{\mathbb{R}^d}\left|\det\begin{pmatrix}1 & \boldsymbol{x}_0\\ \vdots & \vdots\\ 1 & \boldsymbol{x}_d\end{pmatrix}\right| dF(\boldsymbol{x}_0)\cdots dF(\boldsymbol{x}_d)\,. \tag{7.8}$$

With an empirical distribution $F_{\boldsymbol{A}}$ on $\boldsymbol{a}_1, \boldsymbol{a}_2, \ldots, \boldsymbol{a}_n \in \mathbb{R}^d$ obtain

$$\mathrm{vol}_{d+1}\widehat{Z}(F_{\boldsymbol{A}}) = \frac{1}{n^{d+1}}\sum_{\{i_0,\ldots,i_d\}\subset\{1,\ldots,n\}}\left|\det\begin{pmatrix}1 & \boldsymbol{a}_{i_0}\\ \vdots & \vdots\\ 1 & \boldsymbol{a}_{i_d}\end{pmatrix}\right|. \tag{7.9}$$

Consider $d+1$ points $\boldsymbol{x}_0, \ldots, \boldsymbol{x}_d \in \mathbb{R}^d$. They span a simplex $conv(\{\boldsymbol{x}_0, \ldots, \boldsymbol{x}_d\})$ that has volume

$$\mathrm{vol}_d\, conv(\{\boldsymbol{x}_0, \ldots, \boldsymbol{x}_d\}) = \frac{1}{d!}\left|\det\begin{pmatrix}1 & \boldsymbol{x}_0\\ \vdots & \vdots\\ 1 & \boldsymbol{x}_d\end{pmatrix}\right|. \tag{7.10}$$

Therefore,

$$\mathrm{vol}_{d+1}\widehat{Z}(F) = \frac{1}{d+1}\int_{\mathbb{R}^d}\cdots\int_{\mathbb{R}^d}\mathrm{vol}_d\, conv(\{\boldsymbol{x}_0, \ldots, \boldsymbol{x}_d\})\, dF(\boldsymbol{x}_0)\cdots dF(\boldsymbol{x}_d)\,. \tag{7.11}$$

Definition 7.2 (Scaled distribution, Lorenz zonoid) *Let $\mathcal{P}_{1*} \subset \mathcal{P}_1$ be the class of distributions that have finite nonnull expectation in each component.*

(i) *For $F \in \mathcal{P}_{1*}$ with expectation $\boldsymbol{\epsilon}(F) = (\epsilon_1, \ldots, \epsilon_d)$, the* scaled distribution *$\widetilde{F}$ is defined by*

$$\widetilde{F}(x_1, \ldots, x_d) = F(x_1\epsilon_1, \ldots, x_d\epsilon_d)\,, \quad (x_1, \ldots, x_d) \in \mathbb{R}^d\,, \tag{7.12}$$

(ii) *The* Lorenz zonoid *of F is defined as the lift zonoid of the scaled distribution,*

$$LZ(F) = \widehat{Z}(\widetilde{F})\,. \tag{7.13}$$

Consequently, if F is the distribution of a random vector X, $\widetilde{F}$ is the distribution of the random vector

$$\frac{X}{\boldsymbol{\epsilon}(F)} = \left(\frac{X_1}{\epsilon_1}, \ldots, \frac{X_d}{\epsilon_d}\right).$$

The Lorenz zonoid has volume equal to

$$\mathrm{vol}_{d+1} LZ(F) = \frac{1}{\prod_{j=1}^{d} |\epsilon_j|} \mathrm{vol}_{d+1} \widehat{Z}(F)\,. \tag{7.14}$$

Note that the volume of the lift zonoid (and therefore that of the Lorenz zonoid) equals zero if F is concentrated at one point. However, in dimension $d \geq 2$ the reverse is not true. Consider, e.g., an empirical distribution $F_{\boldsymbol{A}}$ where $\boldsymbol{A}$ has rank less than d; then the volume of the lift zonoid (see (7.9)) vanishes. In general, the lift zonoid volume vanishes whenever the support of F is contained in a linear subspace of $\mathbb{R}^d$ that has dimension less than d.

7.3 Expected volume of a random convex hull

In this section the lift zonoid volume is used to derive lower and upper bounds for the expected volume of a random convex hull, that is the convex hull of a random sample. These bounds equal the volume of the lift zonoid multiplied by factors that depend on dimension and sample length only.

Beginning with the work of Rényi and Sulanke (1963), Rényi and Sulanke (1964) and Efron (1965), the convex hull of random points has been studied. Here bounds are provided for the expected volume in terms of the volume of the lift zonoid of the sampling distribution. The only restriction imposed is that the distribution has a finite expectation vector.

Definition 7.3 (Random lift zonotope) *Let $\boldsymbol{X}_1, \ldots, \boldsymbol{X}_n$ be n independent random points in $\mathbb{R}^d$, each having the distribution function F, $F \in \mathcal{P}_1$.*

The random lift zonotope *of $\boldsymbol{X}_1, \ldots, \boldsymbol{X}_n$ is the random zonotope in $\mathbb{R}^{d+1}$ defined by*

$$\widehat{Z}(\boldsymbol{X}_1, \ldots, \boldsymbol{X}_n) = \sum_{i=1}^{n} \left[(0, \boldsymbol{0}), \frac{1}{n}(1, \boldsymbol{X}_i) \right]. \tag{7.15}$$

The next theorem shows that the expected volume of the random lift zonotope equals, up to a constant, the volume of the lift zonoid of the random vector.

Theorem 7.4 (Expected volume) *Let $\boldsymbol{X}_1, \ldots, \boldsymbol{X}_n$ be independent random vectors which all are distributed as μ. Then*

$$E(\mathrm{vol}_{d+1} \widehat{Z}(\boldsymbol{X}_1, \ldots, \boldsymbol{X}_n)) = \frac{(n-1) \cdot \ldots \cdot (n-d)}{n^d} \mathrm{vol}_{d+1} \widehat{Z}(F)\,. \tag{7.16}$$

Proof. For any points $\boldsymbol{y}_1, \ldots, \boldsymbol{y}_n$ in $\mathbb{R}^{d+1}$, by (2.13) holds that

$$\begin{aligned} \operatorname{vol}_{d+1} \sum_{i=1}^{n} [\mathbf{0}, \boldsymbol{y}_i] &= \frac{1}{(d+1)!} \sum_{i_0=1}^{n} \cdots \sum_{i_d=1}^{n} \left|\det[\boldsymbol{y}_{i_0}, \ldots, \boldsymbol{y}_{i_d}]\right| \\ &= \sum_{1 \le i_0 < \ldots < i_d \le n} \left|\det\left[\boldsymbol{y}_{i_0}, \ldots, \boldsymbol{y}_{<} i_d\right]\right| . \qquad (7.17) \end{aligned}$$

In view of (7.17), the $(d+1)$-variate volume of the random lift zonotope (7.15) amounts to

$$\sum_{1 \le i_0 < \ldots i_d \le n} \left| \det \left[\frac{1}{n}(1, \boldsymbol{X}_{i_0}), \ldots, \frac{1}{n}(1, \boldsymbol{X}_{i_d}) \right] \right| , \qquad (7.18)$$

which has expectation

$$\sum_{1 \le i_0 < \ldots < i_d \le n} \mathrm{E} \left(\left| \det \left[\frac{1}{n}(1, \boldsymbol{X}_{i_0}), \ldots, \frac{1}{n}(1, \boldsymbol{X}_{i_d}) \right] \right| \right) . \qquad (7.19)$$

For $1 \le i_0 < \ldots < i_d \le n$ holds

$$\begin{aligned} &\mathrm{E} \left(\left| \det \left[\frac{1}{n}(1, \boldsymbol{X}_{i_0}), \ldots, \frac{1}{n}(1, \boldsymbol{X}_{i_d}) \right] \right| \right) \\ &= \frac{1}{n^{d+1}} \mathrm{E} \left(\left|\det\left[(1, \boldsymbol{X}_{i_0}), \ldots, (1, \boldsymbol{X}_{i_d})\right]\right| \right) \\ &= \frac{1}{n^{d+1}} \mathrm{E}(|\det \widehat{M}(F)|) = \frac{(d+1)!}{n^{d+1}} \operatorname{vol}_{d+1} \widehat{Z}(F) , \end{aligned}$$

where Theorem 2.37 is used in the last equation. We conclude that

$$\begin{aligned} \mathrm{E}(\operatorname{vol}_{d+1} \widehat{Z}(\boldsymbol{X}_1, \ldots, \boldsymbol{X}_n)) &= \frac{(d+1)!}{n^{d+1}} \sum_{1 \le i_0 < \ldots i_d \le n} \operatorname{vol}_{d+1} \widehat{Z}(F) \\ &= \frac{(d+1)!}{n^{d+1}} \binom{n}{d+1} \operatorname{vol}_{d+1} \widehat{Z}(F) \\ &= \operatorname{vol}_{d+1} \widehat{Z}(F) \prod_{j=1}^{d} \frac{n-j}{n} , \qquad (7.20) \end{aligned}$$

which proves (7.16). Q.E.D.

Next, let $\boldsymbol{x}_1, \ldots, \boldsymbol{x}_n$ be points in $\mathbb{R}^d$ and consider the zonotope

$$Z_n = \left[(0, \mathbf{0}), \frac{1}{n}(1, \boldsymbol{x}_1) \right] + \cdots + \left[(0, \mathbf{0}), \frac{1}{n}(1, \boldsymbol{x}_n) \right] .$$

Obviously, Z_n is a convex, compact zonotope in $\mathbb{R}^{d+1}$. We want to establish inequalities between the volume of Z_n and the volume of the convex hull, $C_n = conv\{\boldsymbol{x}_1, \ldots, \boldsymbol{x}_n\}$. While Z_n is a set in $\mathbb{R}^{d+1}$, C_n is one in $\mathbb{R}^d$. In the sequel the volumes are considered with respect to these dimensions. We first provide a lower bound for the volume of C_n.

Theorem 7.5 (Lower bound) *There holds*

$$\text{vol}_d C_n \geq 2^d (d+1) \text{vol}_{d+1} Z_n \,. \tag{7.21}$$

Proof. Consider the hyperplanes $G_\alpha = \{\boldsymbol{x} \in \mathbb{R}^{d+1} : x_0 = \alpha\}$ and the halfspaces $G_{\leq\alpha} = \{\boldsymbol{x} \in \mathbb{R}^{d+1} : x_0 \leq \alpha\}$ and $G_{\geq\alpha} = \{\boldsymbol{x} \in \mathbb{R}^{d+1} : x_0 \geq \alpha\}$. Let $S_n = Z_n \cap G_{1/n}$. Then $C_n = n\, pr_{\{1,\ldots,d\}}(S_n)$, where $pr_{\{1,\ldots,d\}}$ is the projection to the last d coordinates. Therefore the d-variate volume of S_n equals

$$\text{vol}_d S_n = \frac{1}{n^d} \text{vol}_d C_n \,. \tag{7.22}$$

It is easy to check that for any $\alpha \leq 1/n$ holds

$$n\, S_n = n\,(Z_n \cap G_{1/n}) = \frac{1}{\alpha}\,(Z_n \cap G_\alpha). \tag{7.23}$$

Therefore, $Z_n \subset cone(S_n)$, where $cone(S) = \{\lambda \boldsymbol{s} : \lambda \in \mathbb{R}_+, \boldsymbol{s} \in S\}$ for $S \subset \mathbb{R}^d$. Because Z_n is centrally symmetric around $\frac{1}{2}(1, \boldsymbol{\epsilon}(F))$,

$$\text{vol}_{d+1} Z_n \leq 2\, \text{vol}_{d+1}(cone(S_n) \cap G_{\leq 1/2}). \tag{7.24}$$

Further, $cone(S_n) \cap G_{\leq 1/2}$ is a pyramid in $\mathbb{R}^{d+1}$ that has base $cone(S_n) \cap G_{1/2}$ and height $1/2$. As the d-variate volume of the base amounts to $(n/2)^d\, \text{vol}_d S_n$, obtain

$$\text{vol}_{d+1}(cone(S_n) \cap G_{\leq 1/2}) = \frac{1}{d+1} \frac{1}{2} \left(\frac{n}{2}\right)^d \text{vol}_d S_n \,. \tag{7.25}$$

In view of (7.22), (7.24) and (7.25) holds

$$\text{vol}_{d+1} Z_n \leq \left(\frac{n}{2}\right)^d \frac{1}{d+1} \left(\frac{1}{n}\right)^d \text{vol}_d C_n \,. \tag{7.26}$$

Q.E.D.

A reverse bound is established in the following theorem.

Theorem 7.6 (Upper bound) *Let $n \geq d+1$. Then*

$$\mathrm{vol}_d C_n \leq \frac{n^{d+1}(d+1)}{n(d+1)-2d} \mathrm{vol}_{d+1} Z_n \,. \tag{7.27}$$

Proof. First, note that (7.27) can be written as

$$\mathrm{vol}_{d+1} Z_n \geq \frac{1}{n^d}\left(\frac{n-2}{n} + 2\frac{1}{(d+1)n}\right) \mathrm{vol}_d C_n \,. \tag{7.28}$$

Consider the partition of Z_n that consists of the two pyramids $S_n^1 = Z_n \cap G_{\leq 1/n}$, $S_n^2 = Z_n \cap G_{\geq (n-1)/n}$ and the convex body $B_n = Z_n \cap G_{\leq (n-1)/n} \cap G_{\geq 1/n}$. The pyramids have the same volumes of bases, $n^{-d}\,\mathrm{vol}_d C_n$, and the same heights $1/n$. Thus, their volumes add to

$$\frac{2}{n^d}\frac{1}{(d+1)n} \mathrm{vol}_d C_n \,.$$

The body B_n is convex and has centrally, around the point $\frac{1}{2}(1, \epsilon(F))$, symmetric bases, S_n^1 and S_n^2, and the height $(n-2)/n$. Therefore, due to Schwartz symmetrization (Bonnesen and Fenchel, 1934), holds

$$\mathrm{vol}_{d+1} B_n \geq \frac{n-2}{n} \mathrm{vol}_{d+1} S_n^1 = \frac{n-2}{n}\frac{1}{n^d}\, \mathrm{vol}_d C_n \,.$$

That yields the proof. Q.E.D.

As a consequence of the above three theorems we obtain:

Theorem 7.7 (Bounds for the expected volume) *Let $\boldsymbol{X}_1, \ldots, \boldsymbol{X}_n$ be random vectors independent and identically distributed as F, $n \geq d+1$. Then*

$$\begin{aligned} &\left(\frac{n}{2}\right)^{-d} \left(\prod_{j=1}^{d} (n-j)\right) (d+1) \mathrm{vol}_{d+1} \widehat{Z}(F) \\ \leq\; & \mathrm{E}(\mathrm{vol}_d \, conv\{\boldsymbol{X}_1, \ldots, \boldsymbol{X}_n\}) \\ \leq\; & \frac{(d+1)n}{n(d+1)-2d} \left(\prod_{j=1}^{d} (n-j)\right) \mathrm{vol}_{d+1} \widehat{Z}(F) \,. \end{aligned} \tag{7.29}$$

7.4 The multivariate volume-Gini index

The univariate Gini index amounts to twice the area between the Lorenz curve and the diagonal, that is, to the volume of the Lorenz zonoid. We now extend the index to the multivariate case.

Given $F \in \mathcal{P}_1$, let $\boldsymbol{X}, \boldsymbol{X}_1, \ldots, \boldsymbol{X}_d$ be independent random vectors each of which is distributed according to F. $\widehat{\boldsymbol{Q}}$ denotes the $(d+1) \times (d+1)$ matrix having rows $(1, \boldsymbol{X}), (1, \boldsymbol{X}_1), \ldots, (1, \boldsymbol{X}_d)$. The term

$$M_{Wilks} = \frac{1}{d!} \mathrm{E}(|\det \widehat{\boldsymbol{Q}}|) \tag{7.30}$$

was proposed as a multivariate index of dispersion by Wilks (1960); see Oja (1983) and Giovagnoli and Wynn (1995). Oja (1983) has interpreted it via the average volume of random simplexes with vertices $\boldsymbol{X}, \boldsymbol{X}_1, \ldots, \boldsymbol{X}_d$. Recall from Equation (7.11) that $(d+1)^{-1} M_{Wilks}$ equals the volume of the lift zonoid. In Mosler (1994a) the $(d+1)$-dimensional volume of the Lorenz zonoid (7.14) was proposed as a multivariate Gini index, called the *Gini zonoid index*. Although this index shows a number of useful properties (boundedness between 0 and 1, 0 at one-point distributions, vector scale invariance, monotonicity with multivariate dilations), it can be zero also for distributions that are not concentrated at one point. The same applies for M_{Wilks}.

To avoid this drawback, Koshevoy and Mosler (1997a) propose the following definitions of a Gini mean difference and a Gini index: Let

$$C^d = \{(z_0, z_1, \ldots, z_d) \in \mathbb{R}^{d+1} : z_0 = 0,\ 0 \le z_s \le 1,\ s = 1, \ldots, d\},$$

which is a d-dimensional cube in $\mathbb{R}^{d+1}$. Instead of the volume of the lift zonoid, the subsequent definition uses the volume of the lift zonoid 'expanded' by this cube.

Definition 7.4 (Volume-Gini mean difference and index) *For* $F \in \mathcal{P}_1$, *the* volume-Gini mean difference *is defined by*

$$M_V(F) = \frac{1}{2^d - 1}\left(\mathrm{vol}_{d+1}(\widehat{Z}(F) + C^d) - 1\right). \tag{7.31}$$

For $F \in \mathcal{P}_{1*}$, *the* volume-Gini index *is defined by* $R_V(F) = M_V(\widetilde{F})$.

Here, again, $+$ is the Minkowski sum of sets. Equation (7.31) can be rewritten in the form

$$M_V(F) = \frac{1}{2^d - 1}\left(\mathrm{vol}_{d+1}(\widehat{Z}(F) + C^d) - \mathrm{vol}_{d+1}(\widehat{Z}(\delta_{\epsilon(F)}) + C^d)\right). \tag{7.32}$$

Equation (7.32) says that the volume-Gini mean difference is proportional to the volume of the lift zonoid of a distribution, 'expanded' by the d-dimensioned unit cube, minus the volume of the lift zonoid of the one-point distribution at the mean vector, 'expanded' by the same cube. The choice of the constant $1/(2^d - 1)$ in the defining equation (7.31) will be justified later by Proposition 7.9(v).

Figure 7.2 illustrates this definition in the case $d = 1$. GL_F is the generalized Lorenz curve of a univariate distribution F, and $\overline{GL}_F$ is its dual. The segment D is the lift zonoid of the one-point distribution δ_ϵ at the mean ϵ of F. GL_F and $\overline{GL}_F$ form the boundary of the lift zonoid $\widehat{Z}(F)$. The cube in dimension one is the segment C. Thus, $\widehat{Z}(\delta_\epsilon) + C$ is the area bounded by D, C, D', C', and $\widehat{Z}(F) + C$ is the area bounded by GL_F, $\overline{GL}'$, C and C'. By (7.32), the volume-Gini mean difference amounts to the difference of these two areas, which is equal to the area between GL_F and $\overline{GL}_F$.

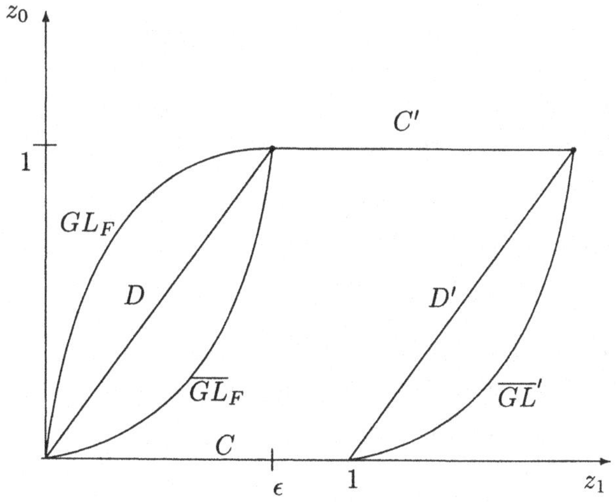

FIGURE 7.2: Illustrating the definition of M_V when $d = 1$.

Remark 1. By the lift zonoid approach we are also able to prove a general result on the univariate Gini mean difference for an arbitrary distribution F in $\mathcal{P}_1$ which is not restricted to the positive axis. For $d = 1$, by Definition 7.4, the volume-Gini mean difference equals the two-dimensional volume of the lift zonoid. With (7.11) obtain

$$M_V(F) = \mathrm{vol}_2(\widehat{Z}(F)) = \frac{1}{2}\int_{\mathbb{R}}\int_{\mathbb{R}}\left|\det\begin{pmatrix} 1 & 1 \\ x & y \end{pmatrix}\right| dF(x)dF(y)$$
$$= \frac{1}{2}\int_{\mathbb{R}}\int_{\mathbb{R}} |y-x|\, dF(x)dF(y)\,. \tag{7.33}$$

Recall that, for $d = 1$, the lift zonoid is the area between the generalized Lorenz curve and its dual and conclude that the usual Gini mean difference equals the area between the generalized Lorenz curve and its dual. This proves that Proposition 7.1 is true not only for distributions on $\mathbb{R}_+$, as previously stated, but also for general ones.

For a nonempty subset $K \subset \{1,\dots,d\}$, F_K denotes the marginal distribution with respect to the coordinates indexed by K.

Theorem 7.8 (Volume-Gini mean difference and index)

$$M_V(F) = \frac{1}{2^d-1}\sum_{\emptyset\neq K\subset\{1,\dots,d\}} \mathrm{vol}_{|K|+1}\widehat{Z}(F_K)\,, \tag{7.34}$$

$$R_V(F) = \frac{1}{2^d-1}\sum_{\emptyset\neq K\subset\{1,\dots,d\}} \mathrm{vol}_{|K|+1}\widehat{Z}(\widetilde{F}_K). \tag{7.35}$$

If $d = 2$, the theorem says that three times $M_V(F)$ is equal to the volume of the lift zonoid plus the Gini mean differences of the two univariate marginals.

Remark 2. By Equation (7.34), the volume-Gini mean difference is the average of the volumes of projections of the lift zonoid over all coordinate subspaces. They are spanned by $(1,0,\dots,0)$ and $(0,\mathbf{e}_r)$, $r \in K$, $K \subset \{1,\dots,d\}$. Here $\mathbf{e}_r$ is the r-th canonical unit vector in $\mathbb{R}^d$. By (7.35) the same holds for the volume-Gini index and the Lorenz zonoid.

Proof of Theorem 7.8 . We prove (7.34) for a discrete distribution F. Then an approximation argument yields (7.34) for a general distribution. (7.35) is an immediate consequence of (7.34).
Let F be a discrete distribution on $\boldsymbol{x}_1,\dots,\boldsymbol{x}_m$ in $\mathbb{R}^d$ with probabilities $q_1,\dots,q_m$. Then

$$\widehat{Z}(F) + C^d = \sum_{i=1}^{m}[(0,\mathbf{0}),(q_i, q_i\boldsymbol{x}_i)] + \sum_{s=1}^{d}[(0,\mathbf{0}),(0,\mathbf{e}_s)].$$

Hence, by (7.17),

$$\begin{aligned}
&\mathrm{vol}_{d+1}(\widehat{Z}(F) + C^d) = \qquad (7.36)\\
&\sum_{1\le i_1<\dots i_{d+1}\le m} |\det[(q_{i_1}, q_{i_1}\boldsymbol{x}_{i_1}), \dots, (q_{i_{d+1}}, q_{i_{d+1}}\boldsymbol{x}_{i_{d+1}})]|\\
&+\sum_{l=1}^{d-1} \sum_{1\le i_1<\dots i_{d+1-l}\le m} \sum_{1\le s_1<\dots s_l\le d}\\
&|\det[(q_{i_1}, q_{i_1}\boldsymbol{x}_{i_1}), \dots, (q_{i_{d+1-l}}, q_{i_{d+1-l}}\boldsymbol{x}_{i_{d+1-l}}), (0, \mathbf{e}_{s_1}), \dots, (0, \mathbf{e}_{s_l})]|\\
&+\sum_{i=1}^{m} |\det[(q_i, q_i\boldsymbol{x}_i), (0, \mathbf{e}_1), \dots, (0, \mathbf{e}_d)]|.
\end{aligned}$$

The first sum is the lift zonoid volume of F. In view of $\sum_{i=1}^m q_i = 1$, the third sum equals

$$\sum_{i=1}^{m} |\det[(q_i, q_i\boldsymbol{x}_i)(0, \mathbf{e}_1), \dots, (0, \mathbf{e}_d)]| = 1\,. \qquad (7.37)$$

To analyze the second sum, let $1 \le l \le d-1$ and $1 \le s_1 < \dots s_l \le d$ be fixed, $K = \{1, \dots, d\} \setminus \{s_1, \dots, s_l\}$. Hence $|K| = d - l$. For any $\boldsymbol{x}_i$, $\boldsymbol{x}_i^K$ denotes the vector composed of those components of $\boldsymbol{x}_i$ whose index belongs to K. Then the lift zonoid of F_K has $(|K| + 1)$-dimensional volume

$$\begin{aligned}
&\mathrm{vol}_{|K|+1}\widehat{Z}(F_K) \qquad (7.38)\\
&= \sum_{1\le i_1<\dots i_{d+1-l}\le m} |\det[(q_{i_1}, q_{i_1}\boldsymbol{x}_{i_1}^K), \dots, (q_{i_{d+1-l}}, q_{i_{d+1-l}}\boldsymbol{x}_{i_{d+1-l}}^K)]|\\
&= \sum_{1\le i_1<\dots i_{d+1-l}\le m}\\
&|\det[(q_{i_1}, q_{i_1}\boldsymbol{x}_{i_1}), \dots, (q_{i_{d+1-l}}, q_{i_{d+1-l}}\boldsymbol{x}_{i_{d+1-l}}), (0, \mathbf{e}_{s_1}), \dots, (0, \mathbf{e}_{s_l})]|\,.
\end{aligned}$$

(7.31), (7.38) and (7.37) together yield (7.34). Q.E.D.

The following three propositions establish properties of R_V and M_V. (Recall that $\boldsymbol{\beta} \cdot \boldsymbol{x}$ denotes the componentwise product of two vectors $\boldsymbol{\beta}$ and $\boldsymbol{x}$.)

Proposition 7.9 (Properties of R_V) *For all $F \in \mathcal{P}_{1*}$,*

(i) $0 \le R_V(F)$,

(ii) $R_V(F) = 0$ *if and only if F is a one-point distribution,*

(iii) $R_V(G) = R_V(F)$ *if* $G(\boldsymbol{x}) = F(\boldsymbol{\beta} \cdot \boldsymbol{x})$ *for* $\boldsymbol{x} \in \mathbb{R}^d$ *and some* $\boldsymbol{\beta} = (\beta_1, \ldots, \beta_d)$ *with* $\beta_1, \ldots, \beta_d > 0$,

(iv) R_V *is continuous w.r.t. weak convergence of distributions,*

(v) *If* $F \in \mathcal{P}_{1+}$, *then* $R_V(F) < 1$ *and the bound is sharp.*

Proof. (i): The volume is a nonnegative function.
(ii): If F is a one-point distribution, then, for every K, $\widehat{Z}(\widetilde{F}_K)$ is the main diagonal of the unit hypercube in $\mathbb{R}^{|K|+1}$ and has volume zero. Therefore $R_V(F) = 0$. If F is no one-point distribution, at least one of its univariate marginals, say $F_{j\bullet}$, has also no singleton support. Then the univariate Gini index $R(F_{j\bullet})$ is positive. Since $\mathrm{vol}_2 \widehat{Z}(\widetilde{F}_{j\bullet}) = R(F_{j\bullet})$, at least one summand in (7.35) does not vanish, and therefore $R_V(F) > 0$.
(iii): The vector scale invariance is obvious from the definition of $R_V(F)$, as it is based on the scaled distribution $\widetilde{F}$ only.
(iv) derives from Theorem 2.37.
(v): For every K, $\widehat{Z}(\widetilde{F}_K)$ is contained in the unit hypercube of $\mathbb{R}^{|K|+1}$, hence $0 \le \mathrm{vol}_{|K|+1} \widehat{Z}(\widetilde{F}_K) < 1$, and, by (7.35), $0 \le R_V(F) < 1$. It is easily seen that the upper bound 1 cannot be improved. For example, consider the distribution $F(\boldsymbol{x}) = \prod_{j=1}^d F_j(x_j)$ where $F_j(x_j) = 0$ if $x_j < 0$, $F_j(x_j) = (n-1)/n$ if $0 \le x_j < 1$, $F_j(x_j) = 1$ if $x_j \ge 1$. Then $R_V(F) \to 1$ for $n \to \infty$. Q.E.D.

Proposition 7.10 (Properties of M_V) *For all* $F \in \mathcal{P}_1$,

(i) $0 \le M_V(F)$,

(ii) $M_V(F) = 0$ *if and only if* F *is a one-point distribution,*

(iii) $M_V(G) = M_V(F)$ *if* $G(\boldsymbol{x}) = F(\boldsymbol{x} + \mathbf{c})$ *for all* $\boldsymbol{x}$ *and some* $\mathbf{c} \in \mathbb{R}^d$,

(iv) M_V *is continuous w.r.t. weak convergence of distributions,*

(v) *If* $F \in \mathcal{P}_{1+}$, *then*

$$M_V(F) < \frac{1}{2^d - 1} \sum_{\emptyset \ne K \subset \{1, \ldots, d\}} \prod_{i \in K} \epsilon_i(F) \le \frac{1}{2^d - 1}\left((\max_i \epsilon_i(F) + 1)^d - 1 \right)$$

and the first inequality cannot be improved.

The proof is similar to that of Proposition 7.9.

So far, we have considered properties for distributions of fixed dimension d. Our next property, the ceteris paribus property, concerns the effect of adding or removing a variable. It says that the dispersion index remains essentially unchanged if a variable is added (or removed) which has dispersion zero.

Let Φ^d be a d-variate *index of dispersion* that is defined for any $d \in \mathbb{N}$. We say that Φ^d has the *ceteris paribus property* if

$$\Phi^{d+1}(F \otimes \delta_\xi) = \gamma(d)\, \Phi^d(F) \tag{7.39}$$

for every d-variate distribution F and every $\xi \in \mathbb{R}$, $d \in \mathbb{N}$. Here δ_ξ is the univariate one-point distribution at ξ, $F \otimes \delta_\xi$ is the product distribution on $\mathbb{R}^{d+1}$, and $\gamma(d)$ is a constant which may depend on d but not on F or ξ.

Theorem 7.11 (Ceteris paribus property) *M_V and R_V have the ceteris paribus property with*

$$\gamma(d) = \frac{2^d - 1}{2^{d+1} - 1}\,.$$

Proof. It is easily seen, that $\mathrm{vol}_{|K|+1}\widehat{Z}((F \cdot E_\xi)_K) = 0$ if $d+1 \in K$. If $d+1 \notin K$, then $F_K = (F \cdot E_\xi)_K$. This and (7.34) yield the proposition. Q.E.D.

7.5 Volume statistics in cluster analysis

Another family of volume statistics is given by the volumes of the zonoid central regions. For $\alpha \in]0,1[$, the d-variate volume of the α-trimmed region D_α measures the dispersion of a certain central part of the data. This dispersion measurement focusses on points that have at least depth α and neglects all points whose depth is less.

The volume of central regions has been used as a measure of inhomogeneity in *cluster analysis.* Here the basic problem is to partition a given data set S into subsets (= 'clusters') $S_1, \dots, S_k$ so that each S_i is relatively homogeneous compared to the remaining data. Many clustering procedures employ an inhomogeneity index for clusters and minimize its sum over all clusters to obtain an 'optimal' partition.

In particular, Hardy and Rasson (1982) evaluate the inhomogeneity of clusters by their *hypervolume*, that is the sum of volumes of their convex hulls; see also Rasson (1996). The hypervolume has a nice property: If the data are uniformly distributed on a given number k of convex sets, minimizing the hypervolume yields a maximum likelihood estimator for these sets. Observe that the convex hull of S_j is the zonoid central region $D_0(F_{S_j})$, where F_{S_j}

denotes the empirical distribution on the data subset S_j. Thus, the Hardy-Rasson hypervolume is the aggregate volume of zonoid central regions.

In order to reduce sensitivity to outliers, Ruts and Rousseeuw (1996) suggest to replace each convex hull by a trimmed region. In particular, they propose to use a halfspace trimmed region.

Hoberg (2000) considers the following compound volume statistic as an inhomogeneity index for clusters,

$$\int_0^1 \mathrm{vol}_d D_\alpha(F_{S_j})\, \pi(d\alpha)\,, \tag{7.40}$$

where π is a weighting measure on $[0, 1]$. E.g., with π concentrated at some fixed α_0 the statistic specializes to the volume of the central region D_{α_0}. With $\pi(d\alpha) = \alpha^d\, d\alpha$ and D_α being the zonoid central region, the lift zonoid volume,

$$\mathrm{vol}_{d+1}\widehat{Z}(F_{S_j}) = \int_0^1 \mathrm{vol}_d D_\alpha(F_{S_j})\, \alpha^d\, d\alpha\,, \tag{7.41}$$

is obtained. Clustering with the inhomogeneity index (7.41) is mentioned as *zonoid clustering*. The idea to use a weighted sum of volumes of trimmed regions can be extended to other notions of depth.

A preliminary simulation study (Hoberg, 2000) demonstrates that the zonoid clustering approach gives good results in the case of noisy observations from uniform distributions on convex sets (if the noise is slight). However, it comes out that searching for the best zonoid clustering is very expensive. Although, if the clusters are assumed to be disjoint, the complexity is polynomial, the computational burden appears to be prohibitive. In order to obtain a solution with real data sets one has to make use of genetic algorithms, simulated annealing and other local search procedures.

7.6 Measuring dependency

This section provides an approach to the measurement of dependence which is based on the volume of central regions. Consider two random vectors $\boldsymbol{X} = (X_1, \ldots, X_d)$ and $\boldsymbol{Y} = (Y_1, \ldots, Y_d)$ that have the same univariate marginal distributions, $X_j =_{st} Y_j$ for all j. We are interested in the question whether $\boldsymbol{X}$ is *more dependent* than $\boldsymbol{Y}$, that is, whether the dependency among the components of $\boldsymbol{X}$ is larger than the dependency among the components of $\boldsymbol{Y}$.

For example, with bivariate normals the dependence is characterized and measured by their correlation coefficients. The dependence structure of a d-variate normal $\boldsymbol{X}$, $d \geq 2$, is fully described by its correlation matrix $\boldsymbol{P}_{\boldsymbol{X}}$.

Two general random vectors can be compared by the Loewner order of their correlation matrices, i.e., $\boldsymbol{X}$ is more dependent than $\boldsymbol{Y}$ if $\boldsymbol{P}_{\boldsymbol{X}} - \boldsymbol{P}_{\boldsymbol{Y}}$ is negative semidefinite. This yields a partial ordering of dependence among random vectors, the *correlation matrix order*. To obtain a complete ordering, the determinant of the correlation matrix can be used as an index and any two random vectors be ordered by decreasing correlation determinant.

The correlation matrix order is just one dependence order among general distributions. In the sequel we introduce further dependence orders that are based on notions of data depth and trimmed regions.

Observe, at least for normal distributions, that the zonoid trimmed regions become 'smaller' with decreasing correlation determinant.

Definition 7.5 (Central regions dependence order) *Let $\mathcal{R}$ be a family of trimmed regions*[1] *and consider two d-variate random vectors $\boldsymbol{X}$ and $\boldsymbol{Y}$ which have the same univariate marginals. Define $\boldsymbol{X} \preceq_{\mathcal{R}} \boldsymbol{Y}$ if*

$\mathrm{vol}_d D_\alpha(F_{\boldsymbol{X}}) \geq \mathrm{vol}_d D_\alpha(F_{\boldsymbol{Y}})$ *holds for every α.*

$\preceq_{\mathcal{R}}$ *is called the* $\mathcal{R}$-dependence order.

In particular, when $\mathcal{R}$ consists of the Mahalanobis central regions, the *Mahalanobis dependence order* is obtained. The *zonoid dependence order* and the *halfspace dependence order* are similarly defined.

The following two propositions relate central region orders to the ordering of certain determinants. Their proofs are left to the reader. The first makes use of the special form of Mahalanobis central regions (see Equation (3.13)), while the second follows from the affine equivariance of central regions. For a definition of elliptical distributions, see Section 2.3.5.

Proposition 7.12 (Mahalanobis dependence order) *Let $\boldsymbol{X}$ and $\boldsymbol{Y}$ have correlation matrices $\boldsymbol{P}_{\boldsymbol{X}}$ and $\boldsymbol{P}_{\boldsymbol{Y}}$. If $\mathcal{R}$ is the family of Mahalanobis regions,*

$$\boldsymbol{X} \preceq_{\mathcal{R}} \boldsymbol{Y} \quad \Leftrightarrow \quad |\det \boldsymbol{P}_{\boldsymbol{X}}| \geq |\det \boldsymbol{P}_{\boldsymbol{Y}}|.$$

[1] I.e., an affine equivariant family of nested sets which are closed, bounded and star-shaped with respect to every point in the smallest set; see Section 5.1.

Proposition 7.13 (Elliptical distributions) *If* $\boldsymbol{X}$ *and* $\boldsymbol{Y}$ *are elliptically distributed,*
$\boldsymbol{X} \sim \mathcal{E}(\mu, \boldsymbol{R}, \psi)$, $\boldsymbol{Y} \sim \mathcal{E}(\mu, \boldsymbol{S}, \psi)$, *then*

$$\boldsymbol{X} \preceq_{\mathcal{R}} \boldsymbol{Y} \quad \Leftrightarrow \quad |\det \boldsymbol{R}| \geq |\det \boldsymbol{S}| .$$

Example 7.1 (Bivariate normal distribution) Table 7.1 contains the volumes of Mahalanobis regions of a bivariate normal distribution with standard normal marginals and correlation $\rho = 0.0, 0.1, \ldots, 0.9$, based on 500 observations. It illustrates how, for given α, the volumes decrease with ρ. Table 7.2 contains the volumes of zonoid regions for the same distribution.

Example 7.2 (Bivariate exponential distribution) A bivariate random vector (X_1, X_2) has a *Marshall-Olkin distribution* if $X_1 = \min\{Y_1, Y_3\}$, $X_2 = \min\{Y_2, Y_3\}$, and Y_1, Y_2, Y_3 are independent exponentially distributed, $F_{Y_j}(t) = 1 - e^{-\lambda_j t}$ with some $\lambda_j > 0$. Then the marginals X_1 and X_2 have, again, exponential distributions with parameters $\lambda_1 + \lambda_3$ and $\lambda_2 + \lambda_3$, respectively, and their correlation is equal to $\rho = \lambda_3/(\lambda_1 + \lambda_2 + \lambda_3)$. Tables 7.3 and 7.4 present the volumes of Mahalanobis and zonoid regions, respectively, for Marshall-Olkin distributions with standard exponential marginals and different correlations.

α	Correlation ρ									
0.00	0.00	0.10	0.20	0.30	0.40	0.50	0.60	0.70	0.80	0.90
0.10	27.20	27.07	26.64	25.94	24.91	23.51	21.69	19.35	16.22	11.76
0.20	12.07	12.01	11.82	11.50	11.05	10.43	9.62	8.58	7.20	5.21
0.30	7.03	7.00	6.89	6.71	6.44	6.07	5.61	5.00	4.19	3.04
0.40	4.51	4.49	4.42	4.30	4.13	3.90	3.60	3.21	2.69	1.95
0.50	3.00	2.99	2.94	2.86	2.75	2.59	2.39	2.13	1.78	1.30
0.60	2.00	1.99	1.96	1.90	1.83	1.72	1.59	1.42	1.19	0.86
0.70	1.28	1.27	1.25	1.22	1.17	1.11	1.02	0.91	0.76	0.55
0.80	0.74	0.74	0.73	0.71	0.68	0.64	0.59	0.53	0.44	0.32
0.90	0.33	0.33	0.32	0.31	0.30	0.28	0.26	0.23	0.19	0.14

TABLE 7.1: Volumes of Mahalanobis trimmed regions for $n = 500$ bivariate normal observations with $\mu_1 = \mu_2 = 0, \sigma_1 = \sigma_2 = 1$ and several ρ and α

Joe (1997) states that a dependence order $\preceq$ should satisfy the following *postulates*:

1. preorder (reflexive and transitive),

α	Correlation ρ									
	0.00	0.10	0.20	0.30	0.40	0.50	0.60	0.70	0.80	0.90
0.10	9.00	8.96	8.82	8.59	8.25	7.80	7.20	6.42	5.38	3.90
0.20	5.65	5.63	5.54	5.39	5.17	4.88	4.51	4.02	3.37	2.45
0.30	3.88	3.87	3.80	3.70	3.56	3.36	3.10	2.76	2.32	1.68
0.40	2.70	2.69	2.64	2.57	2.47	2.33	2.15	1.92	1.61	1.16
0.50	1.83	1.83	1.80	1.75	1.68	1.59	1.47	1.31	1.09	0.79
0.60	1.19	1.19	1.17	1.14	1.09	1.03	0.95	0.85	0.71	0.52
0.70	0.70	0.70	0.70	0.67	0.65	0.61	0.56	0.50	0.42	0.30
0.80	0.35	0.34	0.34	0.33	0.32	0.30	0.28	0.24	0.21	0.15
0.90	0.11	0.11	0.10	0.10	0.10	0.09	0.08	0.07	0.06	0.05

TABLE 7.2: Volumes of Zonoid trimmed regions for $n = 500$ bivariate normal observations with $\mu_1 = \mu_2 = 0, \sigma_1 = \sigma_2 = 1$ and several ρ and α.

α	Correlation ρ									
0.00	0.00	0.10	0.20	0.30	0.40	0.50	0.60	0.70	0.80	0.90
0.10	28.87	29.78	28.73	28.42	26.69	24.57	21.85	18.13	14.56	9.83
0.20	12.83	13.22	12.74	12.61	11.84	10.91	9.69	8.04	6.45	4.36
0.30	7.41	7.70	7.42	7.35	6.89	6.35	5.64	4.68	3.76	2.54
0.40	4.77	4.94	4.77	4.71	4.42	4.08	3.62	3.00	2.41	1.62
0.50	3.21	3.29	3.17	3.14	2.95	2.71	2.41	2.00	1.60	1.08
0.60	2.10	2.19	2.11	2.09	1.96	1.80	1.60	1.33	1.06	0.72
0.70	1.36	1.40	1.35	1.34	1.26	1.15	1.03	0.85	0.68	0.46
0.80	0.80	0.81	0.78	0.78	0.73	0.67	0.60	0.49	0.40	0.27
0.90	0.35	0.36	0.35	0.34	0.32	0.29	0.26	0.22	0.17	0.12

TABLE 7.3: Volumes of Mahalanobis trimmed regions for $n = 500$ dependent exponential observations (Marshall-Olkin distribution) with $X_1, X_2 \sim Exp(1)$ and several α and $\rho = corr(X_1, X_2)$

2. antisymmetric,

3. continuous w.r.t. weak convergence of distributions,

4. invariant to marginalization,

5. invariant to permutation of components,

6. invariant to increasing transform of components,

α	Correlation ρ									
0.00	0.00	0.10	0.20	0.30	0.40	0.50	0.60	0.70	0.80	0.90
0.10	8.60	9.15	9.07	8.80	8.49	7.84	6.89	5.76	4.40	2.39
0.20	5.30	5.42	5.16	4.96	4.63	4.09	3.48	2.78	1.94	0.98
0.30	3.57	3.55	3.32	3.13	2.85	2.46	2.04	1.60	1.08	0.53
0.40	2.46	2.40	2.21	2.05	1.84	1.57	1.29	1.02	0.68	0.33
0.50	1.67	1.62	1.48	1.37	1.22	1.04	0.86	0.67	0.45	0.22
0.60	1.08	1.06	0.97	0.90	0.81	0.69	0.57	0.45	0.30	0.15
0.70	0.65	0.64	0.60	0.56	0.51	0.44	0.37	0.29	0.20	0.10
0.80	0.32	0.33	0.31	0.30	0.28	0.25	0.21	0.17	0.12	0.06
0.90	0.10	0.11	0.11	0.10	0.10	0.09	0.08	0.07	0.05	0.03

TABLE 7.4: Volumes of Zonoid trimmed regions for $n = 500$ dependent exponential observations (Marshall-Olkin distribution) with $X_1, X_2 \sim Exp(1)$ and several α and $\rho = corr(X_1, X_2)$

7. consistent with the bivariate concordance order,[2]

8. maximal at the upper Fréchet bound.[3]

Note that Postulate 8 is implied by 7. While Postulates 1–5 are satisfied with the Mahalanobis dependence order and the zonoid dependence order, Postulates 6–7 are not. Neither the weaker Postulate 8 is fulfilled.

To obtain dependence orderings that are invariant to increasing transforms of the components, we apply the previous dependence orders to the copulae of two distributions instead of the distributions themselves. For given $\boldsymbol{X}$ and $\boldsymbol{Y}$ denote

$$\boldsymbol{X}^0 = (F_1(X_1), \ldots, F_d(X_d)),$$

where F_j is the distribution function of the marginal X_j. The distribution of $\boldsymbol{X}^0$ is mentioned as the *standardized distribution* and its distribution function as the *copula* of $\boldsymbol{X}$. Define $\boldsymbol{Y}^0$ similarly.

Definition 7.6 (Copula regions dependence order) $\boldsymbol{X} \preceq^0_{\mathcal{R}} \boldsymbol{Y}$ *if* $\text{vol}_d D_\alpha(F^0_{\boldsymbol{X}}) \geq \text{vol}_d D_\alpha(F^0_{\boldsymbol{Y}})$ *holds for every* α.

The dependence order $\preceq^0_{\mathcal{R}}$ satisfies Joe's Postulates 1 and 3–6, but not 2 and 8, hence 7.

[2]The *bivariate concordance order* means that all two-dimensional marginals are ordered in the lower orthant order.

[3]Given the univariate marginals $F_1, \ldots, F_d$, the upper Fréchet bound is $F^+(\boldsymbol{x}) = \min_j F_j(x_j)$; see also Example 9.3.

In fact, the above dependence orders measure any departures from independence in either direction and do not distinguish whether the dependence is in the positive or negative.

Example 7.3 (Bivariate exponential distribution) Figures 7.3 exhibits the zonoid regions for 50 and 500 observations simulated from a standardized Marshall-Olkin distribution with standard exponential marginals and correlation $\rho = 1/3$.

Dependence indices. Indices consistent with the dependence order are easily constructed. When the dependent distribution is related to the independent distribution with the same marginals, notions of generalized correlations can be defined.

Let $\boldsymbol{X}$ have correlation matrix $P_{\boldsymbol{X}}$ and choose $\boldsymbol{Y}$ to be independent, that is, be the independent version of $\boldsymbol{X}$. Consider, for some fixed α^*, the ratio

$$GC_{\alpha^*}(\boldsymbol{X}) = \frac{\text{vol}_d D_{\alpha^*}(F_{\boldsymbol{Y}})}{\text{vol}_d D_{\alpha^*}(F_{\boldsymbol{X}})}.$$

If either both random vectors are normal or the regions are Mahalanobis,

$$GC_{\alpha^*}(\boldsymbol{X}) = |\det P_{\boldsymbol{X}}|$$

holds independently of α^* and for any family of starshaped central regions. By this, the ratio $GC_{\alpha^*}(\boldsymbol{X})$ can be introduced as a *generalized correlation* of $\boldsymbol{X}$. Of course, α^* has still to be chosen.

Note that with these generalized correlations neither the existence of second moments is assumed (for zonoid and halfspace regions) nor the dependence is restricted to some linear interrelation among components.

Instead of picking some α^*, volumes over α can be aggregated, e.g., by using the weighted index

$$\int_0^1 \alpha^d \, \text{vol}_d D_\alpha(\boldsymbol{X}) \, d\alpha\,,$$

where each D_α receives weight α^d. With zonoid central regions, this index is equal to the lift zonoid volume. Observe that $\boldsymbol{X} \preceq_{\mathcal{R}} \boldsymbol{Y}$ with zonoid central regions implies that the lift zonoid of $\boldsymbol{X}$ has larger volume than that of $\boldsymbol{Y}$.

An aggregate notion of *generalized correlation* of $\boldsymbol{X}$ is obtained by the ratio of lift zonoid volumes,

$$GC_{lz}(\boldsymbol{X}) = \frac{\text{vol}_d \widehat{Z}(\boldsymbol{Y})}{\text{vol}_d \widehat{Z}(\boldsymbol{X})},$$

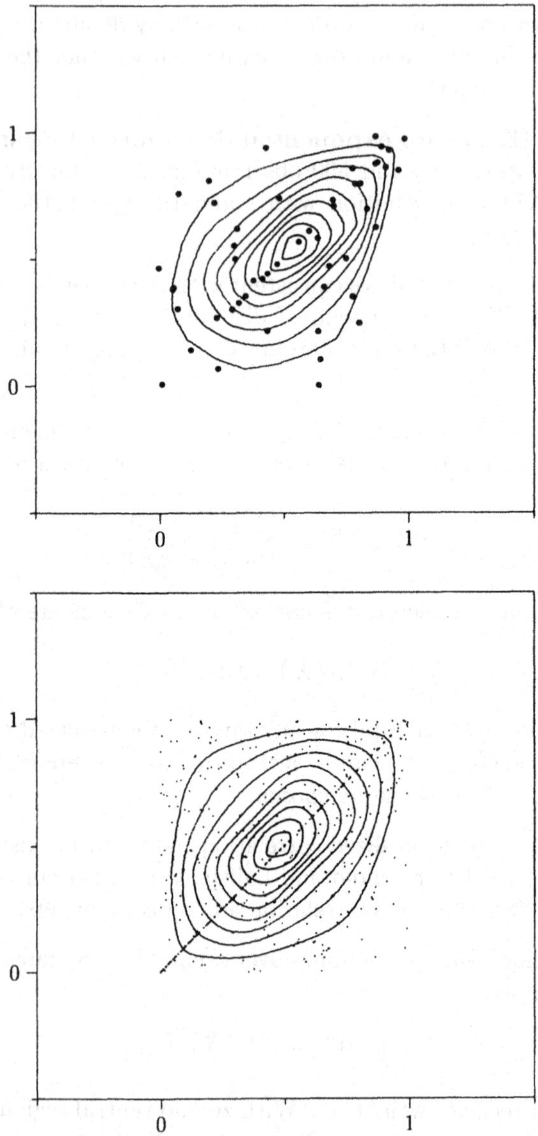

FIGURE 7.3: Zonoid trimmed regions for $n = 50$ and $n = 500$ standardized Marshall-Olkin observations,
$(1 - \exp(-X_1), 1 - \exp(-X_2))$, $X_1, X_2 \sim Exp(1)$, $\rho = 1/3$.

where, again, $\boldsymbol{Y}$ is the independent version of $\boldsymbol{X}$.

Application to data. When the above dependence orders are applied to given data volumes of central regions have to be calculated for empirical distributions. This is most simple for the Mahalanobis ellipsoids, but can also be done in reasonable time for the zonoid regions; see Section 3.9. For the likelihood regions a proper density estimate has to be used.

When determining these central regions from data, that is, from an empirical distribution, $\Sigma_{\boldsymbol{X}}$ and $\mu_{\boldsymbol{X}}$ can be substituted by other parameters which are more robust against outliers. See, e.g., Rousseeuw's minimum volume ellipsoid (Rousseeuw and Leroy, 1987), or the rank correlation matrix and center rank vector by Hettmansperger et al. (1998) and Oja (1999).

In comparing empirical distributions, rather often a dependence order fails because it is violated by a few outliers, that is, 'extreme' data points which lie far away from the 'main bulk' of the data. A natural robust definition of our dependence order is the following restricted order:

Definition 7.7 (Restricted regions dependence order) *Let $\beta \geq 0$ and define $\boldsymbol{X} \preceq_{\mathcal{R}}^{\beta} \boldsymbol{Y}$ if*

$$\mathrm{vol}_d D_\alpha(F_{\boldsymbol{X}}) \geq \mathrm{vol}_d D_\alpha(F_{\boldsymbol{Y}}) \quad \textit{holds for every } \alpha \geq \beta\,.$$

The parameter β controls the robustness of the order.

7.7 Notes

There are several attempts in the literature to define a multivariate Gini mean difference or Gini index as a volume statistic. Wilks (1960) introduces the volume of a convex body associated with F. Oja (1983) shows that the Wilks index is the expected volume of a simplex generated by $d+1$ random vertices which are independent and identically distributed according to F; see also Giovagnoli and Wynn (1995). In our framework the Wilks index amounts to $(d+1)!$ times the volume of the lift zonoid. Torgersen (1991) uses, as a multivariate Gini mean difference, the volume of the zonoid of the distribution, that is the projection of its lift zonoid on the last d coordinates. For a one-point distribution, both the Wilks-Oja and the Torgersen indexes vanish. But they are zero also for many other distributions, which appears to be unsatisfactory. Our notion $M_V(F)$ avoids this drawback; it vanishes if and only if F is a one-point distribution. In addition, the correct scaling factor is provided which makes, for nonnegative data, R_V vary between 0 and 1.

Another multivariate Gini index, associated with a concentration surface, has been introduced by Taguchi (1981). For the relations between Taguchi's concentration surface and the lift zonoid, see Koshevoy and Mosler (1996). Arnold (1987) proposes a multivariate Gini index that is based on the Euclidean distances of data points; see Chapter 8 below.

Section 7.3 has been adapted from Koshevoy and Mosler (1998), Section 7.4 from Koshevoy and Mosler (1997a), and Section 7.6 from Mosler (2002). Section 7.1 is also based on Koshevoy and Mosler (1997a). The notion of the Lorenz zonoid is due to Koshevoy and Mosler (1996).

The approach of Section 7.6 is in the spirit of Liu et al. (1999) who use notions of depth and trimmed regions to describe the *location* and *dispersion* of multivariate distributions. They employ central regions and parameterize them by their probability content, which is different from the approach chosen here. Liu et al. (1999) do not consider dependence.

Here I have introduced a dependence order by *comparing volumes* of all central regions. It suggests itself to consider also the stronger order that is defined by the *set inclusion* of all central regions. However, this order is too strong. With the zonoid regions it amounts to the set inclusion of the lift zonoids and becomes trivial (as only identical distributions are ordered). Dall'Aglio and Scarsini (2000b) define a dependence order by the set inclusion of zonoids (not lift zonoids); this yields a linear dependence order which implies the concordance order.

In fact, the above dependence orders measure *departures from independence* in either direction and do not distinguish whether the dependence is in the positive or negative. When the dependent distribution is related to the independent distribution with the same marginals, sort of generalized correlation is measured. Neither the existence of second moments is assumed (for zonoid and halfspace regions) nor the dependence restricted to some linear interrelation among components.

For distributions with *unequal marginals* the copula dependence order can be used as it is, and the non-copula dependence order after proper standardization with respect to location and scale.

8

Orderings and indices of dispersion

The problem of comparing random vectors with respect to their dispersion is common to many parts of applied probability and statistics, among them estimation problems and the comparison of experiments.

We introduce the lift zonoid order among probability distributions (or random vectors), characterize it in various respects and show that it is a dispersion ordering similar to the convex order but slightly weaker than the latter. The lift zonoid order is a *stochastic order* in $\mathbb{R}^d$, that is, a partial order (transitive, reflexive, and antisymmetric) of probability distributions and a preorder (transitive and reflexive) of random vectors. It has many useful properties, allows the derivation of useful probability inequalities, and can be modified to cope with outlying data.

Section 8.1 presents several equivalent definitions of the lift zonoid order and establishes a number of useful properties: preservation under affine transforms – in particular scale transforms, shifts and projections –, preservation under mixtures, and continuity. Section 8.2 concerns the ordering of marginals and independent components. In Section 8.3 preservation under convolutions and convex combination of independent random vectors is investigated. In Section 8.4 the lift zonoid order is contrasted with the multivariate convex order (= dilation). Section 8.5 gives an example of probability inequalities derived from the lift zonoid order. The lift zonoid order between two distributions implies that their means are equal. In Section 8.6 related lift zonoid orders are introduced by which distributions with unequal means can be compared: an increasing order, a scaled and a centered order. Finally,

Section 8.8 is devoted to multivariate indices of dispersion. An extended Gini index is discussed in detail and its consistency with various orders is shown.

8.1 Lift zonoid order

This section introduces the lift zonoid order as a partial order of distributions that reflects their dispersion. It presents several characterizations and the principal preservation properties of the order.

Definition 8.1 (Lift zonoid order) *For $F, G \in \mathcal{P}_1$, define the* lift zonoid order $\preceq_{lz}$ *by*

$$F \preceq_{lz} G \quad \text{if} \quad \hat{Z}(F) \subset \hat{Z}(G)\,.$$

Similarly, for two random vectors $\boldsymbol{X}$ and $\boldsymbol{Y}$ that are distributed as F and G, respectively, define

$$\boldsymbol{X} \preceq_{lz} \boldsymbol{Y} \quad \text{if} \quad F \preceq_{lz} G\,.$$

In view of Theorem 2.13, $\boldsymbol{X} \preceq_{lz} \boldsymbol{Y}$ means that

$$\mathrm{E}([(0,\mathbf{0}),(1,\boldsymbol{X})]) \subset \mathrm{E}([(0,\mathbf{0}),(1,\boldsymbol{Y})])\,,$$

i.e., the segment $[(0,\mathbf{0}),(1,\boldsymbol{X})]$ has a 'smaller' set-valued expectation than the segment $[(0,\mathbf{0}),(1,\boldsymbol{Y})]$.

Let $\mathcal{X}_1$ denote the class of random vectors that have distributions in $\mathcal{P}_1$. It follows immediately from the definition and Theorem 2.21 that $\preceq_{lz}$ is a partial order on $\mathcal{P}_1$ and a preorder on $\mathcal{X}_1$.

The lift zonoid order can also be defined in terms of central regions. In Chapter 3 the zonoid trimmed regions $D_\alpha(F)$ of a distribution F have been introduced. From Proposition 3.2 follows immediately that the lift zonoid order is equivalent to the set inclusion of zonoid trimmed regions:

Proposition 8.1 (Central regions order) *Let $F, G \in \mathcal{P}_1$. Then*

$$F \preceq_{lz} G \quad \Leftrightarrow \quad D_\alpha(F) \subset D_\alpha(G) \quad \text{for all } \alpha \in [0,1]\,.$$

We start with a discussion of the univariate case.

Proposition 8.2 (Univariate lift zonoid order) *Two univariate distributions F and $G \in \mathcal{P}_1$ are lift-zonoid ordered, $F \preceq_{lz} G$, if and only if $\epsilon(F) = \epsilon(G)$ and*

$$\int_0^t Q_F(s)ds \geq \int_0^t Q_G(s)ds \quad \text{for all } t \in [0,1]\,, \tag{8.1}$$

where Q_F and Q_G denote the quantile functions.

Proof. This follows from Theorem 2.17(ii) and $\epsilon(F) = \int_0^1 Q_F(s)ds = \int_0^1 Q_G(s)ds = \epsilon(G)$. Q.E.D.

Restriction (8.1) and $\epsilon(F) = \epsilon(G)$ define a well-known order of dispersion, the *univariate convex order*, $F \preceq_{cx} G$. The proposition says that, in dimension $d = 1$, the lift zonoid order is the same as the convex order. The next two propositions are well-known for the univariate convex order; see, e.g., Müller and Stoyan (2002, Ch. 1.5) or Shaked and Shanthikumar (1994, Ch. 2). The first provides three additional characterizations of the order.

Proposition 8.3 (Univariate convex order) *Let X and Y be univariate random variables distributed as F and $G \in \mathcal{P}_1$. Equivalent are:*

(i) $X \preceq_{lz} Y$.

(ii) $\mathrm{E}(\psi(X)) \le \mathrm{E}(\psi(Y))$ *if $\psi : \mathbb{R} \to \mathbb{R}$ is convex and the expectations exist.*

(iii) $\mathrm{E}(X) = \mathrm{E}(Y)$ *and* $\mathrm{E}(\psi(X)) \le \mathrm{E}(\psi(Y))$ *if $\psi : \mathbb{R} \to \mathbb{R}$ is increasing convex and the expectations exist.*

(iv) $Y =_{st} X + U$ *for some random variable U with* $\mathrm{E}(U|X) = 0$.

Proposition 8.4 (Univariate dispersion) *Let X be a univariate random variable having finite expectation. Then* $\mathrm{E}(X) \preceq_{lz} X$ *and*

$$X - \mathrm{E}(X) \preceq_{lz} \alpha(X - \mathrm{E}(X)) \quad \textit{if } \alpha > 1 .$$

The following theorem characterizes the lift zonoid order in higher dimensions. For $F \in \mathcal{P}_1$ and $\boldsymbol{p} = (p_1, \ldots, p_d) \in \mathbb{R}^d$, denote

$$F_{\boldsymbol{p}}(t) = \int_{\langle \boldsymbol{x}, \boldsymbol{p}\rangle \le t} dF(\boldsymbol{x}), \quad t \in \mathbb{R}.$$

Thus, if F is the distribution of the random vector $\boldsymbol{X}$ in $\mathbb{R}^d$, $F_{\boldsymbol{p}}$ is the distribution of the random variable $\langle \boldsymbol{X}, \boldsymbol{p}\rangle = p_1 X_1 + \ldots + p_d X_d$ in $\mathbb{R}$.

Theorem 8.5 (Convex-linear order) *Let $\boldsymbol{X}$ and $\boldsymbol{Y} \in \mathcal{X}_1$. The following conditions* (i) – (vi) *are equivalent.*

(i) $\boldsymbol{X} \preceq_{lz} \boldsymbol{Y}$.

(ii) *For all* $\boldsymbol{p} \in \mathbb{R}^d$, $\langle \boldsymbol{X}, \boldsymbol{p}\rangle \preceq_{lz} \langle \boldsymbol{Y}, \boldsymbol{p}\rangle$.

(iii) *For all* $\boldsymbol{p} \in \mathbb{R}^d$ *and all* $t \in [0, 1]$, $\int_0^t Q_{F_{\boldsymbol{p}}}(s)ds \ge \int_0^t Q_{G_{\boldsymbol{p}}}(s)ds$.

(iv) $\mathrm{E}(\psi \circ l(\boldsymbol{X})) \leq \mathrm{E}(\psi \circ l(\boldsymbol{Y}))$ *if* $\psi : \mathbb{R} \to \mathbb{R}$ *is convex,* $l : \mathbb{R}^d \to \mathbb{R}$ *is linear and the expectations exist.*

(v) $\mathrm{E}(\psi \circ l(\boldsymbol{X})) \leq \mathrm{E}(\psi \circ l(\boldsymbol{Y}))$ *if* $\psi : \mathbb{R} \to \mathbb{R}$ *is increasing convex,* $l : \mathbb{R}^d \to \mathbb{R}$ *is linear and the expectations exist.*

(vi) *For all* $\boldsymbol{p} \in \mathbb{R}^d$, $\langle \boldsymbol{Y}, \boldsymbol{p} \rangle =_{st} \langle \boldsymbol{X}, \boldsymbol{p} \rangle + U_{\boldsymbol{p}}$, *where* $U_{\boldsymbol{p}}$ *is a random vector in* $\mathbb{R}^d$ *and* $\mathrm{E}(U_{\boldsymbol{p}} | \langle \boldsymbol{X}, \boldsymbol{p} \rangle) = 0$.

A function $f : \mathbb{R}^d \to \mathbb{R}$ is called *convex-linear* if $f = \psi \circ l$ with some linear $l : \mathbb{R}^d \to \mathbb{R}$ and some convex $\psi : \mathbb{R} \to \mathbb{R}$. Therefore, the lift zonoid order is mentioned as the *convex-linear order* in the literature; see Müller and Stoyan (2002, Ch. 3.5).

Proof. By Theorem 2.20, $\boldsymbol{X} \preceq_{lz} \boldsymbol{Y}$ iff $\widehat{Z}(F_{\boldsymbol{p}}) \subset \widehat{Z}(G_{\boldsymbol{p}})$ for all $\boldsymbol{p} \in \mathbb{R}^d$. $F_{\boldsymbol{p}}$ and $G_{\boldsymbol{p}}$ are the distributions of $\langle \boldsymbol{X}, \boldsymbol{p} \rangle$ and $\langle \boldsymbol{Y}, \boldsymbol{p} \rangle$, respectively; hence (i) $\Leftrightarrow$ (ii). From Proposition 8.2 obtain (ii) $\Leftrightarrow$ (iii). Note that (v) with $\psi(\alpha) = \alpha$ and $l(\boldsymbol{x}) = \pm x_j$, $j = 1, \ldots, d$, implies $\mathrm{E}(\boldsymbol{X}) = \mathrm{E}(\boldsymbol{Y})$; hence, from Proposition 8.3 conclude (v) $\Leftrightarrow$ (iv). All other statements can be immediately derived from Proposition 8.3. Q.E.D.

Corollary 8.6 (Dispersion) *For* $\boldsymbol{X} \in \mathcal{P}_1$,

(i) $\mathrm{E}(\boldsymbol{X}) \preceq_{lz} \boldsymbol{X}$,

(ii) $\boldsymbol{X} - \mathrm{E}(\boldsymbol{X}) \preceq_{lz} \alpha(\boldsymbol{X} - \mathrm{E}(\boldsymbol{X}))$ *if* $\alpha > 1$.

Proof. Use Theorem 8.5(ii) in combination with Proposition 8.4. Q.E.D.

These results clarify what the lift zonoid order actually measures. The order implies equal expectations,

$$\boldsymbol{X} \preceq_{lz} \boldsymbol{Y} \quad \Rightarrow \quad \mathrm{E}(\boldsymbol{X}) = \mathrm{E}(\boldsymbol{Y}) , \tag{8.2}$$

and, consequently, does not indicate differences in location. Instead it is sensitive against changes in scale. The lift zonoid order increases with a positive scale factor, and the one-point distributions form its minimal elements. Thus, it compares distributions with respect to some sort of dispersion. Dispersion in this sense is described by the size and shape of zonoid central regions: $\boldsymbol{Y}$ is *more dispersed* than $\boldsymbol{X}$ if for every α the zonoid α-trimmed region of $\boldsymbol{Y}$ includes that of the $\boldsymbol{X}$.

In a family of elliptical distributions (see Section 2.3.5), the lift zonoid order corresponds to an ordering of parameter matrices. The lift zonoid of an elliptical distribution has a special form which is given in Proposition 2.29. From this conclude:

Proposition 8.7 (Elliptical distributions) *Let* $\boldsymbol{X} \sim \mathcal{E}(\boldsymbol{a}, \boldsymbol{R}, \psi)$ *and* $\boldsymbol{Y} \sim \mathcal{E}(\boldsymbol{b}, \boldsymbol{S}, \psi)$. *Then* $\boldsymbol{X} \preceq_{lz} \boldsymbol{Y}$ *iff*

$$\boldsymbol{a} = \boldsymbol{b} \text{ and } \boldsymbol{S} - \boldsymbol{R} \text{ is positive semidefinite.} \tag{8.3}$$

Particularly, two normal variables $\boldsymbol{X} \sim N(\boldsymbol{a}, \boldsymbol{R})$ and $\boldsymbol{Y} \sim N(\boldsymbol{b}, \boldsymbol{S})$ are lift-zonoid ordered iff (8.3) applies, that is, the means are equal and the covariance matrices are Loewner ordered.

Now, let us turn to preservation properties. First, it will be shown that the lift zonoid order is preserved under affine transformations. It follows that the order is translation and scale invariant. Moreover, the lift zonoid order of two distributions implies the same order of all marginal distributions. Further results concern the preservation under mixture and weak convergence.

Proposition 8.8 (Affine transform) *Let* $\boldsymbol{A} \in \mathbb{R}^{d \times k}$ *and* $\boldsymbol{c} \in \mathbb{R}^k$. *Then*

$$\boldsymbol{X} \preceq_{lz} \boldsymbol{Y} \quad \Rightarrow \quad \boldsymbol{X}\boldsymbol{A} + \boldsymbol{c} \preceq_{lz} \boldsymbol{Y}\boldsymbol{A} + \boldsymbol{c}\,.$$

Proof. We use the formula $\widehat{Z}_t(\boldsymbol{X}\boldsymbol{A} + \boldsymbol{c}) = \widehat{Z}_t(\boldsymbol{X})\boldsymbol{A} + t\boldsymbol{c}$ from Corollary 2.27. $\boldsymbol{X} \preceq_{lz} \boldsymbol{Y}$ is equivalent to $\widehat{Z}_t(\boldsymbol{X}) \subset \widehat{Z}_t(\boldsymbol{Y})$ for all $0 \le t \le 1$. Let $(t, \boldsymbol{z}) \in \widehat{Z}(\boldsymbol{X}\boldsymbol{A} + \boldsymbol{c})$. Then $\boldsymbol{z} = \boldsymbol{u}\boldsymbol{A} + t\boldsymbol{c}$ with $\boldsymbol{u} \in \widehat{Z}_t(\boldsymbol{X})$, hence $\boldsymbol{u} \in \widehat{Z}_t(\boldsymbol{Y})$, $\boldsymbol{z} \in \widehat{Z}_t(\boldsymbol{Y}\boldsymbol{A}+\boldsymbol{c})$ and $(t, \boldsymbol{z}) \in \widehat{Z}(\boldsymbol{Y}\boldsymbol{A}+\boldsymbol{c})$. Conclude $\widehat{Z}(\boldsymbol{X}\boldsymbol{A}+\boldsymbol{c}) \subset \widehat{Z}(\boldsymbol{Y}\boldsymbol{A}+\boldsymbol{c})$. Q.E.D.

The following two corollaries concern important special cases of Proposition 8.8. The lift zonoid order is preserved under marginalization and transformations of location and scale.

Corollary 8.9 (Marginals) *If two random vectors are lift-zonoid ordered, then all their marginals are lift-zonoid ordered in the same direction:*

$$\boldsymbol{X} \preceq_{lz} \boldsymbol{Y}\,, \quad \emptyset \neq J \subset \{1, \ldots, d\} \quad \Rightarrow \quad \boldsymbol{X}_J \preceq_{lz} \boldsymbol{Y}_J\,, \tag{8.4}$$

where $\boldsymbol{X}_J = (X_j)_{j \in J}$ *denotes the vector of components with index in* J.

Corollary 8.10 (Translation and scale transform) *There holds*

(i) $\boldsymbol{X} \preceq_{lz} \boldsymbol{Y}\,, \quad \boldsymbol{c} \in \mathbb{R}^d \quad \Rightarrow \quad \boldsymbol{X} + \boldsymbol{c} \preceq_{lz} \boldsymbol{Y} + \boldsymbol{c}\,,$

(ii) $\boldsymbol{X} \preceq_{lz} \boldsymbol{Y}\,, \quad \alpha > 0 \quad \Rightarrow \quad \alpha\boldsymbol{X} \preceq_{lz} \alpha\boldsymbol{Y}\,.$

Proposition 8.11 (Equal marginals) *If the univariate marginal distributions of two random vectors $\boldsymbol{X}$ and $\boldsymbol{Y}$ coincide, then*

$$\boldsymbol{X} \preceq_{lz} \boldsymbol{Y} \quad \Rightarrow \quad \boldsymbol{X} =_{st} \boldsymbol{Y}.$$

Proof. Let $\boldsymbol{X} \preceq_{lz} \boldsymbol{Y}$, i.e., $\langle \boldsymbol{X}, \boldsymbol{p}\rangle \preceq_{cx} \langle \boldsymbol{Y}, \boldsymbol{p}\rangle$ for all $\boldsymbol{p}$. Under this premise Scarsini and Shaked (1990) have shown that $\boldsymbol{X} =_{st} \boldsymbol{Y}$ iff $\boldsymbol{X}_j =_{st} \boldsymbol{Y}_j$ holds for $j = 1, \ldots, d$. Q.E.D.

Proposition 8.12 (Mixture) *Let F and G be mixed distributions, $F = \int_\Theta F_\theta \, \pi(d\theta)$ and $G = \int_\Theta G_\theta \, \pi(d\theta)$. Then*

$$F_\theta \preceq_{lz} G_\theta \quad \text{for all } \theta \in \Theta \quad \Rightarrow \quad F \preceq_{lz} G.$$

Proof. By assumption and Theorem 8.5(iv),

$$\int_{\mathbb{R}^d} f(\boldsymbol{x}) F_\theta(d\boldsymbol{x}) \le \int_{\mathbb{R}^d} f(\boldsymbol{x}) G_\theta(d\boldsymbol{x})$$

for every convex-linear f and $\theta \in \Theta$. Integrating this inequality with the mixing distribution $\pi(d\theta)$ yields the result. Q.E.D.

From Proposition 8.12 follow useful special results concerning finite mixtures of two components:

Corollary 8.13 (Finite mixture) *Let F^1, F^2, G^1, G^2, and $H \in \mathcal{P}_1$. For every $\alpha \in [0,1]$ holds*

(i) $F^1 \preceq_{lz} G^1$ *and* $F^2 \preceq_{lz} G^2 \quad \Rightarrow \quad \alpha F^1 + (1-\alpha)F^2 \preceq_{lz} \alpha G^1 + (1-\alpha)G^2$,

(ii) $F^1 \preceq_{lz} H$ *and* $F^2 \preceq_{lz} H \quad \Rightarrow \quad \alpha F^1 + (1-\alpha)F^2 \preceq_{lz} H$,

(iii) $H \preceq_{lz} G^1$ *and* $H \preceq_{lz} G^2 \quad \Rightarrow \quad H \preceq_{lz} \alpha G^1 + (1-\alpha)G^2$.

The lift zonoid order is continuous in the following sense.

Proposition 8.14 (Continuity) *Let $F, G \in \mathcal{P}_1$ and consider a sequence $(F^n)_{n\in\mathbb{N}}$ in $\mathcal{P}_1$ that converges weakly to F.*

(i) $F^n \preceq_{lz} G$ *for all* $n \quad \Rightarrow \quad F \preceq_{lz} G$.

(ii) $G \preceq_{lz} F^n$ *for all n and $(F^n)_{n\in\mathbb{N}}$ uniformly integrable* $\quad \Rightarrow \quad G \preceq_{lz} F$.

The theorem follows from Theorem 2.30 and its corollaries in Section 2.4.1. Alternatively, the proposition derives from Theorem 8.5(iv) and the fact that every convex function ψ can be weakly approximated by continuous convex functions.

8.2 Order of marginals and independence

We have seen in Corollary 8.9 that the lift zonoid order of two random vectors implies the same for all their marginals. The reverse is not true in general: Two distributions are not necessarily lift-zonoid ordered if all their marginals are.

For example, let $\boldsymbol{X}$ and $\boldsymbol{Y}$ be bivariate normal, $\boldsymbol{X} \sim N(\boldsymbol{0}, \boldsymbol{R})$, $\boldsymbol{Y} \sim N(\boldsymbol{0}, \boldsymbol{S})$, with

$$\boldsymbol{R} = \begin{pmatrix} 1 & \alpha \\ \alpha & 1 \end{pmatrix}, \qquad \boldsymbol{S} = \begin{pmatrix} 4 & \beta \\ \beta & 1 \end{pmatrix}$$

and $\alpha \neq \beta$. Then $X_j \preceq_{lz} Y_j$ for $j = 1, 2$, but not $\boldsymbol{X} \preceq_{lz} \boldsymbol{Y}$.

However, under independency assumptions a reverse result can be proved. Again, denote $\boldsymbol{X}_J = (X_j)_{j \in J}$ and $\boldsymbol{X}_{-J} = (X_j)_{j \notin J}$. F_J is the marginal distribution of $\boldsymbol{X}_J$, and $F_J \cdot F_{-J}$ the product distribution of F_J and F_{-J}.

Proposition 8.15 (Independent marginals) *Consider* $\boldsymbol{X} = (\boldsymbol{X}_J, \boldsymbol{X}_{-J})$ *and* $\boldsymbol{Y} = (\boldsymbol{Y}_J, \boldsymbol{Y}_{-J}) \in \mathcal{X}_1$, *and let* $\boldsymbol{X}_J$ *be independent of* $\boldsymbol{X}_{-J}$ *and* $\boldsymbol{Y}_J$ *be independent of* $\boldsymbol{Y}_{-J}$. *Then*

$$\boldsymbol{X} \preceq_{lz} \boldsymbol{Y} \quad \Leftrightarrow \quad \boldsymbol{X}_J \preceq_{lz} \boldsymbol{Y}_J \ \textit{and} \ \boldsymbol{X}_{-J} \preceq_{lz} \boldsymbol{Y}_{-J}\,. \tag{8.5}$$

Proof. $F = F_J \cdot F_{-J}$ and $G = G_J \cdot G_{-J}$ are the distributions of $\boldsymbol{X}$ and $\boldsymbol{Y}$. We have to prove that $\widehat{Z}(F) \subset \widehat{Z}(G)$ if and only if $\widehat{Z}(F_J) \subset \widehat{Z}(G_J)$ and $\widehat{Z}(F_{-J}) \subset \widehat{Z}(G_{-J})$. The only if part follows from Corollary 8.9. To prove the reverse, consider the support function of $\widehat{Z}(F_J \cdot F_{-J})$. For given $(p_0, \boldsymbol{p}) \in \mathbb{R}^{d+1}$,

$$\begin{aligned} & h(\widehat{Z}(F_J \cdot F_{-J}), (p_0, \boldsymbol{p})) \\ &= \int_{p_0 + \langle \boldsymbol{x}, \boldsymbol{p} \rangle \geq 0} (p_0 + \langle \boldsymbol{x}, \boldsymbol{p} \rangle) F_J(d\boldsymbol{x}_J) F_{-J}(d\boldsymbol{x}_{-J}) \\ &= \int_{-\infty}^{\infty} \Big[\int_{p_0 + \langle \boldsymbol{x}_{-J}, \boldsymbol{p}_{-J} \rangle = s} \psi(s, F_J) F_{-J}(d\boldsymbol{x}_{-J}) \Big] ds \\ &= \int_{\mathbb{R}^{|-J|}} \psi(p_0 + \langle \boldsymbol{x}_{-J}, \boldsymbol{p}_{-J} \rangle, F_J) F_{-J}(d\boldsymbol{x}_{-J})\,, \end{aligned} \tag{8.6}$$

where the notation

$$\psi(s, F_J) = \int_{s + \langle \boldsymbol{p}_J, \boldsymbol{x}_J \rangle \geq 0} (s + \langle \boldsymbol{p}_J, \boldsymbol{x}_J \rangle) F_J(d\boldsymbol{x}_J)$$

has been introduced. Note that $s \mapsto \psi(s, F_J)$ is a convex function, and $\boldsymbol{x}_{-J} \mapsto \psi(p_0 + \langle \boldsymbol{x}_{-J}, \boldsymbol{p}_{-J}\rangle, F_J)$ is a convex-linear function. Therefore, from $\widehat{Z}(F_{-J}) \subset \widehat{Z}(G_{-J})$, Theorem 8.5(iv), and (8.6) obtain

$$\begin{aligned}
&h(\widehat{Z}(F_J \cdot F_{-J}), (p_0, \boldsymbol{p})) \\
&= \int_{\mathbf{R}^{|-J|}} \psi(p_0 + \langle \boldsymbol{x}_{-J}, \boldsymbol{p}_{-J}\rangle, F_J)\, F_{-J}(d\boldsymbol{x}_{-J}) \\
&\le \int_{\mathbf{R}^{|-J|}} \psi(p_0 + \langle \boldsymbol{x}_{-J}, \boldsymbol{p}_{-J}\rangle, F_J)\, G_{-J}(d\boldsymbol{x}_{-J}) \\
&= \int_{-\infty}^{\infty} \Big[\int_{p_0 + \langle \boldsymbol{x}_{-J}, \boldsymbol{p}_{-J}\rangle = s} \psi(s, F_J)\, G_{-J}(d\boldsymbol{x}_{-J}) \Big] ds\,. \qquad (8.7)
\end{aligned}$$

Further, $\widehat{Z}(F_J) \subset \widehat{Z}(G_J)$ with Theorem 8.5(iv) yields $\psi(s, F_J) \le \psi(s, G_J)$ for all s. Therefore the right-hand side of (8.7) is bounded by

$$\begin{aligned}
&\int_{-\infty}^{\infty} \Big[\int_{p_0 + \langle \boldsymbol{x}_{-J}, \boldsymbol{p}_{-J}\rangle = s} \psi(s, G_J) G_{-J}(d\boldsymbol{x}_{-J}) \Big] ds \\
&= h(\widehat{Z}(G_J \cdot G_{-J}), (p_0, \boldsymbol{p}))\,.
\end{aligned}$$

Conclude $h(\widehat{Z}(F_J \cdot F_{-J}), \cdot) \le h(\widehat{Z}(G_J \cdot G_{-J}), \cdot)$, hence $\widehat{Z}(F_J \cdot F_{-J}) \subset \widehat{Z}(G_J \cdot G_{-J})$. Q.E.D.

By repeated application of (8.5) obtain:

Corollary 8.16 (Independent random vectors) *If all univariate marginals are independent, then*

$$\boldsymbol{X} \preceq_{lz} \boldsymbol{Y} \quad \Leftrightarrow \quad X_j \preceq_{lz} Y_j \quad \text{for all } j\,. \qquad (8.8)$$

8.3 Order of convolutions

The lift zonoid order is preserved under convolutions.

Theorem 8.17 (Convolutions) *Let $\boldsymbol{X}, \boldsymbol{Y}, \boldsymbol{Z}, \boldsymbol{U} \in \mathcal{X}_1$. Assume that $\boldsymbol{X}$ is independent of $\boldsymbol{Z}$ and $\boldsymbol{Y}$ is independent of $\boldsymbol{U}$. Then*

$$\boldsymbol{X} \preceq_{lz} \boldsymbol{Y}, \quad \boldsymbol{Z} \preceq_{lz} \boldsymbol{U} \quad \Rightarrow \quad \boldsymbol{X} + \boldsymbol{Z} \preceq_{lz} \boldsymbol{Y} + \boldsymbol{U}\,. \qquad (8.9)$$

In particular,

$$\boldsymbol{X} \preceq_{lz} \boldsymbol{Y} \quad \Rightarrow \quad \boldsymbol{X} + \boldsymbol{Z} \preceq_{lz} \boldsymbol{Y} + \boldsymbol{Z}\,. \qquad (8.10)$$

Proof. Consider the random vectors $(\boldsymbol{X}, \boldsymbol{Z})$ and $(\boldsymbol{Y}, \boldsymbol{U})$ in $\mathbb{R}^{2d}$. From the assumptions and Proposition 8.15 follows that $(\boldsymbol{X}, \boldsymbol{Z}) \preceq_{lz} (\boldsymbol{Y}, \boldsymbol{U})$. Let I_d denote the $d \times d$ unit matrix and $\boldsymbol{A}' = (I_d, I_d)$. Then $(\boldsymbol{X}, \boldsymbol{Z})\boldsymbol{A} = \boldsymbol{X} + \boldsymbol{Z}$ and $(\boldsymbol{Y}, \boldsymbol{U})\boldsymbol{A} = \boldsymbol{Y} + \boldsymbol{U}$. By Proposition 8.8, $\boldsymbol{X} + \boldsymbol{Z} \preceq_{lz} \boldsymbol{Y} + \boldsymbol{U}$. Q.E.D.

The next proposition concerns *convex convolutions*, that is, convex combinations of independent random vectors.

Proposition 8.18 (Convex convolution) *Let $\boldsymbol{X}, \boldsymbol{Y}, \boldsymbol{Z} \in \mathcal{X}_1$, $\alpha \in [0, 1]$, and assume that $\boldsymbol{X}$ and $\boldsymbol{Z}$ are independent. Denote by $\boldsymbol{V}_\alpha$ the* α-mixture *which has distribution $F_{\boldsymbol{V}_\alpha} = \alpha F_{\boldsymbol{X}} + (1 - \alpha) F_{\boldsymbol{Z}}$. There holds*

(i) $\alpha \boldsymbol{X} + (1 - \alpha)\boldsymbol{Z} \preceq_{lz} \boldsymbol{V}_\alpha$,

(ii) $\boldsymbol{X} \preceq_{lz} \boldsymbol{Y}$ *and* $\boldsymbol{Z} \preceq_{lz} \boldsymbol{Y} \quad \Rightarrow \quad \alpha \boldsymbol{X} + (1 - \alpha)\boldsymbol{Z} \preceq_{lz} \boldsymbol{Y}$,

(iii) $\boldsymbol{Y} \preceq_{lz} \boldsymbol{X}$ *and* $\boldsymbol{Y} \preceq_{lz} \boldsymbol{Z} \quad \Rightarrow \quad \boldsymbol{Y} \preceq_{lz} \alpha \boldsymbol{X} + (1 - \alpha)\boldsymbol{Z}$.

Proposition 8.18(ii) implies that, if several independent random vectors are $\preceq_{lz}$-smaller than a given random vector, every convex combination of them is smaller than the given vector, too. (iii) implies the same for independent vectors that are $\preceq_{lz}$-greater than a given vector.

Proof. Define $\boldsymbol{U} = \alpha \boldsymbol{X} + (1 - \alpha)\boldsymbol{Z}$. Let $F_{\boldsymbol{X}}$, $F_{\boldsymbol{Y}}$, $F_{\boldsymbol{Z}}$, $F_{\boldsymbol{U}}$, and $F_{\boldsymbol{V}_\alpha}$ be the distributions of $\boldsymbol{X}$, $\boldsymbol{Y}$, $\boldsymbol{Z}$, $\boldsymbol{U}$, and $\boldsymbol{V}_\alpha$, respectively, and show that for $0 \le \alpha \le 1$ the inclusion

$$\widehat{Z}(F_{\boldsymbol{U}}) \subset \alpha \widehat{Z}(F_{\boldsymbol{X}}) + (1 - \alpha)\widehat{Z}(F_{\boldsymbol{Z}}) \tag{8.11}$$

holds. Let $\boldsymbol{z} \in \widehat{Z}(F_{\boldsymbol{U}})$. Then, with some $g : \mathbb{R}^d \to [0, 1]$,

$$\begin{aligned} \boldsymbol{z} &= \int_{\mathbb{R}^d} g(\boldsymbol{u})(1, \boldsymbol{u}) F_{\boldsymbol{U}}(d\boldsymbol{u}) \\ &= \int_{\mathbb{R}^d} \int_{\mathbb{R}^d} g(\alpha \boldsymbol{x} + (1 - \alpha)\boldsymbol{z})(1, \alpha \boldsymbol{x} + (1 - \alpha)\boldsymbol{z}) F_{\boldsymbol{X}}(d\boldsymbol{x}) F_{\boldsymbol{Z}}(d\boldsymbol{z}) \\ &= \alpha \int_{\mathbb{R}^d} \left[\int_{\mathbb{R}^d} g(\alpha \boldsymbol{x} + (1 - \alpha)\boldsymbol{z})(1, \boldsymbol{x}) F_{\boldsymbol{X}}(d\boldsymbol{x}) \right] F_{\boldsymbol{Z}}(d\boldsymbol{z}) \\ &\quad + (1 - \alpha) \int_{\mathbb{R}^d} \left[\int_{\mathbb{R}^d} g(\alpha \boldsymbol{x} + (1 - \alpha)\boldsymbol{z})(1, \boldsymbol{z}) F_{\boldsymbol{Z}}(d\boldsymbol{z}) \right] F_{\boldsymbol{X}}(d\boldsymbol{x}). \end{aligned} \tag{8.12}$$

The inner integral of the first summand in (8.12) is an element of $\widehat{Z}(F_{\boldsymbol{X}})$ for every $\boldsymbol{z}$. Integration over $\boldsymbol{z}$ again yields an element of $\widehat{Z}(F_{\boldsymbol{X}})$ because $\widehat{Z}(F_{\boldsymbol{X}})$ is convex. Therefore, the first summand in (8.12) equals $\alpha \boldsymbol{z}^*$ with some $z^* \in \widehat{Z}(F_X)$. Similarly, the second summand amounts to $(1-\alpha)z^{**}$ with some $z^{**} \in \widehat{Z}(F_Z)$. We conclude (8.11). Therefore, $\widehat{Z}(F_U) \subset \widehat{Z}(F_{V_\alpha})$, which proves part (i) of the theorem. From the assumptions of part (ii) and the Corollary 8.13(ii) conclude that $F_{V_\alpha} = \alpha F_X + (1-\alpha)F_Z \preceq_{lz} F_Y$. Hence, with (i), $U \preceq_{lz} V_\alpha$, follows $U \preceq_{lz} Y$. The part (iii) is a consequence of Theorem 8.17. Q.E.D.

8.4 Lift zonoid order vs. convex order

There are many possibilities to extend the univariate convex order to the multivariate case.

The lift zonoid order is just one of them. Another one is the multivariate *convex order*, which is now considered and contrasted with the lift zonoid order.

Theorem 8.19 (Convex order) *For* X *and* $Y \in \mathcal{X}_1$ *the following statements are equivalent:*

(i) $\mathrm{E}(f(X)) \le \mathrm{E}(f(Y))$ *if* $f : \mathbb{R}^d \to \mathbb{R}$ *is convex and the expectations exist,*

(ii) $\mathrm{E}(X) = \mathrm{E}(Y)$, *and* $\mathrm{E}(f(X)) \le \mathrm{E}(f(Y))$ *if* $f : \mathbb{R}^d \to \mathbb{R}$ *is increasing convex and the expectations exist,*

(iii) $Y =_{st} X + U$ *with some* U *and* $\mathrm{E}(U|X) = 0$ *almost surely.*

X is said to be not larger than Y in *convex order*, $X \preceq_{cx} Y$, if one of these equivalent restrictions is satisfied. The convex order is a partial order on $\mathcal{P}_1$ and a preorder on $\mathcal{X}_1$.

Theorem 8.19 is wellknown and its proof due to Strassen (1965); see also Meyer (1966). Many more results on the convex order are found in Shaked and Shanthikumar (1994) and Müller and Stoyan (2002). The random vector U in Theorem 8.19(iii) can be interpreted as 'noise', so that Y is distributed as X plus some noise.

In view of Theorems 8.19(i) and 8.5(iv), as every convex-linear function is convex, the convex order implies the lift zonoid order,

$$X \preceq_{cx} Y \quad \Rightarrow \quad X \preceq_{lz} Y \quad \Rightarrow \quad \mathrm{E}(X) = \mathrm{E}(Y). \tag{8.13}$$

In case $d = 1$ the convex order coincides with the lift zonoid order. But, in general, for $d \ge 2$ the two orders are different. The following counterexample is due Elton and Hill (1992); see also Müller and Stoyan (2002, example 3.5.4).

Example 8.1 (Lift zonoid order is finer than convex order) Consider two empirical distributions F and G in $\mathbb{R}^2$ (see Figure 8.1). F gives probability 1/3 to $(-1,-1)$, $(0,1)$ and $(1,0)$, while G gives probability 1/6 to $(-2,0)$, $(0,-2)$, and $(2,2)$ and probability 1/2 to $(0,0)$. Elton and Hill (1992) demonstrate that $F \preceq_{lz} G$. But with $f(x_1,x_2) = \max\{x_1,x_2,0\}$ obtain

$$\int f(x_1,x_2)dF(x_1,x_2) = \frac{2}{3} > \frac{1}{3} = \int f(x_1,x_2)dG(x_1,x_2).$$

As f is convex, conclude that $F \not\preceq_{cx} G$.

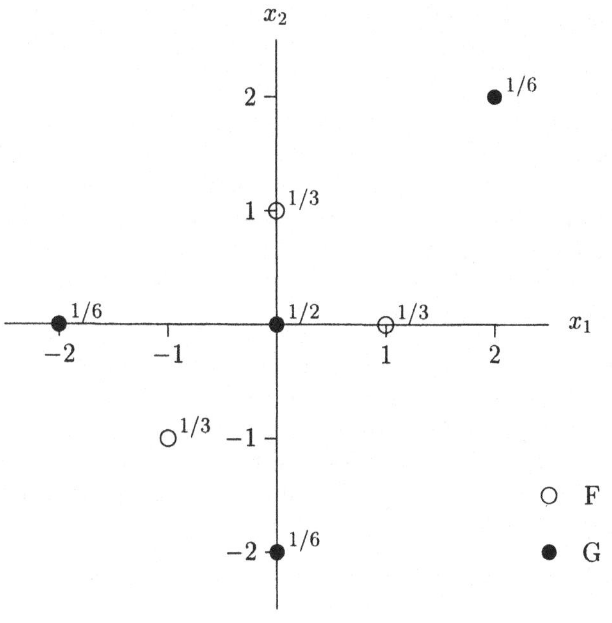

FIGURE 8.1: Example for $F \preceq_{lz} G$, but $F \not\preceq_{cx} G$.

The two orders have many properties in common. The Propositions 8.8 – 8.15 and their corollaries and Theorem 8.17, that have been derived for the lift zonoid order, can be shown for the convex order as well. In particular, the minimal elements of the convex order consist of all one-point distributions and the order increases with scale. Also, the convex order is affine invariant and continuous, it implies the order of all marginals and is preserved under mixtures and convolutions.

If the univariate marginals are independent, the convex order is equivalent to the univariate convex order of all marginals and, by (8.8), to the lift zonoid order. In multivariate normal distributions the two orders coincide; both are equivalent to the equality of means and Loewner order of covariance matrices (8.3).

To summarize, the lift zonoid order as well as the zonoid order compare distributions with respect to their dispersion. Both fulfil virtually all properties which may be postulated for an order of dispersion. They are equal in special cases, but in general the lift zonoid order is weaker than – that is, implied by – the convex order.

Convex order means that the larger random vector is distributed like the smaller one plus some noise. On the other hand, lift zonoid order means that every α-trimmed central region of the larger random vector includes the respective region of the smaller one.

The question remains which of the two orders should be used in applications. From a practical view, comparability of real data sets and computational effort in checking the order are important. Given two empirical distributions, it is quite easy to check whether they are convexly ordered. This amounts to solving a linear program; see, e.g., Mosler and Scarsini (1991). Checking for the lift zonoid order between two given empirical distributions is computationally much more involved.

The convex order is a rather strong ordering of dispersion, which limits its practical use. The same holds for the lift zonoid order. It happens very rarely in applications that two data sets are comparable by $\preceq_{cx}$ or $\preceq_{lz}$. This effect is also apparent with other stochastic orders, like the *increasing stochastic order*[1]. For practical use, we need weaker orders by which many real data sets are comparable.

Often a stochastic order is satisfied for the 'main bulk' of the data, while it fails at more 'outlying' points. In contrast to the convex order, the lift zonoid order can be adapted to cope with such situations. In view of Proposition 8.1, as an 'outlier resistant' version of the lift zonoid order, the following restricted order is proposed:

Definition 8.2 (Lift zonoid β-order)

$$\boldsymbol{X} \preceq_{lz}^{\beta} \boldsymbol{Y} \quad \textit{if} \quad D_\alpha(F_{\boldsymbol{X}}) \subset D_\alpha(F_{\boldsymbol{Y}}) \quad \textit{for} \ 0 < \beta \le \alpha \le 1 - \beta < 1 \,. \tag{8.14}$$

The parameter β signifies the degree by which outlying data are neglected.

[1]The increasing stochastic order $\boldsymbol{X} \preceq_{st} \boldsymbol{Y}$ is defined by $\mathrm{E}(f(\boldsymbol{X})) \le \mathrm{E}(f(\boldsymbol{Y}))$ for all increasing $f : \mathbb{R}^d \to \mathbb{R}$.

8.5 Volume inequalities and random determinants

Like any stochastic order, the lift zonoid order can be used to generate probability inequalities. This section contains an application that is derived from inequalities among the volumes of lift zonoids. From the volume formula comparison results for random determinants are obtained.

Let $F_{\boldsymbol{X}}$ denote the distribution of a random vector $\boldsymbol{X}$ in $\mathbb{R}^d$, $F_{\boldsymbol{X}} \in \mathcal{P}_1$. By Theorem 2.37, the volume of $\widehat{Z}(F_{\boldsymbol{X}})$ equals $((d+1)!)^{-1}\mathrm{E}[|\det \widehat{\boldsymbol{Q}}(F_{\boldsymbol{X}})|]$, where $\widehat{\boldsymbol{Q}}(F_{\boldsymbol{X}})$ is a $(d+1) \times (d+1)$ matrix whose columns are independent and identically distributed copies of $(1, \boldsymbol{X})$.

From the definition of the lift zonoid order, $\boldsymbol{X} \preceq_{lz} \boldsymbol{Y}$ holds if and only if $\widehat{Z}(F_{\boldsymbol{X}}) \subset \widehat{Z}(F_{\boldsymbol{Y}})$. Since the zonoids of $\boldsymbol{X}$ and $\boldsymbol{Y}$ are projections of the corresponding lift zonoids on their last d coordinates, the same inclusion follows for the zonoids, $Z(F_{\boldsymbol{X}}) \subset Z(F_{\boldsymbol{Y}})$. This implies that the volumes of the zonoids are ordered, $\mathrm{vol}_d(Z(F_{\boldsymbol{X}})) \le \mathrm{vol}_d(Z(F_{\boldsymbol{Y}}))$. Hence

$$\boldsymbol{X} \preceq_{lz} \boldsymbol{Y} \quad \Longrightarrow \quad \mathrm{E}(|\det \boldsymbol{Q}(F_{\boldsymbol{X}})|) \le \mathrm{E}(|\det \boldsymbol{Q}(F_{\boldsymbol{Y}})|)\,. \tag{8.15}$$

Here, $\boldsymbol{Q}(F_{\boldsymbol{X}})$ denotes the $d \times d$ matrix whose columns are independent and identically distributed copies of $\boldsymbol{X}$.

For example, let $\boldsymbol{Y} =_{st} \boldsymbol{X} + \boldsymbol{U}$ and $\mathrm{E}[\boldsymbol{U}|\boldsymbol{X}] = \boldsymbol{0}$. Then $\boldsymbol{X} \preceq_{cx} \boldsymbol{X} + \boldsymbol{U}$ and, thus, $\boldsymbol{X} \preceq_{lz} \boldsymbol{X} + \boldsymbol{U}$. Therefore by (8.15)

$$\mathrm{E}(|\det \boldsymbol{Q}(F_{\boldsymbol{X}})|) \le \mathrm{E}(|\det \boldsymbol{Q}(F_{\boldsymbol{X}+\boldsymbol{U}})|)\,. \tag{8.16}$$

Inequalities like these are interesting in stochastic geometry. The implication (8.15) strengthens comparison results that have been given in Vitale (1991a) and Vitale (1991b). In particular, Theorem 5.2 in Vitale (1991b) is extended by inequality (8.16).

8.6 Increasing, scaled, and centered orders

Both the lift zonoid order and the convex order imply equal expectations. Two distributions are comparable only if they have the same mean.

In this section three approaches are presented to compare the dispersion of distributions that have different means. Firstly, dispersion is measured together with monotone shifts in location. Secondly, the previous dispersion orders are applied to distributions scaled down by their means and, thirdly, the same is done with distributions centered by their means.

Stochastic orders of dispersion are sometimes collected under the general heading 'dilations'. In this connection the convex order is mentioned as *dilation* (in the narrow sense) and the lift zonoid order as ***directional dilation***. Here we will consider increasing dilations and dilations of scaled and of centered distributions.

Definition 8.3 (Increasing and decreasing orders) *For* $\boldsymbol{X}$ *and* $\boldsymbol{Y}$ *in* $\mathcal{X}_1$ *define*

(i) *the* increasing lift zonoid order, $\boldsymbol{X} \preceq_{ilz} \boldsymbol{Y}$, *if*
$\mathrm{E}(\psi(\langle \boldsymbol{X}, \boldsymbol{p}\rangle)) \leq \mathrm{E}(\psi(\langle \boldsymbol{Y}, \boldsymbol{p}\rangle))$ *for all increasing convex* $\psi : \mathbb{R} \to \mathbb{R}$ *and all* $\boldsymbol{p} \in \mathbb{R}^d_+$ *as far as the expectations exist,*

(ii) *the* increasing convex order, $\boldsymbol{X} \preceq_{icx} \boldsymbol{Y}$, *if*
$\mathrm{E}(f(\boldsymbol{X})) \leq \mathrm{E}(f(\boldsymbol{Y}))$ *for all increasing convex* $f : \mathbb{R}^d \to \mathbb{R}$ *as far as the expectations exist.*

The decreasing lift zonoid order $\preceq_{dlz}$ *and the* decreasing convex order $\preceq_{dcx}$ *are similarly defined with decreasing functions* ψ *and* f, *respectively.*

Several implications are immediate from the definitions:

$$\begin{array}{ccccccc} & & \boldsymbol{X} \preceq_{cx} \boldsymbol{Y} & \Rightarrow & \boldsymbol{X} \preceq_{lz} \boldsymbol{Y} & & \\ & & \Downarrow & & \Downarrow & & \\ \boldsymbol{X} \preceq_{st} \boldsymbol{Y} & \Rightarrow & \boldsymbol{X} \preceq_{icx} \boldsymbol{Y} & \Rightarrow & \boldsymbol{X} \preceq_{ilz} \boldsymbol{Y} & \Rightarrow & \mathrm{E}(\boldsymbol{X}) \leq \mathrm{E}(\boldsymbol{Y}) \end{array}$$

Here $\boldsymbol{X} \preceq_{st} \boldsymbol{Y}$ means the *multivariate increasing order*, defined by $\mathrm{E}(f(\boldsymbol{X})) \leq \mathrm{E}(f(\boldsymbol{Y}))$ for every increasing function $f : \mathbb{R}^d \to \mathbb{R}$.

If $d = 1$, the increasing lift zonoid order and the increasing convex order coincide, $\preceq_{icx} = \preceq_{ilz}$. For larger dimension d, the increasing lift zonoid order is equivalent to univariate lift zonoid order in every nonnegative direction,

$$\begin{aligned} \boldsymbol{X} \preceq_{ilz} \boldsymbol{Y} \quad &\Leftrightarrow \quad \langle \boldsymbol{X}, \boldsymbol{p}\rangle \preceq_{ilz} \langle \boldsymbol{Y}, \boldsymbol{p}\rangle \quad \text{for all } \boldsymbol{p} \in S^{d-1} \cap \mathbb{R}^d_+ \\ &\Leftrightarrow \quad \langle \boldsymbol{X}, \boldsymbol{p}\rangle \preceq_{icx} \langle \boldsymbol{Y}, \boldsymbol{p}\rangle \quad \text{for all } \boldsymbol{p} \in S^{d-1} \cap \mathbb{R}^d_+ , \end{aligned}$$

while the increasing convex order is characterized by an additive random perturbation $\boldsymbol{U}$ which has nonnegative conditional expectation,

$$\boldsymbol{X} \preceq_{icx} \boldsymbol{Y} \quad \Leftrightarrow \quad \boldsymbol{Y} =_{st} \boldsymbol{X} + \boldsymbol{U} \text{ with } \mathrm{E}(\boldsymbol{U}|\boldsymbol{X}) \geq 0 ;$$

see e.g. Müller and Stoyan (2002, Ch. 3.4)).

Among normal distributions the increasing lift zonoid order can be described by the parameters:

Proposition 8.20 (Increasing lift zonoid order of normals)
Let $\boldsymbol{X} \sim N(\boldsymbol{a}, \boldsymbol{R})$ *and* $\boldsymbol{Y} \sim N(\boldsymbol{b}, \boldsymbol{S})$. *Then*
$\boldsymbol{X} \preceq_{ilz} \boldsymbol{Y}$ *if and only if* $\boldsymbol{a} \leq \boldsymbol{b}$ *and* $\mathbf{p}(\boldsymbol{S} - \boldsymbol{R})\mathbf{p}' \geq 0$ *for all* $\boldsymbol{p} \in S^{d-1} \cap \mathbb{R}^d_+$.

Proof. See Equations (2.59) or (3.12).

Now, let us turn to orders of scaled and centered distributions. Let $\mathcal{P}_{1*}$ be the class of those distributions in $\mathcal{P}_1$ whose mean vector has no component equal to zero. $\mathcal{X}_{1*}$ is the class of random vectors having distribution in $\mathcal{P}_{1*}$.

Definition 8.4 (Scaled orders) *Two random vectors* $\boldsymbol{X}$ *and* $\boldsymbol{Y} \in \mathcal{X}_{1*}$ *are said to be ordered in*

(i) *the* lift-zonoid scaled order, $\boldsymbol{X} \preceq^s_{lz} \boldsymbol{Y}$, *if*

$$\frac{\boldsymbol{X}}{\mathrm{E}(\boldsymbol{X})} \preceq_{lz} \frac{\boldsymbol{Y}}{\mathrm{E}(\boldsymbol{Y})},$$

(ii) *the* convex scaled order, $\boldsymbol{X} \preceq^c_{lz} \boldsymbol{Y}$, *if*

$$\frac{\boldsymbol{X}}{\mathrm{E}(\boldsymbol{X})} \preceq_{cx} \frac{\boldsymbol{Y}}{\mathrm{E}(\boldsymbol{Y})}.$$

The scaled orders relate to the comparison of Lorenz curves of univariate distributions and to the majorization of vectors and matrices. The lift-zonoid scaled order has been named the *Lorenz order* in Section 7.1. If $d = 1$, this order coincides with the convex scaled order; it is the usual, wellknown Lorenz order which corresponds to the pointwise *reverse order of Lorenz curves.* Both the lift-zonoid scaled and the convex scaled order are multivariate extensions of the Lorenz order. Another extension, the *price Lorenz order* will be investigated in Chapter 9. For empirical distributions in $\mathbb{R}$ that have the same mean, the reverse order of Lorenz curves is equivalent to *majorization* of the data vectors. If we compare empirical distributions in $\mathbb{R}^d$ that have the same number, say n, of support points, the convex order and the lift zonoid order correspond to majorization and *directional majorization* of $n \times d$ matrices, respectively (Marshall and Olkin, 1979, Ch. 15).

Definition 8.5 (Centered orders) *Two random vectors* $\boldsymbol{X}$ *and* $\boldsymbol{Y} \in \mathcal{X}_1$ *are said to be ordered in*

(i) *the* lift-zonoid centered order, $\boldsymbol{X} \preceq^c_{lz} \boldsymbol{Y}$, *if*

$$\boldsymbol{X} - \mathrm{E}(\boldsymbol{X}) \preceq_{lz} \boldsymbol{Y} - \mathrm{E}(\boldsymbol{Y}),$$

(ii) *the* convex centered order, $X \preceq_{cx}^{c} Y$, *if*

$$X - \mathrm{E}(X) \preceq_{cx} Y - \mathrm{E}(Y).$$

These scaled and centered orders are partial orders on distributions and preorders on random vectors. The following implications are immediate from the definitions.

$$\begin{array}{ccc} X \preceq_{cx}^{s} Y & \Longrightarrow & X \preceq_{lz}^{s} Y \\ \Uparrow & & \Uparrow \\ X \preceq_{cx} Y & \Longrightarrow & X \preceq_{lz} Y \\ \Downarrow & & \Downarrow \\ X \preceq_{cx}^{c} Y & \Longrightarrow & X \preceq_{lz}^{c} Y \end{array} \tag{8.17}$$

It can be shown by counterexamples that, in general, no reverse implication holds.

The lift-zonoid scaled and centered orders are similarly characterized as the lift-zonoid order. The same applies for the convex scaled and centered orders and the convex order. See Theorems 8.5 and 8.19.

8.7 Properties of dispersion orders

We give a list of properties which a given stochastic order $\preceq$ of dispersion can satisfy or not. It consists of two groups. The properties in the first group (**IS** to **SO**) state that certain pairs of distributions can be ordered, while the properties in the second group (**E** to **C**) say that the order is invariant against certain operations.

IS (Increasing with scale) $\beta_j > 1$ for $j = 1, \ldots, d \Rightarrow X \preceq \beta \cdot X = (\beta_1 X_1, \ldots, \beta_d X_d)$.

ME (Minimal at expectation) $\mathrm{E}(X) \preceq_{lz} X$.

IT (Increasing with shift) $c \in \mathbb{R}^d_+ \Rightarrow X \preceq X + c$.

DO (Deterministic order) $a \leq b \Rightarrow \delta_a \preceq \delta_b$.

SO (Increasing stochastic order) $X \preceq_{st} Y \Rightarrow X \preceq Y$.

E (Expectation) $X \preceq Y \Rightarrow \mathrm{E}(X) = \mathrm{E}(Y)$.

A (Affine transform) $X \preceq Y$, $A \in \mathbb{R}^{d \times d}$ of full rank, $c \in \mathbb{R}^d \quad \Rightarrow \quad XA + c \preceq YA + c$.

S (Scale transform) $X \preceq Y$, $\alpha > 0, \quad \Rightarrow \quad \alpha X \preceq \alpha Y$.

T (Translation) $X \preceq Y$, $c \in \mathbb{R}^d \quad \Rightarrow \quad X + c \preceq Y + c$.

MA (Marginals) $X \preceq Y$, $\emptyset \neq J \subset \{1, \ldots, d\} \quad \Rightarrow \quad X_J \preceq Y_J$.

C (Convolution) $X \preceq Y$, Z independent of X and Y $\Rightarrow \quad X + Z \preceq Y + Z$.

Table 8.1 states whether these properties are satisfied for the eight orders considered in this chapter. The proofs are left to the reader. They are either obvious from the definitions or parallel the proofs given above for the lift zonoid order.

	$\preceq_{lz}$	$\preceq_{cx}$	$\preceq_{ilz}$	$\preceq_{icx}$	$\preceq_{lz}^{s}$	$\preceq_{cx}^{s}$	$\preceq_{lz}^{c}$	$\preceq_{cx}^{c}$
IS	•	•	•	•	•	•	•	•
ME	•	•	•	•	•	•	•	•
IT	-	-	•	•	-	-	-	-
DO	-	-	•	•	•	•	•	•
SO	-	-	•	•	-	-	-	-
E	•	•	-	-	-	-	-	-
A	•	•	-	-	-	-	•	•
S	•	•	•	•	•	•	•	•
T	•	•	•	•	-	-	•	•
MA	•	•	•	•	•	•	•	•
C	•	•	•	•	-	-	•	•

TABLE 8.1: Properties of the lift zonoid order, the convex order, and their increasing, scaled and centered versions.

It is seen from the table that all eight orders measure the comparative dispersion of two distributions in some way: They are minimal at expectation and increasing with scale. The two increasing orders are, in addition, sensitive against positive shifts and they reflect the componentwise order of points in $\mathbb{R}^d$ and the usual increasing stochastic order of distributions. Note also that the table shows no differences in these properties between the various convex orders and their lift zonoid counterparts.

8.8 Multivariate indices of dispersion

The lift zonoid order and the convex order are partial orders of dispersion only. Many pairs of distributions cannot be compared by them. In contrast, a real-valued index of dispersion provides a complete ordering of distributions.

A natural requirement for such an index is that it should be *consistent* with the partial orders, that is, if one distribution is more dispersed in the lift zonoid or in the convex order than another, the index should indicate a higher value with the first distribution.

In Section 7.4 an index of dispersion has been investigated, the multivariate volume-Gini index, which is based on the lift zonoid volume and extends the usual univariate Gini index to many dimensions. For a univariate distribution $F \in \mathcal{P}_{1*}$ the Gini index $R(F)$ amounts to twice the area between the Lorenz curve and the diagonal of the unit square, that is, to the volume of the lift zonoid. Equivalently (see also (7.33)), it equals

$$R(F) = \frac{1}{\epsilon(F)} \int_{\mathbb{R}} \int_{\mathbb{R}} |x - y| \, dF(x) \, dF(y) \,. \tag{8.18}$$

In the sequel another extension of the usual Gini index shall be discussed which is based on a formula corresponding to (8.18) and called the *multivariate distance-Gini index* (Arnold, 1987).

Recall that $\widetilde{F}(\boldsymbol{x}) = F(x_1\epsilon_1, \dots, x_d\epsilon_d)$ if $F \in \mathcal{P}_{1*}$ and $(\epsilon_1, \dots, \epsilon_d) = \epsilon(F)$. If F is the distribution of a random vector $\boldsymbol{X} = (X_1, \dots, X_d)$, $\widetilde{F}$ is the distribution of the random vector

$$\widetilde{\boldsymbol{X}} = \frac{\boldsymbol{X}}{\mathrm{E}(\boldsymbol{X})} = \left(\frac{X_1}{\mathrm{E}(X_1)}, \dots, \frac{X_d}{\mathrm{E}(X_d)} \right).$$

Here and in the sequel, when using $\widetilde{F}$, we tacitly assume that $F \in \mathcal{P}_{1*}$.

Definition 8.6 (Distance-Gini mean difference and index)
(i) *For* $F \in \mathcal{P}_1$

$$M_D(F) = \frac{1}{2d} \int_{\mathbb{R}^d} \int_{\mathbb{R}^d} \|\boldsymbol{x} - \boldsymbol{y}\| \, dF(\boldsymbol{x}) \, dF(\boldsymbol{y}) \tag{8.19}$$

is the distance-Gini mean difference.

(ii) *For* $F \in \mathcal{P}_{1*}$

$$R_D(F) = M_D(\widetilde{F}) \tag{8.20}$$

is the distance-Gini index.

The Gini index equals the Gini mean difference of the scaled data, which are the original data 'scaled down' by their mean.

Several properties of the distance-Gini mean difference and the distance-Gini index follow easily from the definitions.

Proposition 8.21 (Properties of Gini mean difference) *For all* $F \in \mathcal{P}_1$,
(i) $0 \le M_D(F)$,
(ii) $M_D(F) = 0$ *if and only if* F *is a one-point distribution,*
(iii) $M_D(G) = M_D(F)$ *if* $G(\boldsymbol{x}) = F(\boldsymbol{x} + \mathbf{c})$ *for some* $\mathbf{c}$ *and all* $\boldsymbol{x} \in \mathbb{R}^d$,
(iv) M_D *is continuous w.r.t. weak convergence of distributions.*

Proposition 8.22 (Properties of Gini index) *For all* $F \in \mathcal{P}_{1*}$,
(i) $0 \le R_D(F)$,
(ii) $R_D(F) = 0$ *if and only if* F *is a one-point distribution,*
(iii) $R_D(G) = R_D(F)$ *if* $G(x_1, \dots, x_d) = F(x_1\beta_1, \dots, x_d\beta_d)$ *and* $\beta_1, \dots, \beta_d > 0$,
(iv) R_D *is continuous w.r.t. weak convergence of distributions.*

Proposition 8.22(iii) says that R_D is *vector scale invariant,* while Proposition 8.21(iii) states that M_D is *translation invariant.* Regarding upper bounds the following result is obtained.

Recall that $\mathcal{P}_{1+} \subset \mathcal{P}_{1*}$ is the subclass that has, in addition, support in $\mathbb{R}^d_+$.

Proposition 8.23 (Bounds) *If* $F \in \mathcal{P}_{1*}$ *has support in* $\mathbb{R}^d_+$, *then the inequalities*

$$R_D(F) < 1 \quad \textit{and} \quad M_D(F) < \frac{1}{d}\sum_{j=1}^{d} \epsilon_j(F)$$

hold and the bounds are sharp.

Proof. See Koshevoy and Mosler (1997a).

The distance-Gini index and also the mean difference are independent of irrelevant attributes. Both possess the ceteris paribus property (7.39), which means that, if an attribute is added which is constant in the population, the index multiplies by a factor that depends only on dimension.

Theorem 8.24 (Ceteris paribus property) M_D *and* R_D *have the ceteris paribus property with factor*

$$\gamma(d) = \frac{d}{d+1}.$$

The proof is obvious from the definition of M_D. In Koshevoy and Mosler (1997a) more properties of M_D and R_D are found, among them the following two propositions. The first says that the multivariate Gini mean difference is, up to a constant factor, the integral of the directed, univariate Gini mean differences in each direction. Recall that $F_{\boldsymbol{p}}$ is the distribution of $\langle \boldsymbol{X}, \boldsymbol{p} \rangle$ if F is the distribution of $\boldsymbol{X}$.

Proposition 8.25 (Integral of directed indices) *Let $d \geq 2$ and $F \in \mathcal{P}_1$. Then*

$$M_D(F) = \frac{\Gamma(\frac{d+1}{2})}{4d\,\pi^{\frac{d-1}{2}}} \int_{\boldsymbol{p} \in S^{d-1}} \int_{-\infty}^{+\infty} \int_{-\infty}^{+\infty} |u - v| \, dF_{\boldsymbol{p}}(u) \, dF_{\boldsymbol{p}}(v) d\boldsymbol{p}, \qquad (8.21)$$

where $d\boldsymbol{p}$ denotes the Lebesgue measure on the sphere S^{d-1}.

Proof. See Theorem 4.3 in Koshevoy and Mosler (1997a).

Proposition 8.26 (Specific direction) *For every F there exist some $\boldsymbol{p}$ and $\tilde{\boldsymbol{p}} \in S^{d-1}$ such that*[2]

$$M_D(F) = \frac{B(\frac{d+1}{2}, \frac{1}{2})}{2} M_D(F_{\boldsymbol{p}}) \quad \textit{and}$$

$$R_D(F) = \frac{B(\frac{d+1}{2}, \frac{1}{2})}{2} R_D(F_{\tilde{\boldsymbol{p}}}).$$

Proof. Follows from Proposition 8.25 by the mean value theorem.

The Proposition 8.26 says that, for every distribution F, there are directions $\boldsymbol{p}$ and $\tilde{\boldsymbol{p}}$ for the distance-Gini mean difference and index, respectively, that reflect the dependence structure of F so that it can be expressed by a univariate mean difference and index.

There are other multivariate indices of dispersion. The *volume-Gini index* and *mean difference* have been treated in detail in Section 7.4. Next is demonstrated that both multivariate Gini indices are consistent with previously defined dispersion orders.

Theorem 8.27 (Consistency of index with orders) *The distance-Gini index R_D and the volume-Gini index R_V are increasing with*

(i) *convex order,*

(ii) *lift zonoid order,*

(iii) *convex scaled order,*

(iv) *lift zonoid scaled order.*

[2]Note that $B(\rho, \sigma) = \int_0^1 t^{\rho-1}(1-t)^{\sigma-1} dt$ is the Beta function.

Proof. In view of the implications (8.17), only (iv) has to be shown. Suppose $F \preceq_{lz}^{s} G$. For every $p \in S^{d-1}$, follows $F_p \preceq_{lz}^{s} G_p$, hence $M(\widetilde{F}_p) \leq M(\widetilde{G}_p)$,

$$\int_{-\infty}^{+\infty}\int_{-\infty}^{+\infty} |u-v|\, d\widetilde{F}_p(u)\, d\widetilde{F}_p(v) \leq \int_{-\infty}^{+\infty}\int_{-\infty}^{+\infty} |u-v|\, d\widetilde{G}_p(u)\, d\widetilde{G}_p(v)\,.$$

Therefore,

$$\int_{p\in S^{d-1}}\int_{-\infty}^{+\infty}\int_{-\infty}^{+\infty} |u-v|\, d\widetilde{F}_p(u)\, d\widetilde{F}_p(v) dp$$

$$\leq \int_{p\in S^{d-1}}\int_{-\infty}^{+\infty}\int_{-\infty}^{+\infty} |u-v|\, d\widetilde{G}_p(u)\, d\widetilde{G}_p(v) dp$$

for all p. This yields, according to Proposition 8.25, $M_D(\widetilde{F}) \leq M_D(\widetilde{G})$, that is $R_D(F) \leq R_D(G)$.
To demonstrate the same result for R_V, note first that $F \preceq_{lz}^{s} G$ implies $F_K \preceq_{lz}^{s} G_K$ for all K, $\emptyset \neq K \subset \{1, \dots, d\}$. By Theorem 7.8 follows $R_V(F) \leq R_V(G)$. Q.E.D.

For the distance-Gini and the volume-Gini mean differences, we have an analogous theorem.

Theorem 8.28 (Consistency of mean difference with orders) *The distance-Gini mean difference M_D and volume-Gini mean difference M_V are increasing with*
(i) *convex order,*
(ii) *lift zonoid order,*
(iii) *convex centered order,*
(iv) *lift zonoid centered order.*

Proof. The proof is similar to that of Theorem 8.27. Again, because of (8.17), only (iv) has to be proven. Use Propositions 8.21 and 7.10.

Besides these extended Gini notions many other indices can be defined which are consistent with the dispersion orders. For example, a class of such indices is given by

$$\Phi(F) = \int_{S^{d-1}} \Psi(F_p) d\pi(p)\,,$$

where $\Psi : \mathcal{P}_1 \to \mathbb{R}$ is a given univariate dispersion index and π is some measure on the sphere. If Ψ is consistent with the univariate convex order, then Φ is, obviously, consistent with the lift zonoid order.

8.9 Notes

Stochastic orders are widely used in probability and statistics, including applications to economics and operations research. Many problems in these fields boil down to the comparison of two probability distributions regarding location, dispersion, or dependency; see, e.g., Mosler and Scarsini (1991), Mosler (1994a). A great variety of stochastic orders has been designed for the different tasks of comparison. The location, e.g., of multivariate distributions can be compared by the usual increasing stochastic order, the variability by the convex order, and the degree of dependence by the supermodular order; see, e.g., Müller and Stoyan (2002, Ch. 3.9).

Sections 8.1 to 8.6 of this chapter draw on Koshevoy and Mosler (1998), Section 8.8 on Koshevoy and Mosler (1997a).

What I call the (increasing) lift zonoid order is mentioned in Müller and Stoyan (2002, Ch. 3.6) as the *(increasing positive) linear convex order*. $\preceq^s_{lz}$ has been named the *multivariate Lorenz order* in Mosler (1994a); see also Koshevoy and Mosler (1996).

The *zonoid order* $\preceq_z$ between measures F and $G \in \mathcal{P}_1$ is defined by the inclusion of the measures' zonoids, $F \preceq_z G$ if $Z(F) \subset Z(G)$. Corollary 2.12 and Example 2.1 show that there exist many probability measures which are equivalent with respect to the zonoid order. The lift zonoid order avoids this drawback.

9

Measuring economic disparity and concentration

This chapter treats orders of multivariate dispersion which are particularly designed for economic applications, namely to measure economic inequality and industrial concentration.

Two random vectors are lift zonoid ordered if and only if any linear combination of their components is ordered in the univariate convex order. If the components represent commodities, the coefficients of a linear combination can be interpreted as prices, provided they are not negative; then the linear combination can be seen as an expenditure. It suggests itself to define a new order by restricting the coefficients to have positive sign. This leads to the *price majorization order* and its scaled version, the *price Lorenz order*. Both will be investigated in the sequel.

Section 9.1 addresses the problem of economic inequality measurement in a finite population, say, of households and, in particular, its multivariate aspects. Classical univariate inequality measurement focusses on the question how a single attribute of well-being, usually income, is distributed in the population, that is, which shares of the population receive which shares of the total income. Basic tools of univariate measurement are the Lorenz order and indices, like the Gini index, that are consistent with it.

In Section 9.2 the multivariate inverse Lorenz function (ILF) is defined and its properties are investigated. The pointwise ordering of ILFs is a multivariate analogue to the usual Lorenz order. In Section 9.3 the price Lorenz order is introduced as the Lorenz order of expenditures for all possible prices. As a main result, it is demonstrated that the price Lorenz order and the ILF order are the same. Section 9.4 concerns distributions of absolute endowments (instead of shares) and their ordering; weak price majorizations are defined in order to compare distributions with unequal totals. Section 9.5 contrasts our approach with some of the literature on multivariate inequality measurement. A second application area in economics is the measurement of industrial concentration. Section 9.6 gives an introduction into this field. An order of multivariate concentration functions is defined and investigated in Section 9.7, while in Section 9.8 multivariate concentration indices are treated which rest on the concentration function, such as a Rosenbluth index and concentration rates.

9.1 Measuring economic inequality

In the classical measurement of income inequality it is asked which share of the population receives which share of the total income.

The usual Lorenz curve (Lorenz, 1905) depicts the disparity of income in a given population[1]. It is the basic notion for modelling and measuring economic inequality with respect to a single attribute of well-being. The Gini index, which is twice the area between the Lorenz curve and the diagonal of the unit square, serves as the most popular index of income inequality.

But economic inequality does not arise from the distribution of income alone. Other attributes of affluence and well-being appear to be of similar interest in economic analysis. Households vary in income and assets, individuals in earning and education, countries in per capita income and mineral resources, etc. In modern theories of social choice the specific distributional inequality of attributes like these is considered; see Tobin (1970), Sen (1973), Kolm (1977). If inequality in two or more attributes is treated simultaneously, we face the problem of modelling and measuring multidimensional economic disparity.

Kolm (1969) and Atkinson (1970) have considered multivariate endowments of well-being and used matrix majorization and related orders in the analysis of economic inequality. Special multidimensional indexes and orderings have been proposed by various authors (Atkinson and Bourguignon, 1982, 1989; Kolm, 1977; Maasoumi, 1986; Tsui, 1995; Mosler, 1994b, and others).

Let us start with an example.

Example 9.1 Consider three households that have different endowments of wealth and income. Their (wealth, income) vectors be $(1000, 0), (0, 80)$, and

[1] The Lorenz curve is the graph of the Lorenz function; see (2.36).

(0, 20), respectively. For income (wealth) alone, the usual univariate Lorenz curve plots the percentage $L(t)$ of total income (wealth) which is received by the percentage t of lower income (wealth) households; see Figure 9.1.

A surface is proposed in the sequel that reflects the whole bivariate distribution, say of income and wealth. Figure 9.2 exhibits the solution for Example 9.1, which will be developed below.

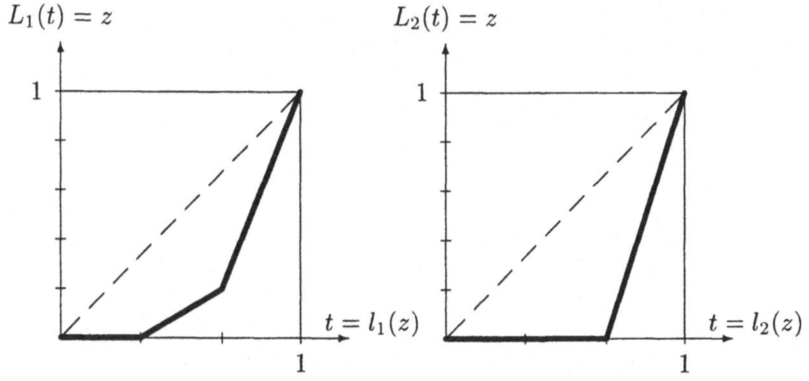

FIGURE 9.1: Lorenz functions, $L_1(t)$ and $L_2(t)$, and inverse Lorenz functions, $l_1(z)$ and $l_2(z)$, of the univariate empirical distributions having support $\{0, 80, 20\}$ and $\{1000, 0, 0\}$.

More formally, consider a population of economic units $i \in \{1, \dots, n\}$ and a set of attributes $k \in \{1, \dots, d\}$. We shall speak of households i and commodities k. Let $\boldsymbol{A} = [a_{ik}] = [\boldsymbol{a}_1, \dots, \boldsymbol{a}_n]$ be a nonnegative $n \times d$ matrix that describes the distribution of d commodities in the population. Its row $\boldsymbol{a}_i$ is the endowment vector of the i-th household, and its column $\boldsymbol{a}^k$ is the distribution of the k-th commodity. The total commodity vector equals $\sum_{i=1}^n \boldsymbol{a}_i = (\sum_i a_{i1}, \dots, \sum_i a_{id})$ and is assumed to be positive in each component. Every such $\boldsymbol{A}$ is called a *distribution matrix*. Let $\breve{\boldsymbol{A}} = [\breve{a}_{ik}]$ be the matrix of shares, $\breve{a}_{ik} = a_{ik}/\sum_i a_{ik}$. In Example 9.1 we have

$$\boldsymbol{A} = \begin{pmatrix} 1000 & 0 \\ 0 & 80 \\ 0 & 20 \end{pmatrix}, \qquad \breve{\boldsymbol{A}} = \begin{pmatrix} 1 & 0 \\ 0 & .8 \\ 0 & .2 \end{pmatrix}.$$

Our basic idea is briefly outlined as follows. We introduce a special view on the usual Lorenz curve and extend this view to the multivariate situation:

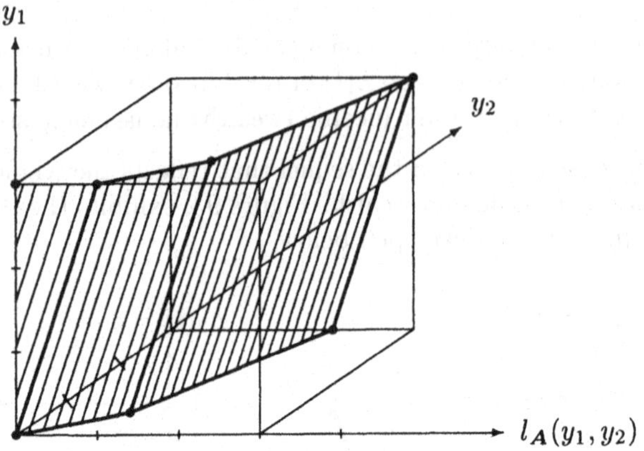

FIGURE 9.2: Lorenz hypersurface of a bivariate empirical distribution (Example 9.1). The hypersurface (= graph of the bivariate ILF) is the hatched surface 'spanned' by $(0,0,0)$, $(0,1,0)$, $(\frac{1}{3},1,0)$, $(\frac{2}{3},0,1)$, $(1,1,1)$, $(\frac{1}{3},0,\frac{1}{5})$ and $(\frac{2}{3},1,\frac{1}{5})$. Note that the axes correspond to the axes in Figure 9.1, that is the value of the inverse Lorenz function is indicated on the abszissa.

The usual Lorenz curve can also be seen as the graph of the *inverse Lorenz function*: Given a portion of the total income, the inverse Lorenz function indicates the maximum percentage of the population that receives this portion. In the multivariate case we consider a share of the total endowment in each commodity and determine the largest percentage of the population by which these shares or less are held. This percentage, depending on the vector of shares, will be called the d-variate inverse Lorenz function. Its graph is the *Lorenz hypersurface*. The Lorenz hypersurface for Example 9.1 is depicted in Figure 9.2. The formal definition will be given in Section 9.2.

9.2 Inverse Lorenz function (ILF)

In this section the inverse Lorenz function is defined for general probability distributions and its main properties are investigated.

Consider a random vector $\boldsymbol{X}$ in $\mathbb{R}^d_+$ which is defined on a probability space $(\Omega, \mathcal{S}, P)$. We interpret the space Ω as a space of economic agents. An element $S \in \mathcal{S}$ is seen as a coalition of agents and the probability $P(S)$ as the relative size of the coalition S with respect to the size of the total space of agents. A value of the random vector, $\boldsymbol{X}(\omega)$, is the endowment of a household $\omega \in \Omega$. The distribution function $F(\boldsymbol{y})$ indicates the relative size of the coalition whose agents have endowments less or equal to $\boldsymbol{y} \in \mathbb{R}_+^d$. In such a model a comparison of two distributions is equivalent to a comparison of distributions of endowments $\boldsymbol{X}$ and $\boldsymbol{Y}$ in two societies $(\Omega, \mathcal{S}, P)$ and $(\Omega', \mathcal{S}', P')$.

Recall that $\mathcal{P}_{1+}$ is the class of d-variate probability distributions on $\mathbb{R}_+^d$ that have a finite and, in each component, positive expectation. For $F \in \mathcal{P}_{1+}$ notate $\boldsymbol{\epsilon}(F) = \int \boldsymbol{x} dF(\boldsymbol{x})$ and $\widetilde{F}(\boldsymbol{x}) = F(\boldsymbol{x} \cdot \boldsymbol{\epsilon}(F)) = F(x_1\epsilon_1(F), \ldots, x_d\epsilon_d(F))$. If $\boldsymbol{X}$ is distributed as F, then

$$\widetilde{\boldsymbol{X}} = \frac{\boldsymbol{X}}{\boldsymbol{\epsilon}(F)} = \left(\frac{X_1}{\epsilon_1(F)}, \ldots, \frac{X_d}{\epsilon_d(F)} \right)$$

is distributed as $\widetilde{F}$. Our central definition is the following:

Definition 9.1 (Inverse Lorenz function) *For $F \in \mathcal{P}_{1+}$, the* inverse Lorenz function (ILF) *is given by*

$$l_F(\boldsymbol{y}) = \max \int_{\mathbb{R}_+^d} g(\boldsymbol{x}) dF(\boldsymbol{x}), \quad \boldsymbol{y} \in [0,1]^d, \tag{9.1}$$

where the maximum extends over all measurable functions $g : \mathbb{R}_+^d \to [0,1]$ for which

$$\int_{\mathbb{R}_+^d} g(\boldsymbol{x}) \frac{\boldsymbol{x}}{\boldsymbol{\epsilon}(F)} dF(\boldsymbol{x}) \leq \boldsymbol{y}.$$

The graph $\{(z_0, \boldsymbol{y}) : z_0 = l_F(\boldsymbol{y}), \boldsymbol{y} \in [0,1]^d\}$ of the ILF is named the Lorenz hypersurface.

An $n \times d$ distribution matrix $\boldsymbol{A} = [\boldsymbol{a}_1, \ldots, \boldsymbol{a}_n]$ corresponds to the n-point empirical distribution $F_{\boldsymbol{A}}$ in $\mathbb{R}_+^d$. The share vector of the i-th household is

$$\breve{\boldsymbol{a}}_i = \frac{\boldsymbol{a}_i}{n\, \boldsymbol{\epsilon}(F_{\boldsymbol{A}})},$$

and from Definition 9.1 we obtain the inverse Lorenz function of $\boldsymbol{A}$,

$$l_{\boldsymbol{A}}(\boldsymbol{y}) = \max \left\{ \frac{1}{n} \sum_{i=1}^n \theta_i : \sum_{i=1}^n \theta_i \breve{\boldsymbol{a}}_i \leq \boldsymbol{y}, 0 \leq \theta_i \leq 1 \right\}, \quad \boldsymbol{y} \in [0,1]^d. \tag{9.2}$$

Here θ_i is a weight by which the endowment of the i-th household enters $\boldsymbol{y}$. The ILF is defined on the unit cube of $\mathbb{R}^d$. Its argument $\boldsymbol{y}$ represents a vector of relative endowments with respect to mean endowments in the d attributes.

In case $d = 1$ the ILF is the inverse function of the Lorenz function. The Lorenz order between univariate distributions consists in the pointwise ordering of their Lorenz functions or, equivalently, of their ILFs.

The ILF can be given the following economic interpretation: $l_F(\boldsymbol{y})$ equals the maximum percentage of people who hold a vector of shares less than or equal to $\boldsymbol{y} = (y_1, \ldots, y_d)$. Indeed, to every household ω of the population Ω the vector $\boldsymbol{X}(\omega)$ of endowments in d commodities is assigned. Then household ω holds the vector $\widetilde{\boldsymbol{X}}(\omega)$ of shares with respect to mean endowment.

A given measurable function $g : \mathbb{R}_+^d \to [0, 1]$ can be considered as a *selection* of some part of the population: Of all those households which have endowment vector $\boldsymbol{x}$ the percentage $g(\boldsymbol{x})$ is selected. Thus, $\int_{\mathbb{R}_+^d} \frac{\boldsymbol{x}}{\epsilon(F)} g(\boldsymbol{x}) dF(\boldsymbol{x})$ amounts to the total portion vector held by this subpopulation and $\int_{\mathbb{R}_+^d} g(\boldsymbol{x}) dF(\boldsymbol{x})$ is the size of the subpopulation selected by g. Given a vector $\boldsymbol{y} \in [0, 1]^d$, we maximize $\int g(\boldsymbol{x}) dF(\boldsymbol{x})$ over all measurable g such that $\int \frac{\boldsymbol{x}}{\epsilon(F)} g(\boldsymbol{x}) dF(\boldsymbol{x}) \le \boldsymbol{y}$. Then $\max \int g(\boldsymbol{x}) dF(\boldsymbol{x})$ is the maximum size of a part of the population that holds total portion vector $\boldsymbol{y}$ or less.

The remainder of this section is devoted to properties of the ILF and examples.

Proposition 9.1 (Properties of the ILF) *For any $F \in \mathcal{P}_1$ holds:*

(i) *l_F is monotone increasing, concave, and has values in $[0, 1]$,*

(ii) *$l_F(\boldsymbol{1}) = 1$ and $l_F(\boldsymbol{0}) = F(\boldsymbol{0})$,*

(iii) *$l_F(\boldsymbol{y}_J, \boldsymbol{1}_{-J}) = l_{F_J}(\boldsymbol{y}_J)$.*

Recall that F_J is the marginal distribution with respect to coordinates in $J \subset \{1, \ldots, d\}$ and $(\boldsymbol{y}_J, \boldsymbol{1}_{-J})$ is a vector in $\mathbb{R}^d$ with components $y_j, j \in J$, and remaining components equal to 1.

Proof. Obvious from the definition of the ILF. Q.E.D.

Example 9.2 (Egalitarian distribution) As a special case, consider the *egalitarian distribution* δ_ξ that poses unit mass to some point $\xi \in \mathbb{R}_+^d$ having positive components. This means that all agents have the same endowment vector ξ. We get $l_{\delta_\xi}(\boldsymbol{y}) = \min_{i=1,\ldots,d} y_i$ for all $\boldsymbol{y}$, which does not depend on ξ. Figure 9.3 shows the graph of l_{δ_ξ} when $d = 2$.

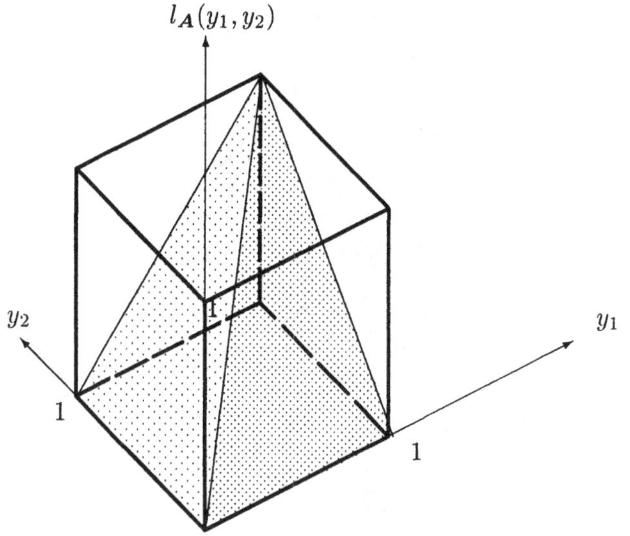

FIGURE 9.3: Lorenz hypersurface of an egalitarian distribution; $d = 2$.

Example 9.3 (Upper Fréchet bound) Let F_1 and F_2 be given univariate distribution functions. The bivariate distribution function $F^+(x_1, x_2) = \min\{F_1(x_1), F_2(x_2)\}$ is the *upper Fréchet bound* of F_1 and F_2. Then there exist real random variables X_1 and X_2 that have joint distribution F and marginals F_1 and F_2 such that X_2 is an increasing function of X_1, *viz.* $X_2 = (F_2)^{-1}(F_1(X_1))$. The ILF is given by

$$l_F^+(y) = \begin{cases} l_{F_2}(y_2) & \text{if } \int_0^t (F_1)^{-1}(s)ds \geq \int_0^t (F_2)^{-1}(s)ds\,, \\ l_{F_1}(y_1) & \text{if } \int_0^t (F_1)^{-1}(s)ds \leq \int_0^t (F_2)^{-1}(s)ds\,, \end{cases}$$

where l_{F_1} and l_{F_2} are the ILFs of the marginals; see Figure 9.4.

The inverse Lorenz function l_F is closely related with a properly modified lift zonoid, the Lorenz zonoid of F. Proposition 9.2 states that the graph of l_F is part of its boundary.

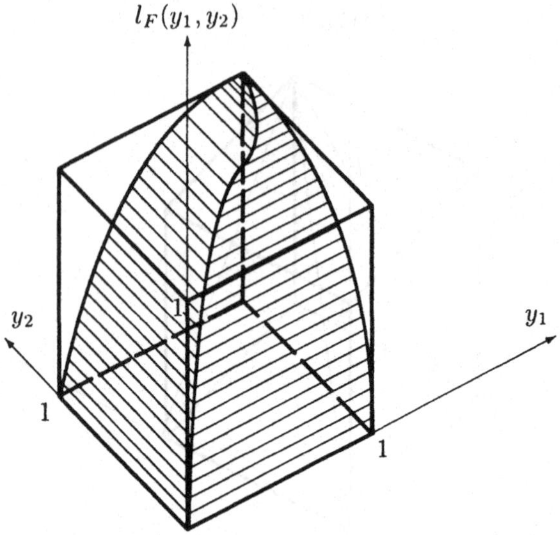

FIGURE 9.4: Lorenz hypersurface of an upper Fréchet bound; $d = 2$.

Definition 9.2 (Lorenz zonoid) *The* Lorenz zonoid $LZ(F)$ *of a distribution* $F \in \mathcal{P}_{1+}$ *is the lift zonoid of the scaled distribution* $\widetilde{F}$, *where* $\widetilde{F}(\boldsymbol{x}) = F(\boldsymbol{\epsilon}(F) \cdot \boldsymbol{x})$. *In symbols,*

$$LZ(F) = \widehat{Z}(\widetilde{F}).$$

Proposition 9.2 (Lorenz zonoid boundary) *The graph of* l_F *is the intersection of the unit cube* $[0,1]^{d+1}$ *with the boundary of the set* $LZ(F) + (\mathbb{R}_- \times \mathbb{R}_+^d)$.

Equivalently, Proposition 9.2 says that the set

$$[0,1]^{d+1} \cap \left(LZ(F) + (\mathbb{R}_- \times \mathbb{R}_+^d)\right)$$

is the hypograph of l_F.

Proof. It has to be shown that, for any $z_0 \in [0,1]$ and $\boldsymbol{y} \in [0,1]^d$, $z_0 = l_F(\boldsymbol{y})$ holds iff the point $(z_0, \boldsymbol{y})$ supports $LZ(F)$ in some direction $(p_0, \boldsymbol{p})$ with $p_0 \geq 0$ and $\boldsymbol{p} \leq \mathbf{0}$. As l_F is a concave function and $LZ(F)$ is a convex set, the proof can be restricted to points which are extreme in $LZ(F)$.

Now, let $(p_0, \boldsymbol{p}) \in \mathbb{R}_+ \times \mathbb{R}_-$ and $(z_0, \boldsymbol{y})$ be an extreme point to $LZ(F)$ in direction $(p_0, \boldsymbol{p})$. Notate

$$H_{p_0,\boldsymbol{p}}(F) = \{\boldsymbol{x} \in \mathbb{R}^d \ : \ \left\langle \frac{\boldsymbol{x}}{\epsilon(F)}, \boldsymbol{p} \right\rangle \geq -p_0\}\,. \tag{9.3}$$

By Corollary 2.5,

$$z_0 = \int_{\mathbb{R}^d_+} g(\boldsymbol{x})dF(\boldsymbol{x})\,, \quad \boldsymbol{y} = \int_{\mathbb{R}^d_+} \frac{\boldsymbol{x}}{\epsilon(F)} g(\boldsymbol{x})dF(\boldsymbol{x})\,,$$

and g is the indicator function of $H_{0,\boldsymbol{p}}(F) \setminus B$ with some set

$$B \subset \{\boldsymbol{x} : \left\langle \frac{\boldsymbol{x}}{\epsilon(F)}, \boldsymbol{p} \right\rangle = 0\}\,.$$

Then it can be shown that $z_0 = l_F(\boldsymbol{y})$ solves the optimization problem (9.1) and, on the reverse, that every solution of (9.1) is a convex combination of points that are extreme to $LZ(F)$ in directions $(p_0, \boldsymbol{p}) \in \mathbb{R}_+ \times \mathbb{R}_-$. The details are left to the reader. Q.E.D.

Let us continue with properties of the ILF. Their proofs make use of Proposition 9.2.

Proposition 9.3 (Uniqueness) *The ILF defines the underlying distribution uniquely, up to a vector of scale factors.*

Proof. Follows from the fact that the lift zonoid determines the underlying distribution uniquely (Theorem 2.21) and from Proposition 9.2. Q.E.D.

Proposition 9.4 (Continuity) *Let a sequence F^n in $\mathcal{P}_{1+}$ weakly converge to F. Then l_{F^n} converges uniformly to l_F if F^n is uniformly integrable.*

In particular, F^n is uniformly integrable if one of the following three restrictions holds:
(i) *There exists a compact set that includes the supports of all F^n,*
(ii) *there exists a distribution G such that $l_{F^n} \leq l_G$ for all n,*
(iii) *there exists a distribution G such that l_{F^n} converges uniformly to l_G.*

Proof. This is a consequence of Theorem 2.30 and Proposition 9.2. The sufficient restrictions for uniform integrability are derived from Corollaries 2.33 – 2.35. Q.E.D.

Proposition 9.5 (Law of large numbers) *Let $F \in \mathcal{P}_{1+}$ and F^n_{emp} be the empirical distribution function of a random sample $\boldsymbol{X}_1, \dots, \boldsymbol{X}_n$ from F. Then the ILF of F^n_{emp} converges uniformly to the ILF of F with probability one, that is,*

$$P\left(\lim_{n\to\infty} \sup_{\boldsymbol{y}} |l_{F^n_{emp}}(\boldsymbol{y}) - l_F(\boldsymbol{y})| = 0\right) = 1 .$$

Proof. Use the law of large numbers for lift zonoids (Theorem 2.38) and Proposition 9.2. Q.E.D.

The case of an empirical distribution deserves particular attention. Then the ILF is a piecewise linear function. It can be seen from (9.2) that the Lorenz hypersurface of an empirical distribution $F_{\boldsymbol{A}}$ is a piecewise linear hypersurface passing through a subset of

$$\left\{\left(\frac{|I|}{n}, \sum_{i\in I} \breve{\boldsymbol{a}}_{i.J}, \boldsymbol{1}_{-J}\right) : I \subset \{1, \dots, n\}, \emptyset \neq J \subset \{1, \dots, d\}\right\} .$$

Here $\boldsymbol{a}_{i,J}$ is the projection of the vector $\boldsymbol{a}_i \in \mathbb{R}^d$ on the coordinate subspace belonging to J. The notation $(\frac{|I|}{n}, \sum_{i\in I} \breve{\boldsymbol{a}}_{i,J}, \boldsymbol{1}_{-J})$ means that the first coordinate (indexed 0) of this point is the cardinality of I divided by n and the remaining coordinates are $\sum_{i\in I} \breve{a}_{ij}$ for $j \in J$ and 1 for $j \notin J$. The following proposition characterizes the extreme points of this Lorenz hypersurface:

Proposition 9.6 (Extreme points) *Let $\boldsymbol{A}$ be a distribution matrix. Then a point $(\frac{|I|}{n}, \sum_{i\in I} \breve{\boldsymbol{a}}_{i,J}, \boldsymbol{1}_{-J})$ is an extreme point of the Lorenz hypersurface if and only if the vector $\boldsymbol{\theta} = (\boldsymbol{1}_I, \boldsymbol{0}_{-I})$ provides a solution to the following maximization problem:*

$$\max\left\{\frac{1}{n}\sum_{i=1}^{n} \theta_i : \sum_{i=1}^{n} \theta_i \breve{a}_{i,J} \leq \sum_{i\in I} \breve{a}_{i,J}, \ \theta_i \in [0,1]\right\}. \tag{9.4}$$

Obviously, in the case $d = 1$, such points have the form $(\frac{k}{n}, \sum_{i=1}^{k} \breve{a}_{(i)})$, $k = 0, \dots, n$, where $\breve{a}_{(1)}, \dots, \breve{a}_{(n)}$ are the shares ordered from below.

9.3 Price Lorenz order

There are many possibilities to extend the univariate Lorenz order to a multivariate setting[2]. An economically meaningful notion is the price Lorenz order, which will be considered now.

[2]See Section 8.6.

For given $F \in \mathcal{P}_{1+}$ and $\boldsymbol{p} \in \mathbb{R}^d_+$, let

$$\widetilde{F}_{\boldsymbol{p}}(t) = \int_{\langle \frac{\boldsymbol{x}}{\boldsymbol{\epsilon}(F)}, \boldsymbol{p} \rangle \le t} dF(\boldsymbol{x})\,. \tag{9.5}$$

If $\boldsymbol{X}$ is a random vector having distribution F, then $\widetilde{F}_{\boldsymbol{p}}(t)$ is the distribution of the random variable $\widetilde{\langle \boldsymbol{X}, \boldsymbol{p} \rangle}$ in $\mathbb{R}$. $\widetilde{F}_{\boldsymbol{p}}$ can be interpreted as the distribution of relative expenditures, i.e. shares of total expenditure, under the price vector $\boldsymbol{p}$.

Definition 9.3 (Price Lorenz order) *Of two distributions F and G in $\mathcal{P}_{1+}$, F is said to be larger than G in the* price Lorenz order, *$F \succeq_{PL} G$, if*

$$\widetilde{F}_{\boldsymbol{p}} \succeq_{lz} \widetilde{G}_{\boldsymbol{p}} \quad \textit{for every } \boldsymbol{p} \in \mathbb{R}^d_+\,.$$

If $\boldsymbol{X}$ and $\boldsymbol{Y}$ are random vectors with distributions F and G, the price Lorenz order means that, for all prices $\boldsymbol{p} \ge \boldsymbol{0}$, the distribution of $\widetilde{\langle \boldsymbol{X}, \boldsymbol{p} \rangle}$ is larger than that of $\widetilde{\langle \boldsymbol{Y}, \boldsymbol{p} \rangle}$ in the univariate lift zonoid order or, what is the same, the distribution of $\langle \boldsymbol{X}, \boldsymbol{p} \rangle$ is larger than that of $\langle \boldsymbol{Y}, \boldsymbol{p} \rangle$ in the univariate Lorenz order.

When $d = 1$, the price Lorenz order is the usual Lorenz order or, in other words, the pointwise order of ILFs. With two univariate empirical distributions $F_{\boldsymbol{A}}$ on $a_1, \dots, a_n$ and $F_{\boldsymbol{B}}$ on $b_1, \dots, b_n$, $F_{\boldsymbol{A}} \succeq_{PL} F_{\boldsymbol{B}}$ amounts to

$$\sum_{i=1}^{k} \breve{a}_{(i)} \le \sum_{i=1}^{k} \breve{b}_{(i)}\,, \quad k = 1, \dots, n-1\,. \tag{9.6}$$

The following theorem, which is the main result of this section, states that, also in higher dimensions and for general distributions, the price Lorenz order coincides with the pointwise ordering of ILFs.

Theorem 9.7 (Equivalence to the ILF order) *For F and $G \in \mathcal{P}_{1+}$, $F \succeq_{PL} G$ if and only if*

$$l_F(\boldsymbol{y}) \ge l_G(\boldsymbol{y}) \quad \textit{for all } \boldsymbol{y} \in [0,1]^d\,. \tag{9.7}$$

Proof. By Proposition 9.2, the ILF order (9.7) holds if and only if

$$\left(LZ(G) + (\mathbb{R}_- \times \mathbb{R}^d_+)\right) \subset \left(LZ(F) + (\mathbb{R}_- \times \mathbb{R}^d_+)\right)\,. \tag{9.8}$$

The set inclusion (9.8) can be characterized by the pointwise order of their support functions[3]. First, consider the support function of $LZ(F)$,

$$h(LZ(F), p_0, \boldsymbol{p}) = \int_{H_{p_0, \boldsymbol{p}}(F)} p_0 + \left\langle \frac{\boldsymbol{x}}{\epsilon(F)}, \boldsymbol{p} \right\rangle dF(\boldsymbol{x}),$$

where the Proposition 2.16 and the abbreviation (9.3) are used. Note that, in directions $(p_0, \boldsymbol{p}) \in \mathbb{R}_+ \times \mathbb{R}^d_-$, the support function of the set $LZ(F) + (\mathbb{R}_- \times \mathbb{R}^d_+)$ is equal to the support function of $LZ(F)$; in all other directions it equals infinity. Therefore, (9.8) if and only if $h(LZ(G), p_0, \boldsymbol{p}) \le h(LZ(F), p_0, \boldsymbol{p})$ for every $p_0 \in \mathbb{R}_+, \boldsymbol{p} \in \mathbb{R}^d_-$, if and only if

$$\int_{H_{p_0, \boldsymbol{p}}(G)} \left(p_0 + \left\langle \frac{\boldsymbol{x}}{\epsilon(G)}, \boldsymbol{p} \right\rangle \right) dG(\boldsymbol{x}) \tag{9.9}$$
$$\le \int_{H_{p_0, \boldsymbol{p}}(F)} \left(p_0 + \left\langle \frac{\boldsymbol{x}}{\epsilon(F)}, \boldsymbol{p} \right\rangle \right) dF(\boldsymbol{x})$$

for all $p_0 \ge 0$ and $\boldsymbol{p} \le \boldsymbol{0}$. Substituting $t = -\langle \boldsymbol{x}/\epsilon(F), \boldsymbol{p} \rangle$ yields an equivalent form of (9.9),

$$\int_{p_0 \ge t} (p_0 - t)\, d\widetilde{G}_{\boldsymbol{p}}(-t) \le \int_{p_0 \ge t} (p_0 - t)\, d\widetilde{F}_{\boldsymbol{p}}(-t). \tag{9.10}$$

Now consider the lift zonoid of $\widetilde{F}_{\boldsymbol{p}}$ and its support function at $(p_0, -1)$. There holds

$$h(\widehat{Z}(\widetilde{F}_{\boldsymbol{p}}), p_0, -1) = \int_{p_0 - t \ge 0} (p_0 - t)\, d\widetilde{F}_{\boldsymbol{p}}(t) = \int_{p_0 - t \ge 0} (p_0 - t)\, d\widetilde{F}_{-\boldsymbol{p}}(-t). \tag{9.11}$$

The last equation holds due to $\widetilde{F}_{\boldsymbol{p}}(-t) = \widetilde{F}_{-\boldsymbol{p}}(t)$. From (9.11) we see that (9.10) and, hence, (9.7) are equivalent to

$$h(\widehat{Z}(\widetilde{G}_{-\boldsymbol{p}}), p_0, -1) \le h(\widehat{Z}(\widetilde{F}_{-\boldsymbol{p}}), p_0, -1)$$

for all $\boldsymbol{p} \le \boldsymbol{0}$ and $p_0 \ge 0$, that is, equivalent to the set inclusion $(\widehat{Z}(\widetilde{G}_{-\boldsymbol{p}}) + (\mathbb{R}_- \times \mathbb{R}_+)) \subset (\widehat{Z}(\widetilde{F}_{-\boldsymbol{p}}) + (\mathbb{R}_- \times \mathbb{R}_+))$. This means that $\widetilde{F}_{-\boldsymbol{p}}$ is larger in lift zonoid order than $\widetilde{G}_{-\boldsymbol{p}}$, for every $\boldsymbol{p} \le \boldsymbol{0}$. The latter says that $F \succeq_{PL} G$ holds, which concludes the proof. Q.E.D.

From the definition of the price Lorenz order and its characterization as an ordering of inverse Lorenz functions (Theorem 9.7), many properties derive. They are collected in the next proposition.

[3]For the definition and properties of support functions, see Appendix A 3.

Proposition 9.8 (Properties of the price Lorenz order)
For the price Lorenz order holds:

(i) *There exists a set of smallest elements containing all one-point distributions.*

(ii) *The order is scale invariant.*

(iii) *If two distribution are ordered, all their marginals are ordered in the same way.*

(iv) *The order is continuous in the following sense:*
If the sequence G^n converges weakly to G and $F \succeq_{PL} G^n$ holds for all n, then $F \succeq_{PL} G$.
If the sequence G^n is uniformly integrable and converges weakly to G and $G^n \succeq_{PL} F$ holds for all n, then $G \succeq_{PL} F$.

Proof. Parts (i) and (ii) are obvious from the definition. (iii) is a consequence of Proposition 9.1(iii) and Proposition 9.7. Part (iv) follows from Propositions 9.4 and 9.7. Q.E.D.

It is seen from Proposition 9.8 and the definition that, of the postulates mentioned in Section 8.7, the price Lorenz order satisfies **IS**, **ME**, **S**, and **MA**, but violates **IT**, **DO**, **E**, **A**, **T**, and **C**. With stochastically independent attributes, the price Lorenz order is equivalent to the usual Lorenz order of all univariate marginals:

Proposition 9.9 (Independent marginals) *Let $F, G \in \mathcal{P}_{1+}$, $F(x) = \Pi_{k=1}^d F_k(x_k)$, $G(x) = \Pi_{k=1}^d G_k(x_k)$. Then*

$$F \succeq_{PL} G \quad \text{if and only if} \quad F_k \succeq_L G_k\,, \quad k = 1, \dots, d. \tag{9.12}$$

Note that there are also other instances where the Lorenz order of all univariate marginals is sufficient for two distributions to be price Lorenz ordered. For example, if F^+ and G^+ are upper Fréchet bounds, $F^+(x) = \min_k F_k(x_k)$ and $G^+(x) = \min_k G_k(x_k)$, then (9.12) holds as well.

The ILF of an *empirical distribution* can be described by the extreme points of its graph (Proposition 9.6). From Theorem 9.7 the following characterization of the price Lorenz order among empirical distributions is obtained. It extends the system of univariate inequalities (9.6).

Proposition 9.10 (Price Lorenz order of empiricals) *For empirical distributions F_A and F_B, the following two restrictions are equivalent:*

(i) *Assume that*

$$x = \left(\frac{|I|}{n}, \sum_{i \in I} \breve{b}_{i,J}, \mathbf{1}_{-J} \right)$$

is an extreme point of the Lorenz hypersurface of F_B *and let*

$$z_0 = \max_{\theta_1, \dots, \theta_n} \sum_{i=1}^{n} \theta_i \quad s.t. \quad \sum_{i=1}^{n} \theta_i \breve{a}_{i,J} \leq \sum_{i \in I} \theta_i \breve{b}_{i,J}, \quad 0 \leq \theta_i \leq 1;$$

then $z_0 \geq |I|$.

(ii) $F_A \succeq_{PL} F_B$.

Proof. According to Theorem 9.7, $F_A \succeq_{PL} F_B$ iff $l_A \geq l_B$. From the convexity of the subgraph of the ILF follows that $l_A \geq l_B$ holds iff $l_A(x) \geq l_B(x)$ for every extreme point x to the Lorenz hypersurface of F_B. In view of this, Proposition 9.6 yields the proof. Q.E.D.

9.4 Majorizations of absolute endowments

So far, we have compared distributions of shares, or relative endowments. Distributions of absolute endowments can be treated in a similar way. This is done in the present section.

Several authors, among them Moyes (1987), have investigated Lorenz orders between univariate distributions of absolute endowments. For a probability distribution F on $\mathbb{R}_+$, the *generalized Lorenz function* is given by

$$GL_F(t) = \int_0^t F^{-1}(s)ds, \quad 0 \leq t \leq 1.$$

Hence $GL_F(0) = 0$, $GL_F(1) = \epsilon(F)$, and $GL_F(t) = \epsilon(F)L_F(t)$ if $\epsilon(F)$ is finite. The *dual generalized Lorenz function* of F is

$$DGL_F(t) = \epsilon(F) - GL_F(1-t), \quad 0 \leq t \leq 1.$$

For two univariate distributions F and G, consider the pointwise ordering of these functions,

$$F \succeq^w G \quad \text{if} \quad GL_F(t) \leq GL_G(t) \text{ for all } t,$$
$$F \succeq_w G \quad \text{if} \quad DGL_F(t) \geq DGL_G(t) \text{ for all } t.$$

$F \succeq^w G$ is mentioned as *supermajorization*, and $F \succeq_w G$ as *submajorization* of F over G. Obviously $F \succeq^w G$ implies that $\epsilon(F) \leq \epsilon(G)$, while $F \succeq_w G$ implies that $\epsilon(F) \geq \epsilon(G)$. If both $F \succeq^w G$ and $F \succeq_w G$ hold, then $F \succeq_{PL} G$ and $\epsilon(F) = \epsilon(G)$. On the other hand, if $\epsilon(F) = \epsilon(G)$, submajorization as well as supermajorization coincide with the usual Lorenz order.

Now, multivariate versions of submajorization and supermajorization are defined. We use the notation $F_{\mathbf{p}}(t) = \int_{\langle \boldsymbol{x}, \mathbf{p}\rangle \leq t} dF(\boldsymbol{x})$.

Definition 9.4 (Price supermajorization, price submajorization) *For F and $G \in \mathcal{P}_{1+}$, define*

$$F \succeq^w_P G \quad \text{if } F_{\mathbf{p}} \succeq^w G_{\mathbf{p}} \text{ for all } \mathbf{p} \in \mathbb{R}^d_+,$$

$$F \succeq_{wP} G \quad \text{if } F_{\mathbf{p}} \succeq_w G_{\mathbf{p}} \text{ for all } \mathbf{p} \in \mathbb{R}^d_+.$$

$\succeq^w_P$ *is called* price supermajorization, $\succeq_{wP}$ price submajorization. *Both* $\succeq^w_P$ *and* $\succeq_{wP}$ *are called* weak price majorizations.

The next definition concerns the intersection of price sub- and supermajorization.

Definition 9.5 (Price majorization) *If both $F \succeq^w_P G$ and $F \succeq_{wP} G$ hold, we say that F* price majorizes *G, in symbols, $F \succeq_P G$.*

Note that, with univariate empirical distributions, these three orders specialize to the usual *majorization*, *submajorization*, and *supermajorization* of vectors; see Marshall and Olkin (1979). That is the reason to name them 'majorizations'. The following corollary is obvious.

Corollary 9.11 (Equivalent definitions)
(i)

$$F \succeq_P G \quad \Leftrightarrow \quad F \succeq_{PL} G \text{ and } \epsilon(F) = \epsilon(G).$$

(ii) *If* $\epsilon(F) = \epsilon(G)$,

$$F \succeq_P G \quad \Leftrightarrow \quad F \succeq_{wP} G \quad \Leftrightarrow \quad F \succeq^w_P G.$$

For the weak majorizations we have another characterization which involves a generalized version of the ILF and its dual.

Definition 9.6 (Inverse generalized Lorenz function)
(i) *For* $y \in \mathbb{R}_+^d$ *consider*

$$gl_F(y) = \max \int_{\mathbb{R}_+^d} g(x)\, dF(x), \tag{9.13}$$

where the maximum is taken over all measurable $g : \mathbb{R}_+^d \to [0,1]$ *with*

$$\int_{\mathbb{R}_+^d} g(x)\, x\, dF(x) \leq y\,.$$

The function $gl_F : \mathbb{R}_+^d \to [0,1]$ *is the* inverse generalized Lorenz function (IGLF) *of* F.
(ii) *The function* dgl_F *is called the* dual inverse generalized Lorenz function (DIGLF) *of* F,

$$dgl_F(y) = \min_g \int_{\mathbb{R}_+^d} g(x)\, dF(x), \quad y \in \mathbb{R}_+^d, \tag{9.14}$$

where the minimum is taken over all measurable functions $g : \mathbb{R}_+^d \to [0,1]$ *for which*

$$\int_{\mathbb{R}_+^d} g(x)\, x\, dF(x) \geq y\,.$$

price super-majorization	$F \succ_P^w G$: $\forall p \geq 0\ F_p \succ^w G_p$	$\Leftrightarrow$	ordering of IGLFs
	$\Uparrow$		$\Uparrow$
price majorization	$F \succ_P G$: $\epsilon(F) = \epsilon(G)$ and $\forall p \geq 0\ \widetilde{F}_p \succ \widetilde{G}_p$	$\Leftrightarrow$	$F \succ_P^w G$ and $F \succ_{wP} G$
	$\Downarrow$		$\Downarrow$
price sub-majorization	$F \succ_{wP} G$: $\forall p \geq 0\ F_p \succ_w G_p$	$\Leftrightarrow$	ordering of DIGLFs

TABLE 9.1: Price majorization and weak price majorizations and their characterization by inverse generalized Lorenz functions (IGLFs) and their duals (DIGLFs).

Theorem 9.12 (Weak majorizations and the ILF)
(i) $F \succeq_P^w G$ *if and only if*

$$gl_F(y) \geq gl_G(y) \quad \text{for all } y \in [0, \epsilon(F)]\,.$$

(ii) $F \succeq_{wP} G$ *if and only if*

$$dgl_F(y) \leq dgl_G(y) \quad \text{for all } y \in [\mathbf{0}, \epsilon(G)] .$$

The proof of Theorem 9.12 is similar to that of Theorem 9.7 and, therefore, omitted. Table 9.1 displays the main definitions and results of this section.

9.5 Other inequality orderings

In this section the price Lorenz order is contrasted with several other multicommodity orderings that are found in the literature. Further, indices are considered which are consistent with the orders.

Atkinson and Bourguignon (1982) suggested two other orders to rank distributions of commodity vectors. Their first order ranks $F \succeq_{lo} G$ if $F(\boldsymbol{x}) \leq G(\boldsymbol{x})$ for every $\boldsymbol{x} \in \mathbb{R}^d$, and their second order ranks $F \succeq_{locc} G$ if $\int_{\boldsymbol{y} \leq \boldsymbol{x}} F(\boldsymbol{y}) d\boldsymbol{y} \leq \int_{\boldsymbol{y} \leq \boldsymbol{x}} G(\boldsymbol{y}) d\boldsymbol{y}$ for every $\boldsymbol{x} \in \mathbb{R}^d$. The orderings are known in the literature as the lower orthant ordering and the lower orthant concave ordering, respectively; see, e.g., Dyckerhoff and Mosler (1997).

Obviously, in dimension one, the two Atkinson-Bourgouignon orders amount to usual first and second degree stochastic dominance. For $d \geq 1$, the first order implies the second one, and both can be characterized by inequalities on expected utilities. In particular, (see, for example, Scarsini (1985)), $F \succeq_{lo} G$ if and only if $\int_{\mathbb{R}^d} u(\boldsymbol{x}) dF(\boldsymbol{x}) \geq \int_{\mathbb{R}^d} u(\boldsymbol{x}) dG(\boldsymbol{x})$ for any utility function u from the class

$$\mathcal{U}_{lo} = \left\{ u : \mathbb{R}^d \to \mathbb{R} \, : \, (-1)^{|J|} \frac{\partial^J u}{\partial \boldsymbol{x}_J} \leq \mathbf{0} \ \text{ for any } J \subset \{1, \dots, d\} \right\}, \ J \neq \emptyset .$$

Here $\frac{\partial^J u}{\partial \boldsymbol{x}_J}$ is the vector of partial derivatives with respect to components of $\boldsymbol{x}_J$.

The first Atkinson-Bourgouignon order, in dimension $d = 2$, implies the order of IGLFs or, equivalently, the price supermajorization. This can be seen from the characterizations of the rankings via expected utilities. A class of utility functions that generates the price supermajorization is

$$\mathcal{U}_{wP} = \left\{ u : u(\boldsymbol{x}) = v(\langle \boldsymbol{p}, \boldsymbol{x} \rangle), \boldsymbol{x} \in \mathbb{R}^d, \text{ with } v : \mathbb{R} \to \mathbb{R} \text{ concave}, \ \boldsymbol{p} \in \mathbb{R}^d_+ \right\} .$$

When $d = 2$, obtain

$$\mathcal{U}_{lo} = \left\{ u : \mathbb{R}^2 \to \mathbb{R} \, : \, \frac{\partial u}{\partial x_1} \geq 0, \frac{\partial u}{\partial x_2} \geq 0, \frac{\partial^2 u}{\partial x_1 \partial x_2} \leq 0 \right\} ,$$

and the inclusion $\mathcal{U}_{wP} \subset \mathcal{U}_{lo}$ holds. Then $F \succeq_{lo} G$ implies $F \succeq_{wP} G$. For $d > 2$ the first Atkinson-Bourgouignon order does not imply the weak price majorization.

The second Atkinson-Bourgouignon order, in dimension $d = 1$, equals price supermajorization. But for $d \geq 2$ it neither implies nor is implied by price supermajorization.

Many other majorizations have been used in the analysis of economic disparity; see the survey by Mosler (1994a). E.g., Kolm (1977) and Russell and Seo (1978) employ the multivariate dilation order which is generated by expected utility inequalities for all concave utility functions. Dilation order implies lift zonoid order, and the latter implies price majorization.

Transfers. In dimension one the univariate price majorization, that is the Lorenz order, is closely connected with *Pigou-Dalton transfers.* It is interesting how the above multivariate majorizations relate to Pigou-Dalton transfers. Consider a finite number, n, of agents. Since price majorization is equivalent to univariate Lorenz order of expenditures, whatever the prices, it is tantamount saying that for any prices there exists a series of usual Pigou-Dalton transfers of expenditures.

A *multivariate Pigou-Dalton transfer* is a transfer of commodity vectors among two agents such that the resulting two endowments are in the convex hull of the previous. A distribution matrix $\boldsymbol{A}$ is *chain-majorized* by another distribution matrix $\boldsymbol{B}$ if $\boldsymbol{A}$ is obtained by a finite number of multivariate Pigou-Dalton transfers from $\boldsymbol{B}$. Chain-majorization implies matrix majorization (= multivariate dilation), but for $n \geq 3$ and $d \geq 2$ the reverse is not true (Marshall and Olkin, 1979, p. 431). Thus, the ordering induced by multivariate Pigou-Dalton transfers is stronger than dilation, lift zonoid order and price majorization.

The univariate Lorenz curve has three principal properties which are the reason for its almost universal use in the measurement of economic inequality:

- The pointwise ordering by Lorenz functions is the Lorenz order.
- The Lorenz curve is visual: it indicates the degree of disparity as the bow is bent.
- Up to a scale parameter, the Lorenz curve fully describes the underlying distribution. No information is lost when we look at the Lorenz curve instead of the distribution function or the density.

All these properties are shared by the multivariate inverse Lorenz function:

- The pointwise ordering of inverse Lorenz functions is equivalent to the price Lorenz order.
- The Lorenz hypersurface is visual in the same way as the Lorenz curve is: increasing deviation from the egalitarian hypersurface indicates more disparity.
- The Lorenz hypersurface determines the underlying distribution uniquely, up to a vector of scaling constants.

Indices. The disparity of distributions can be measured by partial orderings, as presented so far, or by real-valued indices. An important question is which multivariate dispersion indices are consistent, that is, increasing with the price Lorenz order, the price majorization, and the weak price majorizations, respectively.

Kolm (1977, Th. 5) suggests a class of indices of the form

$$\Phi_{Kolm}(\boldsymbol{A}) = \Psi(f(\boldsymbol{a}_1), \dots, f(\boldsymbol{a}_n)),$$

where f is a convex function on $\mathbb{R}^d$ (a utility function) and $\Psi : \mathbb{R}^n \to \mathbb{R}$ is an index consistent with the convex (= lift zonoid) order of univariate empirical distributions. He shows that such an index Φ_{Kolm} is consistent with the dilation order. Tsui (1995) investigates similar indices.

The following indices are consistent with the price Lorenz order: Let Ψ be a univariate index that is consistent with the univariate Lorenz order, e.g., Ψ may be the Gini index. Define an 'average index' by

$$\Phi(F) = \int_{\boldsymbol{p} \in S^d_+} \Psi(F_{\boldsymbol{p}})\pi(d\boldsymbol{p}), \tag{9.15}$$

where $S^d_+ = \{\boldsymbol{p} \in S^d : \boldsymbol{p} \geq \boldsymbol{0}\}$ and integration is done with respect to some measure π on the nonnegative part of the sphere. Obviously, Φ increases with the price Lorenz order. The measure π reflects the information about uncertain prices. In particular, if some restrictions on prices are known which form a subset $R \subset S^d_+$, the Lebesgue measure on R can be used for π.

It can be shown that the distance-Gini index R_D defined in Section 8.8 is consistent with the lift zonoid order and the convex order (Theorem 8.27) but not with the price Lorenz order. A different, particularly visual index is given by

$$\Phi_{ILF}(F) = \frac{d+1}{d} \times (\text{volume between the graphs of } l_F \text{ and } l_{\delta_1}). \tag{9.16}$$

Φ_{ILF} forms another multivariate generalization of the univariate Gini index. It is bounded, $0 \leq \Phi_{ILF}(F) < 1$ for all F, and consistent with the price Lorenz order.

9.6 Measuring industrial concentration

Concentration of firms within an industry is commonly measured by the concentration curve of a size characteristic and by indices derived from it. Usually a single characteristic of size is chosen and its concentration measured by means of the concentration curve and indices like those by Rosenbluth and Herfindahl. This choice is guided by the economic question being analyzed and, to a great extent, by the data available. Most authors (e.g. Hannah and Kay (1977), Marfels (1977)) argue in that way, adding that in empirical applications the various size characteristics appear to measure the same and that their concentration, while not being equal, at least varies in the same direction when considered at different periods.

But, in general, different size characteristics yield different results in the measurement of concentration. This is not only theoretically obvious but also empirically relevant.

In the sequel, for measuring the concentration of a given industry, it is assumed that several characteristics of size appear to be specific and that no prior hierarchy or weights exist for them. Methods are proposed to measure the concentration when many, say d, characteristics are considered simultaneously. A geometric notion, the ***concentration hypersurface***(CH), is introduced which extends the usual concentration curve to more than one attributes. The CH is the graph of a function that maps the d-dimensioned unit cube into the reals. This function, which resembles the inverse Lorenz function, is named the *concentration function.*

In economic terms an element of the unit cube represents a vector of market shares in the attributes, and the concentration function indicates the minimal number – properly interpolated – of firms that together own not less than a given vector of market shares.

The usual univariate concentration curve plots the aggregate market share of the biggest firms against the number of such firms, which is tantamount saying that it plots the minimal number of firms owning a given market share. It comes out that, when $d = 1$, our notion of CH reduces to the usual concentration curve where the ordinate and abscissa axes are interchanged.

Two industries are compared by looking at their CHs or, equivalently, at their concentration functions. The CHs can be checked whether one lies above the other (i.e. whether the first concentration function exceeds the second) either at all points or at selected points.

However, as two CHs may intersect (and they often do) the overall pointwise ordering of CHs provides no unambigous ranking of industrial concentration.

A complete ranking is obtained by a real-valued index. Based on our notion of CH, the classic univariate Rosenbluth index, which amounts to 1/2 divided by the area above the usual univariate concentration curve, allows for a straightforward generalization.

A number of statistical and economic postulates has been proposed in the literature for a unidimensional measure of concentration (Hannah and Kay, 1977; Piesch, 1975; Adelman, 1951, 1969; Rosenbluth, 1955, and others). The principal ones are that any concentration measurement should be anonymous, scale invariant, merger increasing and zero entry invariant, decreasing with progressive sales transfers, and decreasing with 'anti-Gibrat effects'. A real-valued index of concentration should, in addition, be consistent with the ordering of concentration curves, normalized and reciprocal to the concentration equivalent number of equally sized firms.

In Section 9.8 a multivariate Rosenbluth index is defined as $d/(d+1)$ divided by the volume below the CH, which for $d = 1$ equals the usual notion. This index satisfies all postulates. For multidimensional extensions of the Gini index, see Sections 7.4 and 8.8.

The setting is notated as follows. Let n be the number of firms which supply a homogeneous good in a market and let d be the number of characteristics that are considered relevant to measure market concentration. a_{ij} denotes the value of characteristic j with firm i, $a_{ij} \geq 0$. Then $\boldsymbol{A} = [a_{ij}]$ is an $n \times d$ matrix whose row $\boldsymbol{a}_i$ lists the characteristics of the i-th firm and whose column $\boldsymbol{a}^j$ contains the distribution of the characteristic j among the firms. Given the number of characteristics d, the set of all such matrices, nonnegative and $n \times d$, is denoted by $\mathcal{A}_d$. Note that matrices in $\mathcal{A}_d$ have a fixed number, d, of columns (characteristics) but an arbitrary number, n, of rows (firms). The subset of *column stochastic matrices*, where $\sum_{i=1}^n a_{ij} = 1$ for every i, is denoted by $\mathcal{R}^d$.

First, a short treatment of unidimensional measurement is presented. For $\boldsymbol{A} \in \mathcal{A}_1$, $\boldsymbol{A} = (a_1, \dots, a_n)'$, the *k-th concentration rate* is given by

$$CR_{\boldsymbol{A}}(k) = \frac{\sum_{i=1}^k a_{[i]}}{\sum_{i=1}^n a_{[i]}}, \qquad k = 0, \dots, n,$$

where the $a_{[i]}$ are ordered from above. For $k \in \mathbb{N}, k > n$, define $CR_{\boldsymbol{A}}(k) = 1$. The piecewise linear curve that connects the points $(k, CR_{\boldsymbol{A}}(k))$, $k = 0, \dots, n$, in $\mathbb{R}^2$ is the usual *concentration curve.* Figure 9.5 exhibits a univariate concentration curve and its inverse.

We say that $\boldsymbol{A}$ is more concentrated than $\boldsymbol{B}$ if $CR_{\boldsymbol{A}}(k) \geq CR_{\boldsymbol{B}}(k)$ holds at every k or, equivalently, if the usual concentration curve of $\boldsymbol{A}$ lies above

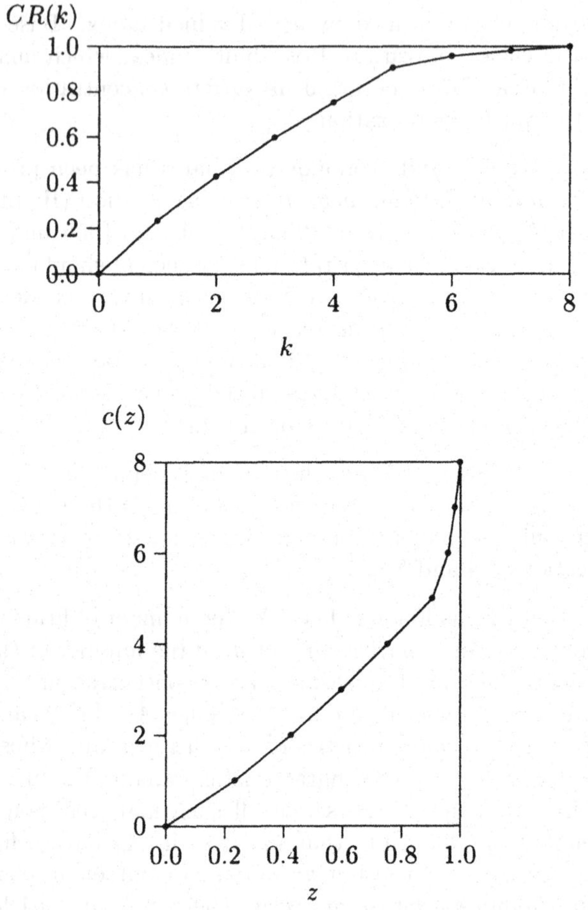

FIGURE 9.5: The concentration curve and the inverse concentration curve. Sales in the German mineral oil industry, 1990; from Koshevoy and Mosler (1999b).

that of $\boldsymbol{B}$. This induces an ordering of $\mathcal{A}_1$, which is called the *concentration order.* The concentration order is reflexive and transitive, but not complete, since concentration curves may cross. Note that the concentration order is unaffected by an arbitrary permutation of the firms and also by a scale transformation of the size characteristic.

Concentration as measured by the concentration order satisfies a number of statistical and economic postulates. The most important are:

- **Anonymity.** Concentration does not change under permutations of the firms.

- **Scale.** Concentration does not change under transformations of the scale.

- **Size transfer.** Concentration decreases with a progressive size transfer. A progressive size transfer means that a firm gives some percentage of its size, e.g. sales, to another firm and receives the same percentage of the second firm in exchange, while the sizes of all other firms remain the same.

- **Zero entry.** Concentration does not change if a firm of size 0 enters the market.

- **Merger.** Concentration increases if two firms merge.

- **Split.** Concentration decreases if every firm is split into k firms of equal size.

The properties of size transfer and zero entry together imply that of merger since a merger can be seen as the reverse of a zero entry and a size transfer.

Further postulates are found in the literature (Hannah and Kay, 1977; Hall and Tideman, 1967, and others). which are closely related to the above and also satisfied by the concentration order:

- **Small entry.** There is some $s \in]0,1[$, such that, if a new firm enters and gains market share $\leq s$ while the relative shares of all existing firms are unchanged, concentration is reduced.

- **Anti-Gibrat.** The 'anti-Gibrat' effect reduces concentration: $\boldsymbol{A}$ has less concentration than $\boldsymbol{B}$ if $\sum_{i=1}^{n} a_i = \sum_{k=1}^{m} b_k$ and each $\breve{a}_i$ is a weighted average of $\breve{b}_1, \ldots, \breve{b}_m$, that is the market size is fixed and each market share a_i is replaced by a weighted average of all market shares. Formally, $\boldsymbol{A}$ is obtained from $\boldsymbol{B}$ via a doubly stochastic transformation. A doubly stochastic transformation is a linear transformation by a doubly stochastic matrix, that is a matrix $\boldsymbol{T} = [t_{ik}]$ with $\sum_i t_{ik} = \sum_k t_{ik} = 1$ for all i, k.

- **Continuous at entrance.** Let s be the market share of a new entrant. Then, as s tends to zero, so does its effect on a the concentration curve.

Observe that, in general, the anti-Gibrat property implies the size transfer property, since any size transfer is a doubly stochastic transformation (Marshall and Olkin, 1979). The reverse is not true.

A complete order is obtained from a real-valued index. A function $\Psi : \mathcal{A}_1 \to \mathbb{R}$. is a *concentration index* if it is consistent (i.e. increasing) with the concentration order. Then the concentration measured by Ψ satisfies the above mentioned postulates and properties as well. For example, the concentration rates (for every k) and the classic indices by Rosenbluth (Rosenbluth, 1955) and Herfindahl are increasing with the concentration order and thus satisfy all these postulates. See Piesch (1975) for these and many other indices of concentration. Disparity indices like the Gini index do not fulfill the postulate of zero entry.

The following additional postulates are often imposed on Ψ.

- **Continuity.** Ψ is continuous.
- **Normalization.** Ψ is always nonnegative. Ψ achieves its minimum value $1/n$ when n firms of equal size are given.
- **Cardinal split.** Ψ decreases by the factor $1/k$ if if every firm is split into k firms of equal size.

To cardinalize an index in this way has been suggested by Adelman (1969) and by Hall and Tideman (1967). For a discussion of these properties, see Hannah and Kay (1977). The Rosenbluth and Herfindahl indices are normalized and satisfy cardinal split, but the concentration rates, besides CR_1, do not.

9.7 Multivariate concentration function

Given a matrix $\boldsymbol{A} = [a_{ij}] \in \mathcal{A}_d$, notate $\breve{\boldsymbol{A}} = [\breve{a}_{ij}]$ with $\breve{a}_{ij} = a_{ij}/(\sum_i a_{ij})$ for all i and j. Then $\breve{\boldsymbol{A}} \in \mathcal{R}^d$.

Definition 9.7 (Concentration function) *We define the multivariate* concentration function $c_{\boldsymbol{A}} : [0,1]^d \to \mathbb{R}_+$ *by*

$$c_{\boldsymbol{A}}(\boldsymbol{y}) = \min_{(\theta_1,\ldots,\theta_n)\in\Theta(\boldsymbol{A})} \sum_{i=1}^{n} \theta_i, \quad \boldsymbol{y} \in [0,1]^d, \tag{9.17}$$

$$\textit{where} \quad \Theta(\boldsymbol{A}) = \{(\theta_1,\ldots,\theta_n) : \sum_i \theta_i \breve{\boldsymbol{a}}_i \geq \boldsymbol{y},\, \theta_i \in [0,1],\, i = 1,\ldots,n\}.$$

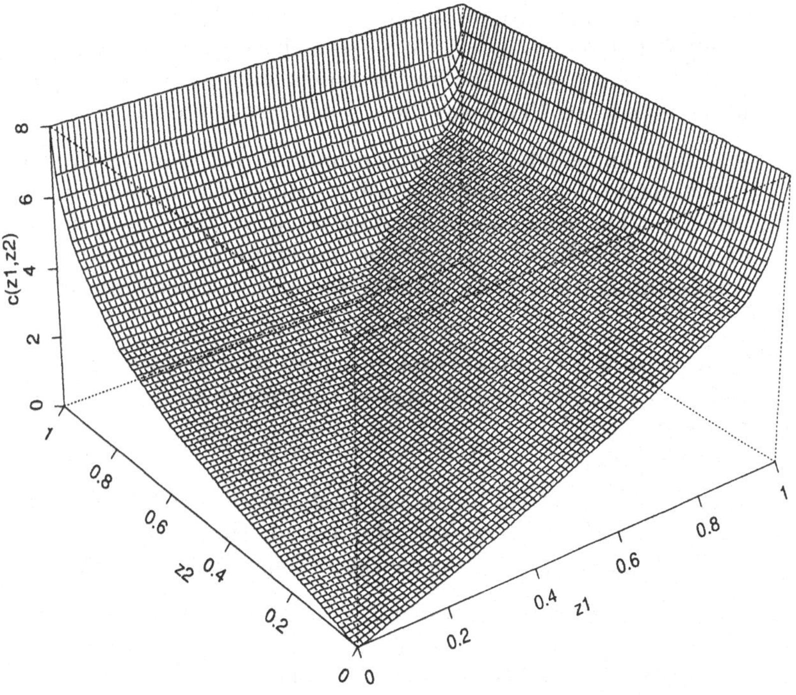

FIGURE 9.6: The concentration hypersurface for two dimensions of size (Sales and employment in the German mineral oil industry, 1990; from Koshevoy and Mosler (1999b)).

The graph of the function $c_{\boldsymbol{A}}$ *is named the* concentration hypersurface *of* $\boldsymbol{A}$ *and denoted by* $CH_{\boldsymbol{A}}$.

The concentration function and the hypersurface are anonymous and scale invariant. An economic interpretation of the concentration function is the following: For a given vector of characteristics (= shares), $\boldsymbol{y} \in [0,1]^d$, determine the minimal number of firms by which these shares or more are held. (Note that this number is no integer and that also parts of the firms enter it in contributing to the aggregate vector of shares.) This number, depending on the vector of shares, is the d-variate concentration function.

Observe that both the definition and the interpretation of the concentration function closely resemble those of the inverse Lorenz function. The difference is that, for a given vector of shares, the concentration function indicates the *smallest number* of units (=firms) that possesses these shares (=attributes of relative size), while the inverse Lorenz function indicates the *largest part* of the population that owns these shares (attributes of relative well-being).

It follows immediately from the definition that the concentration function is continuous, convex and increasing with minimum value 0 and maximum value n. The minimum is attained at $y = (0, \ldots, 0)$ and the maximum at $(1, \ldots, 1)$. Observe that in the univariate case, when $d = 1$, the usual concentration curve is the graph of the inverse concentration function.

Definition 9.8 (Concentration order) *Given two distribution matrices* A *and* $B \in \mathcal{A}_d$*, say that* A has less concentration than B *if* $c_A(y) \geq c_B(y)$ *for all* $y \in [0,1]^d$.

This definition induces a relation on $\mathcal{A}_d$ which is reflexive and transitive. If we restrict the preorder to the set of column stochastic matrices, $\mathcal{R}^d = \{A \in \mathcal{A}_d : \sum_{i=1}^n a_{ij} = 1\}$, then the concentration order is also antisymmetric.

Most of the principles suggested in the literature on univariate concentration measurement extend immediately to the multivariate case. The postulates of anonymity, scale, size transfer, zero entry, merger, anti-Gibrat, and split, carry over *verbatim* if the size of a firm is taken as a vector. Also the postulates of small entry and continuity at entrance extend in the same way if s is replaced by a vector.

We show that our notion of decreasing concentration satisfies all these postulates. Anti-Gibrat (and hence size transfer) is derived from the following proposition.

Proposition 9.13 (Characterization of the concentration order) $A = [a_{ij}]$ *has less concentration than* $B = [b_{lj}]$ *if and only if, with any choice of weights* $p_j \geq 0$*, the same univariate relation holds for the weighted sums of the firms' characteristics,* $\sum_{j=1}^d a_{1j}p_j, \ldots, \sum_{j=1}^d a_{nj}p_j$ *and* $\sum_{j=1}^d b_{1j}p_j, \ldots, \sum_{j=1}^d b_{mj}p_j$.

Formally, the Proposition states that

$$c_A \geq c_B \quad \text{if and only if} \quad c_{pA} \geq c_{pB} \quad \text{for any } p \geq 0\,.$$

Proposition 9.13 corresponds to Theorem 9.7. For proof, see Koshevoy and Mosler (1999b, Appendix).

From the proposition conclude that, if $\boldsymbol{A}$ is obtained from $\boldsymbol{B}$ via a doubly stochastic transformation, then the same holds for $\langle \boldsymbol{A}, \boldsymbol{p}\rangle$ and $\langle \boldsymbol{B}, \boldsymbol{p}\rangle$, for any $\boldsymbol{p} = (p_1, \dots, p_d) \in \mathbb{R}^d_+$. Thus the properties of anti-Gibrat and size transfer are established.

Continuity follows from the continuity, in the Hausdorff distance, of the Lorenz zonoid with respect to weak convergence of distributions that have a compact support; see Corollary 2.33. To see the small entry property, consider

$$\boldsymbol{B} = \begin{pmatrix} \breve{\boldsymbol{A}} \\ \boldsymbol{s} \end{pmatrix} \quad \text{with } \boldsymbol{s} = (s_1, \dots, s_d)\,, \quad 0 \le s_j \le \min_{i:\breve{a}_{ij}>0} \breve{a}_{ij}\,.$$

Then $c_{\boldsymbol{pA}} \ge c_{\boldsymbol{pB}}$ for every $\boldsymbol{p} \ge \boldsymbol{0}$ and, hence, $c_{\boldsymbol{A}} \ge c_{\boldsymbol{B}}$.

The concentration function has also the zero entry property:

$$\text{If } \boldsymbol{B} = \begin{pmatrix} \boldsymbol{A} \\ 0 \end{pmatrix}, \quad \text{then} \quad c_{\boldsymbol{A}}(\boldsymbol{y}) = c_{\boldsymbol{B}}(\boldsymbol{y}) \ \text{ for all } \ \boldsymbol{y} \in [0,1]^d.$$

The split property is satisfied since, if $\boldsymbol{B}$ is the $kn \times d$ matrix obtained from k replications of some $\boldsymbol{A} \in \mathcal{A}_d$, then $c_{\boldsymbol{B}}(\boldsymbol{y}) = k\, c_{\boldsymbol{A}}(\boldsymbol{y})$ holds for all $\boldsymbol{y} \in [0,1]^d$. To summarize:

Theorem 9.14 (Postulates) *Concentration as measured by the multivariate concentration order satisfies the postulates of anonymity, scale, size transfer, zero entry, merger, split, small entry, anti-Gibrat and continuity at entrance.*

9.8 Multivariate concentration indices

Frequently, concentration hypersurfaces will intersect. In such a case the dominance property sheds no light on which industry is more concentrated than the other. In particular, assume that each firm has positive size in at least one attribute. Then, if $n < m$, the concentration ordering will show either less concentration in the distribution with m firms, or the concentrations hypersurfaces will intersect.

In this section we turn to the construction of multidimensioned indices of concentration.

Definition 9.9 (Concentration index) *A function $\Phi : \mathcal{A}_d \to \mathbb{R}$ is a multivariate* concentration index *if it increases with the multivariate concentration order.*

It follows from Theorem 9.14 that every concentration index satisfies the postulates of anonymity, scale, size transfer, zero entry, merger, split, small entry, anti-Gibrat and continuity at entrance.

Also the postulates of normalization, continuity and cardinal split are easily formulated if the size of a firm is taken as a vector. Now a multivariate version of the famous Rosenbluth index is defined.

Definition 9.10 (Multivariate Rosenbluth index) *For $\boldsymbol{A} \in \mathcal{A}_d$,*

$$\Phi_{Ros}(\boldsymbol{A}) = \frac{d}{d+1} \frac{1}{\textit{area below } C_{\boldsymbol{A}}} \tag{9.18}$$

is the multivariate Rosenbluth index *of $\boldsymbol{A}$.*

By the area below $C_{\boldsymbol{A}}$ the $(d+1)$-dimensional volume is meant of the set $\{(c_{\boldsymbol{A}}(\boldsymbol{y}), \boldsymbol{y}) \in \mathbb{R}^{d+1} : \boldsymbol{y} \in [0,1]^d\}$. In the univariate case $\Phi_{Ros}(\boldsymbol{A})$ comes out to be the usual Rosenbluth index. For general d, the index is consistent with the multivariate concentration order and therefore possesses all properties which the concentration order has. Obviously, Φ_{Ros} is continuous in the data.

Lemma 9.15 (Equal shares, replication)
(i) *Let $\boldsymbol{A}$ be a distribution matrix in $\mathcal{A}_d$ with $\boldsymbol{a}_i = \boldsymbol{a}_j$ for all $i, j = 1, \ldots, n$. Then $\Phi_{Ros}(\boldsymbol{A}) = \frac{1}{n}$.*

(ii) *Let $\boldsymbol{B}$ be the k-replication of a given matrix $\boldsymbol{A} \in \mathcal{A}_d$. Then $\Phi_{Ros}(\boldsymbol{B}) = \frac{1}{k}\Phi_{Ros}(\boldsymbol{A})$.*

Proof. (i): Let $\boldsymbol{a}_i = \boldsymbol{a}_j$ for all $i.j = 1, \ldots, n$. Then the area below the concentration hypersurface is a pyramid with base $[0,1]^d$ and height n. The pyramid has volume $n/(d+1)$. Hence the area below the concentration hypersurface equals $n/(d+1)$, and so, $R(\boldsymbol{A}) = 1/n$.
(ii): From the definition of the concentration function, we have $c_{\boldsymbol{B}}(\boldsymbol{y}) = kc_{\boldsymbol{A}}(\boldsymbol{y})$. That yields the proposition. Q.E.D.

It has been shown that, when there are n identical firms, the Rosenbluth index equals $1/n$. If every firm of an industry is split into k equal parts, then Φ_{Ros} reduces by a factor $1/k$. To conclude:

Theorem 9.16 (Postulates) *The multivariate Rosenbluth index satisfies all postulates of Theorem 9.14 and, in addition, continuity, normalization and cardinal split.*

The unidimensioned concentration of a market is often described by particular concentration rates, i.e., the aggregate shares that are held by the 3, 5 or 10 largest firms. In the multivariate case we similarly consider a portion of the concentration function instead of the whole one and define an analogon of the usual concentration rate.

Definition 9.11 (Multivariate concentration rates) *For* $\kappa \in \mathbb{N}$,

$$CR_\kappa = \begin{cases} Vol_d\{y \in [0,1]^d : c_A(y) \leq \kappa\} & \text{if } 0 < \kappa \leq n, \\ 1 & \text{if } \kappa > n, \end{cases}$$

is the multivariate concentration rate *of order* κ.

It is easily seen that for $d = 1$ and integer κ these concentration rates coincide with the usual ones. Observe that, when $d > 1$ and $c_A(y) \leq \kappa$ holds for some κ, different firms can be involved in the aggregate size vector y, for different y. The multivariate concentration rates inherit the properties of the concentration hypersurface (Theorem 9.14). They are continuous, but not normalized nor do they satisfy cardinal split.

So far, we have considered the distribution of relative sizes of the firms and measured its concentration. By similar methods it is possible to consider and measure the distribution of absolute sizes of firms. This is sometimes called the measurement of *absolute concentration* in the literature (Hannah and Kay, 1977). The above notions and results modify in an obvious way.

9.9 Notes

Since some hundred years, economists have been interested in the quantitative description and statistical estimation of economic disparity. To measure the inequality of incomes in a population of households (or individuals), Lorenz (1905), Gini (1912), Pigou (1912) and Dalton (1920) have put forth the Lorenz order and the Gini index and interpreted them in terms of income distributions and redistributions. These ideas are connected with the concept of majorization introduced by Hardy et al. (1929, 1934): For two data vectors that have the same mean, majorization is equivalent to the ordering of their Lorenz curves.

Price majorization has been introduced by several authors under different names. Kolm (1977) terms it income more equal with nonnegative prices. The same is named second degree stochastic dominance by Muliere and Scarsini (1989) and majorization by positive linear combinations by Joe and Verducci (1992). The latter address the problem of checking for price majorization among two given empirical distributions. Arnold (1987) gives an interpretation of it in terms of n individuals with money in d different currencies. Pyatt et al. (1980), Rietveld (1990), and Flückiger and Silber (1994) present attempts to aggregate the univariate Lorenz curves into a univariate quasi-Lorenz curve. In contrast to their approaches, here a Lorenz surface is proposed that reflects the whole bivariate distribution.

This chapter is mainly based on two papers, Koshevoy and Mosler (1999c,b). The notion of the Lorenz zonoid (Definition 9.2) is due to Koshevoy and Mosler (1996). Hildenbrand (1981) has used a function of the form (9.13) for an aggregate production function. Observe that Definition 9.3 with arbitrary vectors $\boldsymbol{p} \in \mathbb{R}^d$ instead of nonnegative ones yields the lift zonoid scaled order, which is treated in detail in Section 8.6. This order implies price majorization, but not vice versa, as can be seen from simple examples. The lift zonoid scaled order has been used by Kolm (1977) in an economic context. Note that the definitions of the inverse generalized Lorenz function and of the price majorization order carry over to distributions on $\mathbb{R}^d_+$ that have no finite mean.

The multivariate concentration order and concentration rates have been applied to data from the German mineral oil market; see Koshevoy and Mosler (1999b).

In practical applications these measures have to be combined and contrasted with other, univariate and multivariate methods of concentration measurement. In particular, each dimension should be first investigated separately. Then several multivariate indices can be employed and checked whether they behave similarly with the data. Finally it must be stressed that the whole approach, following the tradition of the statistical measurement of concentration, does not go beyond a description of the firm size distribution and its changes, under the special view of industrial concentration. No attempt is made to model the causes and dynamics of the concentration process.

Another field of application is the measurement of α-diversity in ecology. This problem corresponds to the measurement of industrial concentration so that similar orders and indices can be used. See Mosler (2001), where also a multivariate version of the classical Herfindahl index of concentration is introduced.

Appendix A: Basic notions

This appendix collects some basic notations and facts from Euclidean spaces, measure theory, and convex analysis which are used in the book.

A 1 Points and sets in Euclidean space

A point $\boldsymbol{x} = (x_1, \ldots, x_d)$ in Euclidean space $\mathbb{R}^d$ is a row, $\boldsymbol{x}'$ denotes the transpose of $\boldsymbol{x}$ and $||\boldsymbol{x}|| = \sqrt{\sum x_i^2}$ the Euclidean norm. The projection of $\boldsymbol{x}$ to the coordinates with indices in $J = \{i_1, \ldots, i_k\} \subset \{1, \ldots, d\}$ is denoted by $pr_J(x_1, \ldots, x_d) = \boldsymbol{x}_J = (x_{i_1}, \ldots, x_{i_k})$.

For two points $\boldsymbol{x}$ and $\boldsymbol{y} \in \mathbb{R}^d$, $\langle \boldsymbol{x}, \boldsymbol{y} \rangle$ is the inner product, $\boldsymbol{x} \cdot \boldsymbol{y} = (x_1 y_1, \ldots, x_d y_d)$ is the componentwise product and $\boldsymbol{x}/\boldsymbol{y} = (x_1/y_1, \ldots, x_d/y_d)$ is the componentwise quotient of the points. The componentwise maximum and minimum are $\boldsymbol{x} \vee \boldsymbol{y} = (\max\{x_1, y_1\}, \ldots, \max\{x_d, y_d\})$ and $\boldsymbol{x} \wedge \boldsymbol{y} = (\min\{x_1, y_1\}, \ldots, \min\{x_d, y_d\})$, respectively. The componentwise order between $\boldsymbol{x}$ and $\boldsymbol{y}$ in $\mathbb{R}^d$ is indicated by $\boldsymbol{x} \leq \boldsymbol{y}$. For any ordering the words "decreasing" and "increasing" are meant in the weak sense.

The indicator function of a set S is signified by 1_S or $1(\cdot|S)$, the number of its elements by $\# S$ or $|S|$.

$\mathbb{R}^d_+$ ($\mathbb{R}^d_-$) is the subset of points in $\mathbb{R}^d$ that have nonnegative (nonpositive) coordinates, and S^{d-1} denotes the unit sphere in $\mathbb{R}^d$.

In $\mathbb{R}^d$, $\mathbf{0}$ is the origin, $\mathbf{1} = (1, \ldots, 1)$, $B(\alpha)$ the α-ball around the origin. $[\mathbf{0}, \boldsymbol{x}] = \{\boldsymbol{y} : \boldsymbol{y} = \alpha \cdot \boldsymbol{x}, \alpha \in [0,1]\}$ stands for the *line segment* from the origin to $\boldsymbol{x}$.

The set $\boldsymbol{y} + \mathbb{R}^d_- = \{\boldsymbol{x} \in \mathbb{R}^d : \boldsymbol{x} \leq \boldsymbol{y}\}$ is the lower orthant of $\boldsymbol{y} \in \mathbb{R}^d$. For $p_0 \in \mathbb{R}$ and $\boldsymbol{p} \in \mathbb{R}^d$, $H_{p_0,\boldsymbol{p}}$ means the halfspace $\{\boldsymbol{x} \in \mathbb{R}^d : p_0 + \langle \boldsymbol{p}, \boldsymbol{x} \rangle \geq 0\}$

in $\mathbb{R}^d$. For $S \subset \mathbb{R}^d$ and $\beta \in \mathbb{R}$, define $\beta \cdot S = \{\beta \cdot \boldsymbol{x} : \boldsymbol{x} \in S\}$. For two sets C and D, $C + D = \{\boldsymbol{x} + \boldsymbol{y} : \boldsymbol{x} \in C, \boldsymbol{y} \in D\}$ denotes their *Minkowski sum* and

$$\delta_H(C, D) = \inf\{\delta : C \subset D + B(\delta), D \subset C + B(\delta), \delta > 0\}$$

their *Hausdorff distance.*

The space $\mathbb{R}^d$, the sphere S^{d-1}, and any other subset S of $\mathbb{R}^d$ are endowed with the usual topology induced by the Euclidean norm. The *closure, interior,* and *boundary* of S are denoted by clS, $intS$, and ∂S, respectively. Convergence of sets is in the Hausdorff distance and indicated by $\xrightarrow{H}$.

The components of a vector $\boldsymbol{z} \in \mathbb{R}^{d+1}$ are indexed from 0 to d, $\boldsymbol{z} = (z_0, z_1, \dots, z_d)$. The hyperplane at a fixed z_0 is written $G_\alpha = \{\boldsymbol{z} \in \mathbb{R}^{d+1} : z_0 = \alpha\}$ for $\alpha \in \mathbb{R}$. Similarly, $G_{\leq\alpha} = \{\boldsymbol{z} \in \mathbb{R}^{d+1} : z_0 \leq \alpha\}$. For a set S in $\mathbb{R}^{d+1}$ and $\alpha \in \mathbb{R}$, $\text{proj}_\alpha(S) \in \mathbb{R}^d$ stands for the projection $pr_{\{1,\dots,d\}}$ of the set $S \cap G_\alpha$ to the last d coordinates.

$\mathbb{R}^{n\times d}$ is the space of $n \times d$ matrices. $[a_{ik}]$ denotes the matrix with general element a_{ik} and $[\boldsymbol{a}_1, \dots, \boldsymbol{a}_n]$ stands for the $n \times d$ matrix with rows $\boldsymbol{a}_1, \dots, \boldsymbol{a}_n \in \mathbb{R}^d$.

A 2 Measure and probability

$\mathcal{B}^d$ denotes the class of Borel sets in $\mathbb{R}^d$, $\mathcal{M}$ the set of all nonnegative measures defined on $\mathcal{B}^d$, $\mathcal{P}$ the set of all probability measures on $\mathcal{B}^d$. Let $\mathcal{M}_k$ be the class of nonnegative measures μ on $(\mathbb{R}^d, \mathcal{B}^d)$ that have finite moments up to order k, $k \in \mathbb{N}$. Define $\alpha(\mu) = \int_{\mathbb{R}^d} \mu(d\boldsymbol{x})$, $\boldsymbol{\epsilon}(\mu) = \int_{\mathbb{R}^d} \boldsymbol{x}\mu(d\boldsymbol{x})$, and let $conv(\mu)$ denote the convex hull of the support of μ. $\mathcal{P}_k$ is the class of probability measures in $\mathcal{M}_k$ and $\mathcal{P}_{1*} \subset \mathcal{P}_1$ is the class of probability measures that have finite nonnull expectation in each component. $\mathcal{P}_{1+} \subset \mathcal{P}_{1*}$ is the subclass that has, in addition, support in $\mathbb{R}^d_+$.

The *Dirac measure* at some $\boldsymbol{b} \in \mathbb{R}^d$ is denoted by $\delta_{\boldsymbol{b}}$ and defined as $\delta_{\boldsymbol{b}}(B) = 1$ if $\boldsymbol{b} \in B$, $\delta_{\boldsymbol{b}}(B) = 0$ otherwise.

An *empirical measure* is a measure that gives equal mass to a finite number of, not necessarily different, points. An *empirical probability measure* is notated

$$\mu_{\boldsymbol{S}} = \sum_{i=1}^{n} \delta_{\boldsymbol{s}_i},$$

where $\boldsymbol{S} = [\boldsymbol{s}_1, \dots, \boldsymbol{s}_n]$. If the $\boldsymbol{s}_i$ are all different points, we notate $\mu_{\boldsymbol{S}} = \mu_S$ with $S = \{\boldsymbol{s}_1, \dots, \boldsymbol{s}_n\}$ as well.

Let $\mathcal{M}^n$ denote those empirical measures in $\mathcal{M}$ which give equal mass to n points in $\mathbb{R}^d$, and let $\mathcal{P}^n = \mathcal{M}^n \cap \mathcal{P}$ be the probability measures among them.

L-continuity of a measure refers to the Lebesgue measure. $\mathcal{M}_{cont}$ contains all L-continuous measures in $\mathcal{M}_1$, while $\mathcal{M}_{disc}$ contains those measures in $\mathcal{M}_1$ that have a density with respect to the counting measure. Let $\mathcal{M}_{comp}$ be the set of measures in $\mathcal{M}_1$ that have a compact support and define $\mathcal{M}_{cc} = \mathcal{M}_{cont} \cap \mathcal{M}_{comp}$. The sets $\mathcal{P}_{cont}$, $\mathcal{P}_{disc}$, $\mathcal{P}_{comp}$, and $\mathcal{P}_{cc}$ are similarly defined.

A measure μ on $\mathcal{B}^d$ has *distribution function* $F(\boldsymbol{x}) = \mu(\boldsymbol{x} - \mathbb{R}_+), \boldsymbol{x} \in \mathbb{R}^d$. We write also $F \in \mathcal{P}_k$ if $\mu \in \mathcal{P}_k$, et cetera. If $\boldsymbol{X}$ is distributed as μ, $\mu_{\boldsymbol{p}}$ indicates the distribution of the random variable $\langle \boldsymbol{X}, \boldsymbol{p} \rangle$. For random vectors $\boldsymbol{X}$ and $\boldsymbol{Y}$, $\boldsymbol{X} =_{st} \boldsymbol{Y}$ means equality in distribution.

The *Lebesgue measure* λ_{sph} *on the sphere* S^{d-1} is constructed as follows: For any $R \in \mathcal{B}^d \cap S^{d-1}$ consider the set $C(R) = \{\beta \boldsymbol{p} : \boldsymbol{p} \in R, \beta \in \mathbb{R}\}$ and let

$$\lambda_{sph}(R) = \lim_{\varepsilon \to 0} \frac{1}{\varepsilon} \lambda \left(C(R) \cap (B(1+\varepsilon) \setminus B(1)) \right),$$

where λ denotes the Lebesgue measure on $\mathbb{R}^d$. Thus, $\lambda_{sph}(S^{d-1})$ equals the surface of the sphere.

A 3 Convex analysis

Consider a set S in $\mathbb{R}^d$.

$$aff\, S = \left\{ \sum_{i=1}^{n} \alpha_i \boldsymbol{x}_i : \boldsymbol{x}_i \in S, \alpha_i \in \mathbb{R}, n \in \mathbb{N} \right\}$$

is the *affine hull* of S. For any $\boldsymbol{x}_0 \in aff\, S$, $lin S = \{\boldsymbol{x} - \boldsymbol{x}_0 : \boldsymbol{x} \in aff\, S\}$ is a linear space that does not depend on $\boldsymbol{x}_0$; its dimension is mentioned as the *affine dimension* of the set S. Endow $aff\, S$ with the Euclidean norm and notate $rint S$ and $r\partial S$ for the *relative interior* and the *relative boundary* of S as subsets of $aff\, S$, respectively. The set

$$conv S = \left\{ \sum_{i=1}^{n} \alpha_i \boldsymbol{x}_i : \boldsymbol{x}_i \in S, \alpha_i \in [0,1], \sum_{i=1}^{n} \alpha_i = 1, n \in \mathbb{N} \right\}$$

denotes the *convex hull* of S. The convex hull of a finite set is a *polyhedron*. If the convex hull of $S = \{\boldsymbol{x}_1, \ldots, \boldsymbol{x}_{k+1}\}$ has the affine dimension k, the points $\boldsymbol{x}_1, \ldots, \boldsymbol{x}_{k+1}$ are said to be in *general position* and $conv S$ is called a *k-simplex*. Let $\Delta(\boldsymbol{x}_1, \ldots, \boldsymbol{x}_{k+1})$ denote the k-simplex generated by the points

$x_1, \ldots, x_{k+1}$, which form its *vertices*. For $k = 2$ a *(line) segment* is obtained, that is the convex hull of two distinct points.

A set $K \subset \mathbb{R}^d$ is *convex* if it equals its convex hull. A function $f : \mathbb{R}^d \to \mathbb{R}$ is *convex* if its epigraph $\{(x_0, x) \in \mathbb{R}^{d+1} : x_0 \geq f(x), x \in \mathbb{R}^d\}$ is a convex set. Let $\mathcal{R}$ denote the set of nonempty convex sets in $\mathbb{R}^d$. For any $K \in \mathcal{R}$ the *support function* $h(K, \cdot) : \mathbb{R}^d \to \mathbb{R} \cup \{\infty\}$ of K is given by

$$h(K, p) = \sup\{\langle x, p\rangle : x \in K\}, \quad p \in \mathbb{R}^d .$$

For example, the support function of the δ-ball $B_\delta = \{x \in \mathbb{R}^d : ||x|| \leq \delta\}$ is $h(B_\delta, p) = \sup_{||x|| \leq \delta} \langle x, p\rangle = \delta ||p||$, $p \in \mathbb{R}^d$. The following list collects basic properties of support functions. (For a proof of these and other properties, the reader is referred e.g. to Bonnesen and Fenchel (1934) and Rockafellar (1970).) For any $K, R \in \mathcal{R}$ holds:

- The function $p \mapsto h(K, p)$, $p \in \mathbb{R}^d$, is convex and positive homogeneous. (A function $f : \mathbb{R}^d \to \mathbb{R}$ is *positive homogeneous* if $f(\lambda x) = \lambda f(x)$ holds for any $x \in \mathbb{R}^d$ and $\lambda > 0$.)
- $h(K, \cdot) \geq 0$ if and only if $\mathbf{0} \in clK$.
- $h(\beta K + \gamma R, \cdot) = \beta h(K, \cdot) + \gamma h(R, \cdot)$ holds for any $K, R \in \mathcal{R}$ and $\beta, \gamma \in \mathbb{R}$.
- $clK \subset clR$ if and only if $h(K, p) \leq h(R, p)$ for all $p \in S^{d-1}$.
- $clK = clR$ if and only if $h(K, p) = h(R, p)$ for all $p \in S^{d-1}$.

Observe that a closed convex set is uniquely determined by its support function. Moreover, as the support function is positive homogeneous in p, a closed convex set is uniquely determined by the restriction of its support function to the sphere.

For any fixed p, the function $h(\cdot, p) : \mathcal{R} \to \mathbb{R}$ is monotone increasing on sets. Further, it is linear, that is homogeneous w.r.t. scalar multiplication and additive w.r.t. Minkowski addition of sets. Note that the support function of a bounded set is finite and continuous. If we restrict ourselves to compact sets in $\mathcal{R}$, the function $h(\cdot, p)$ is also continuous in the Hausdorff distance.

Moreover, there exists a one-to-one relation between convex compact sets and convex functions which are finite and positive homogeneous: Let $\mathcal{K}$ denote the class of nonempty convex compacts.

- For $K, R \in \mathcal{K}$ the Hausdorff distance is equal to

$$\delta_H(K, R) = \max_{p \in S^{d-1}} |h(K, p) - h(R, p)| .$$

- Let $K \in \mathcal{K}$ and $K^n \in \mathcal{K}$ for all n. The sequence $(K^n)_{n \in \mathbb{N}}$ converges, in the Hausdorff distance, to K if and only if the sequence of support functions $h(K^n, \cdot)$ converges uniformly on S^{d-1} to $h(K, \cdot)$.

- Any positive homogeneous finite convex function $\mathbb{R}^d \to \mathbb{R}$ is the support function of a set $K \in \mathcal{K}$, and any such set has a support function which is finite, convex and positive homogeneous.

Let $K \in \mathcal{R}$. A point $\boldsymbol{z} \in clK$ is an *extreme point* of K if $\boldsymbol{z} = \alpha \boldsymbol{x} + (1 - \alpha)\boldsymbol{y}$ with some $\boldsymbol{x}$ and $\boldsymbol{y} \in K$ and $\alpha \in [0, 1]$ implies $\boldsymbol{z} = \boldsymbol{x} = \boldsymbol{y}$. Given $\boldsymbol{p} \in S^{d-1}$, a point $\boldsymbol{z} \in \mathbb{R}^d$ *supports* K in direction $\boldsymbol{p}$ if $\boldsymbol{z} \in clK$ and $\langle \boldsymbol{z}, \boldsymbol{p} \rangle = h(K, \boldsymbol{p})$. Obviously, if $\boldsymbol{z}$ is an extreme point of K it supports K in some direction $\boldsymbol{p}$. A point $\boldsymbol{z}$ is *extreme in direction* $\boldsymbol{p}$ if and only if $\boldsymbol{z}$ is extreme among all points that support K in direction $\boldsymbol{p}$. For almost all $\boldsymbol{p} \in S^{d-1}$ there is exactly one extreme point which supports K in direction $\boldsymbol{p}$. More precisely, it can be shown (Reidemeister, 1921) that the set of directions in which more than one extreme point exists has Lebesgue measure zero on the sphere.

Given some $S \subset \mathbb{R}^d$, $coneS = \{\lambda \boldsymbol{x} : \boldsymbol{x} \in S, \lambda > 0\}$ is the *cone generated by* S. A set C is a *cone* if $C = coneC$ holds. A *simplicial cone* is a cone generated by a finite set. A *solid cone* is a cone which has a nonempty interior.

A finite Minkowski sum of line segments in $\mathbb{R}^d$ is called a *zonotope.* A *zonoid* is a limit, in the Hausdorff sense, of zonotopes. Obviously, every zonotope is a polytope. Its boundary is composed of *faces*, that is polytopes having affine dimensions smaller than d. A face is named *facet* if it has affine dimension equal to $d - 1$. For a set S in $\mathbb{R}^d$ the following restrictions are equivalent (Bolker (1969); Schneider and Weil (1983)):

- S is a zonotope.

- S is a polytope and each facet of S is zonotope.

- S is a polytope and each two-dimensional face of S is centrally symmetric.

- S is a linear image of the unit cube in $\mathbb{R}^k$ for some k.

For a set $S \subset \mathbb{R}^d$ having affine dimension l, $\text{vol}_k S$ signifies its k-dimensional volume if $k = l$; otherwise, $\text{vol}_k S = 0$ if $k \neq l$. The volume of a k-simplex is

$$\text{vol}_d \Delta(\boldsymbol{x}_1, \dots, \boldsymbol{x}_{d+1}) = \frac{1}{d!} |\det[(1, \boldsymbol{x}_1), \dots, (1, \boldsymbol{x}_{d+1})]| \quad \text{if } k = d. \qquad (19)$$

The volume is

$$\mathrm{vol}_k \Delta(\boldsymbol{x}_1, \ldots, \boldsymbol{x}_{k+1}) = \frac{1}{k!} |\det[(1, \boldsymbol{x}_1), \ldots, (1, \boldsymbol{x}_{k+1}), (0, \boldsymbol{b}_1), \ldots, (0, \boldsymbol{b}_{d-k})]| \tag{20}$$

if $k < d$. Here, $\boldsymbol{b}_1, \ldots, \boldsymbol{b}_{d-k}$ denote unit vectors that are orthogonal to the vectors $\boldsymbol{x}_2 - \boldsymbol{x}_1, \ldots, \boldsymbol{x}_{k+1} - \boldsymbol{x}_1$.

A *random compact set* C in $\mathbb{R}^d$ is a measurable map from a probability space $(\Omega, \mathcal{B}, P)$ to the space of nonempty, compact subsets of $\mathbb{R}^d$ endowed with the Hausdorff distance and the pertaining Borel σ-algebra. For details see, e.g., Matheron (1975) and Kendall (1974). The expectation of a random compact set forms a set which is defined by

$$\mathrm{E}(C) = \{\mathrm{E}(\boldsymbol{X}) \;:\; \boldsymbol{X} : \Omega \to \mathbb{R}^d \text{ measurable with } \boldsymbol{X}(\omega) \in C(\omega) \;\; P\text{-a.s.}\}. \tag{21}$$

Equivalently, $\mathrm{E}(C)$ can be defined via support functions,

$$h(\mathrm{E}(C), \boldsymbol{p}) = \mathrm{E}(h(C, \boldsymbol{p})), \quad \boldsymbol{p} \in \mathbb{R}^d. \tag{22}$$

This set-valued expectation has appeared in different connections, see, for example, Weil and Wieacker (1993). From the linearity of support functions follows that the set-valued expectation is a linear operator,

$$\mathrm{E}(\beta C_1 + \gamma C_2) = \beta \mathrm{E}(C_1) + \gamma \mathrm{E}(C_2) \quad \text{for any } \beta, \gamma \in \mathbb{R}.$$

The norm of C,

$$||C|| = \max\{||\boldsymbol{x}|| \;:\; \boldsymbol{x} \in C\} = \delta_H(\{\boldsymbol{0}\}, C),$$

is a random variable in $\mathbb{R}$. $\mathrm{E}(C)$ is a compact set if and only if its norm has a finite expected value.

Appendix B: Lift zonoids of bivariate normals

Consider a bivariate normal distribution $N(\mathbf{a}, \Sigma)$ having the expectation vector $\mathbf{a} = (a_1, a_2)$ and the covariance matrix

$$\Sigma = \begin{pmatrix} \sigma_1^2 & \rho\sigma_1\sigma_2 \\ \rho\sigma_1\sigma_2 & \sigma_2^2 \end{pmatrix}.$$

Its lift zonoid consists of all points (z_0, z_1, z_2) in $\mathbb{R}^3$ which satisfy $0 \leq z_0 \leq 1$ and

$$\frac{z_1 - z_0 a_1}{\sigma_1^2(1-\rho^2)} + \frac{z_2 - z_0 a_2}{\sigma_2^2(1-\rho^2)} - 2\frac{\rho(z_1 - z_0 a_1)(z_2 - z_0 a_2)}{\sigma_1\sigma_2(1-\rho^2)} \leq \frac{1}{2\pi} e^{-u_{z_0}^2},$$

where u_α is the α-quantile of the standard normal distribution, $u_0 = -\infty$, $u_1 = \infty$. Observe that for any $z_0 \in]0,1[$ the slice of $\widehat{Z}(\mu)$ at z_0 is an ellipsis.

In this appendix we depict a number of lift zonoids of bivariate normal distributions that differ in location, scale and correlation. The pictures give a rough impression of their size and form. They provide also an example of misleading visual perception: The lift zonoids look to have two peaks, but in fact they have not, as it follows immediately from the unboundedness of the distributions' support. What is apparent from the pictures is that the gradient appoaches 0 very slowly as z_0 goes to 0 or 1.

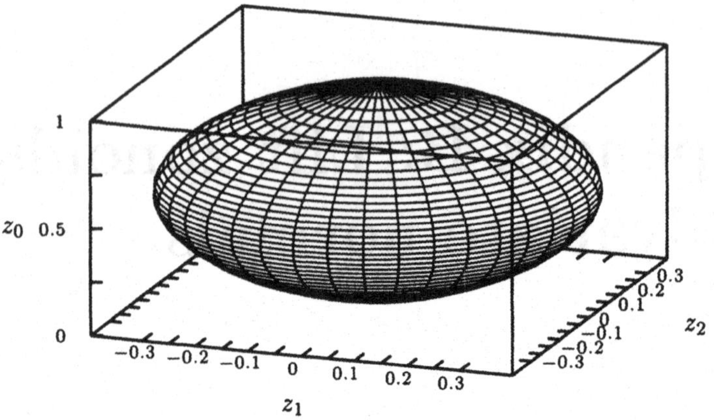

FIGURE A.1: Bivariate normal lift zonoid with $a_1 = a_2 = 0$, $\sigma_1^2 = \sigma_2^2 = 1$, $\rho = 0$.

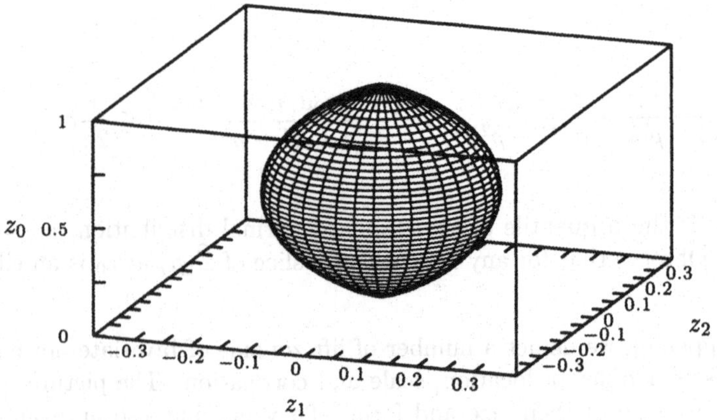

FIGURE A.2: Bivariate normal lift zonoid with $a_1 = a_2 = 0$, $\sigma_1^2 = \sigma_2^2 = 0.5$, $\rho = 0$.

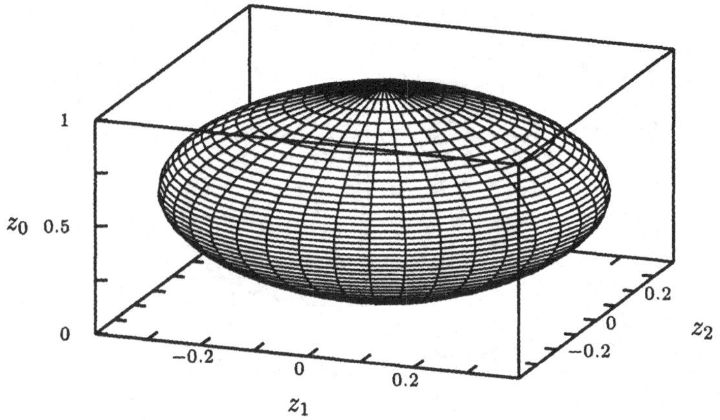

FIGURE A.3: Bivariate normal lift zonoid with
$a_1 = a_2 = 0$, $\sigma_1^2 = \sigma_2^2 = 1$, $\rho = 0$.

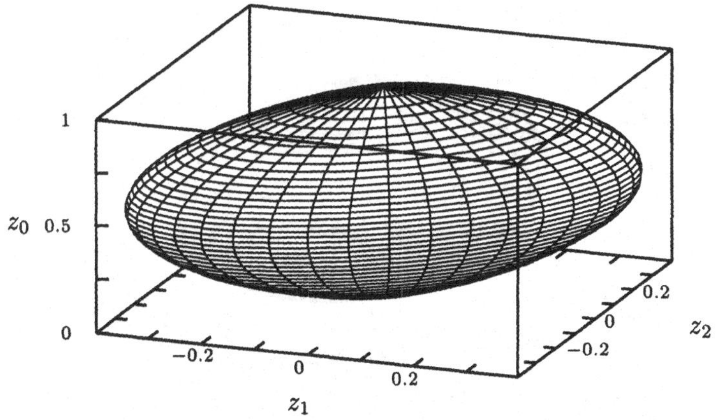

FIGURE A.4: Bivariate normal lift zonoid with
$a_1 = a_2 = 0$, $\sigma_1^2 = \sigma_2^2 = 1$, $\rho = 0.5$.

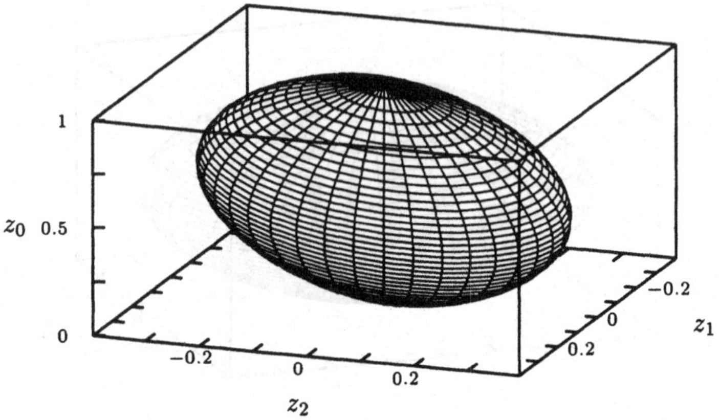

FIGURE A.5: Bivariate normal lift zonoid (rotated) with
$a_1 = a_2 = 0$, $\sigma_1^2 = \sigma_2^2 = 1$, $\rho = 0.5$.

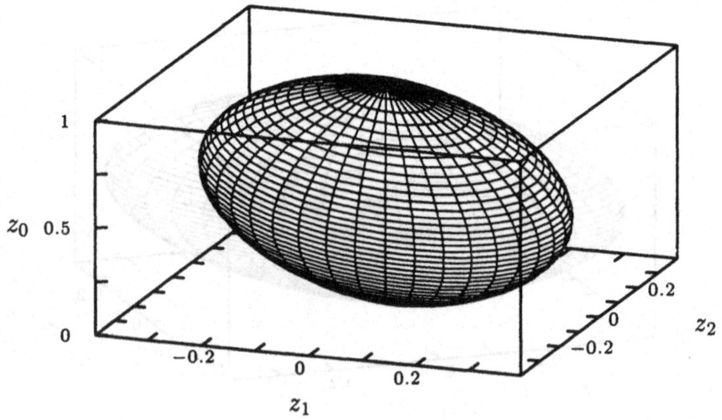

FIGURE A.6: Bivariate normal lift zonoid with
$a_1 = a_2 = 0$, $\sigma_1^2 = \sigma_2^2 = 1$, $\rho = -0.5$.

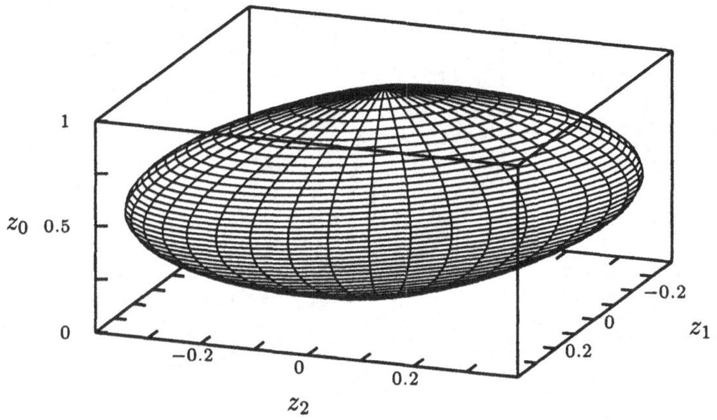

FIGURE A.7: Bivariate normal lift zonoid (rotated) with
$a_1 = a_2 = 0$, $\sigma_1^2 = \sigma_2^2 = 1$, $\rho = -0.5$.

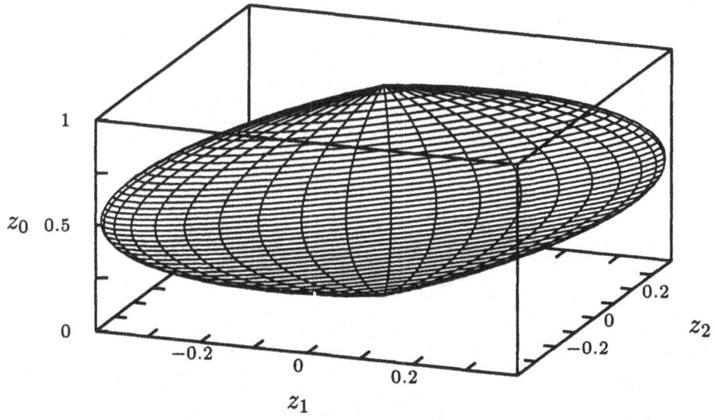

FIGURE A.8: Bivariate normal lift zonoid with
$a_1 = a_2 = 0$, $\sigma_1^2 = \sigma_2^2 = 1$, $\rho = 0.9$.

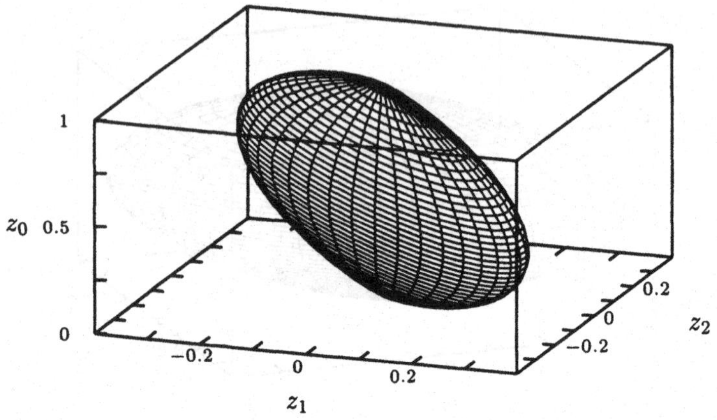

FIGURE A.9: Bivariate normal lift zonoid with
$a_1 = a_2 = 0\ ,\ \sigma_1^2 = \sigma_2^2 = 1\ ,\ \rho = -0.9\,.$

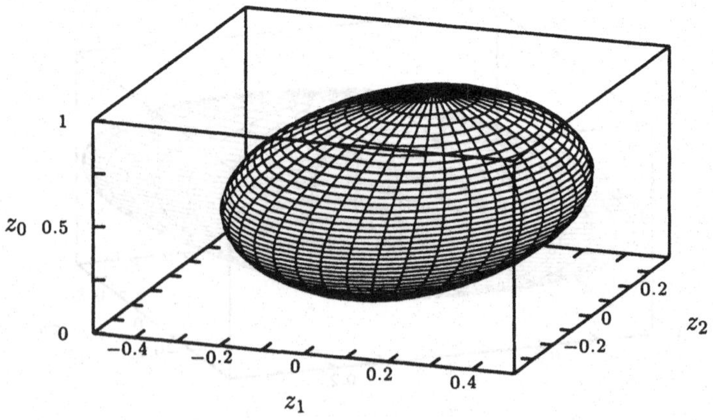

FIGURE A.10: Bivariate normal lift zonoid with
$a_1 = 0.125\ ,\ a_2 = 0\ ,\ \sigma_1^2 = \sigma_2^2 = 1\ ,\ \rho = 0\,.$

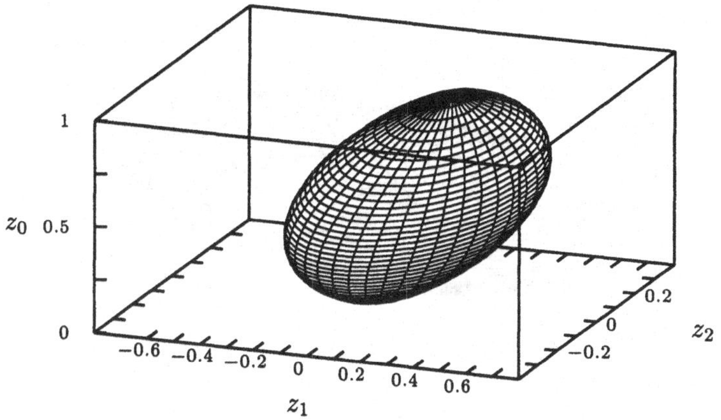

FIGURE A.11: Bivariate normal lift zonoid with
$a_1 = 0.25$, $a_2 = 0$, $\sigma_1^2 = \sigma_2^2 = 1$, $\rho = 0$.

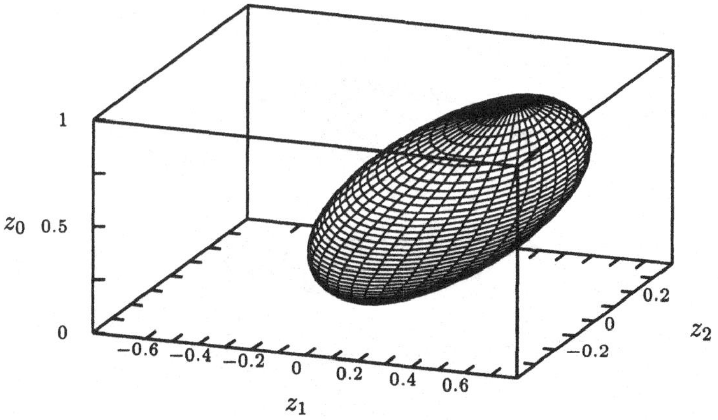

FIGURE A.12: Bivariate normal lift zonoid with
$a_1 = 0.5$, $a_2 = 0$, $\sigma_1^2 = \sigma_2^2 = 1$, $\rho = 0$.

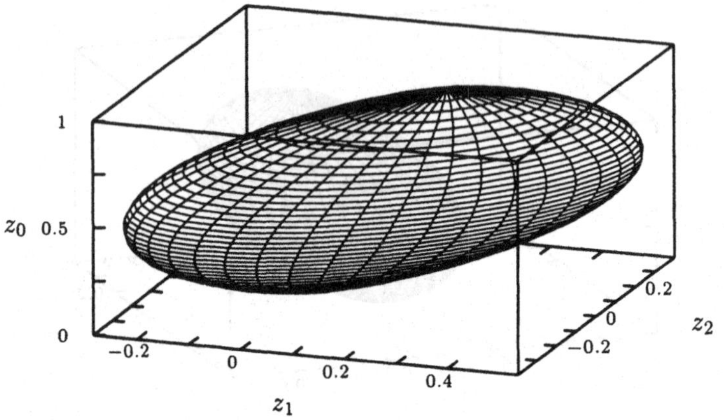

FIGURE A.13: Bivariate normal lift zonoid with $a_1 = 0.25$, $a_2 = 0$, $\sigma_1^2 = \sigma_2^2 = 1$, $\rho = 0.5$.

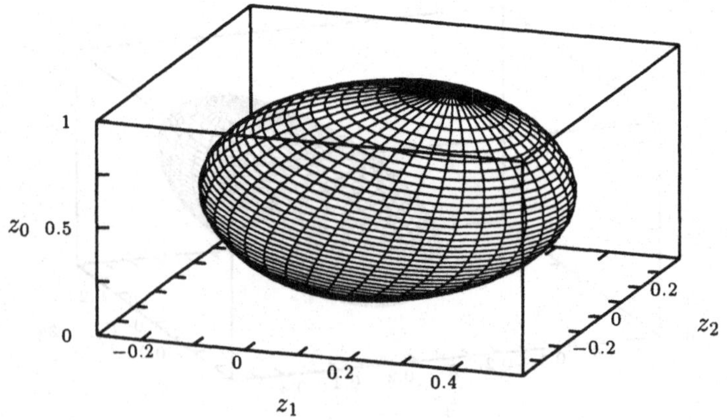

FIGURE A.14: Bivariate normal lift zonoid with $a_1 = 0.25$, $a_2 = 0$, $\sigma_1^2 = \sigma_2^2 = 1$, $\rho = -0.5$.

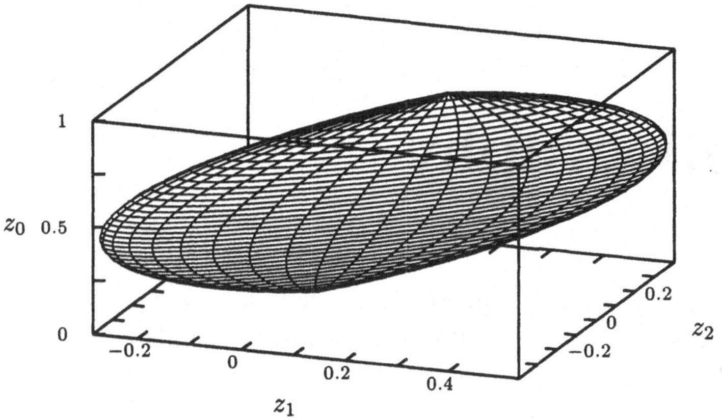

FIGURE A.15: Bivariate normal lift zonoid with
$a_1 = 0.25$, $a_2 = 0$, $\sigma_1^2 = \sigma_2^2 = 1$, $\rho = 0.9$.

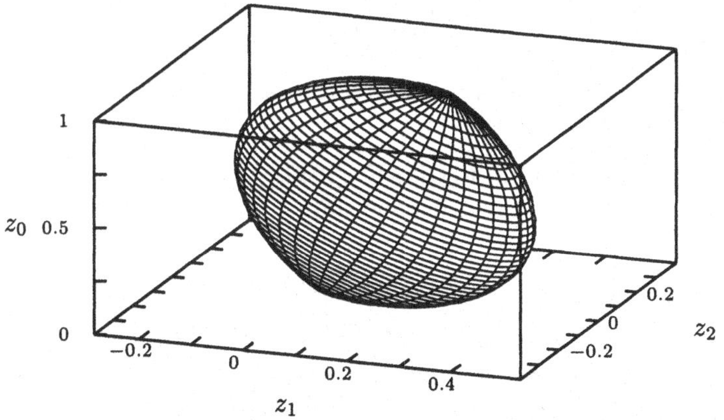

FIGURE A.16: Bivariate normal lift zonoid with
$a_1 = 0.25$, $a_2 = 0$, $\sigma_1^2 = \sigma_2^2 = 1$, $\rho = -0.9$.

Bibliography

ABDOUS, B. and REMILLARD, B. (1995). Relating quantiles and expectiles under weighted-symmetry. *Annals of the Institute of Statistical Mathematics* **47**, 371–384.

ADELMAN, M. (1951). The measurement of industrial concentration. *Review of Economics and Statistics* **33**, 269–296.

ADELMAN, M. (1969). Comment on the "H" concentration measure as a numbers equivalent. *Review of Economics and Statistics* .

ARNOLD, B. (1987). *Majorization and the Lorenz Order: A Brief Introduction.* Springer, Berlin.

ARTSTEIN, Z. and VITALE, R. A. (1975). A strong law of large numbers for random compact sets. *Annals of Probability* **3**, 879–882.

ATKINSON, A. (1970). On the measurement of inequality. *Journal of Economic Theory* **2**, 244–263.

ATKINSON, A. and BOURGUIGNON, F. (1982). The comparison of multidimensioned distributions of economic status. *Review of Economic Studies* **49**, 183–201.

ATKINSON, A. and BOURGUIGNON, F. (1989). The design of direct taxation and family benefits. *Journal of Public Economics* **41**, 3–29.

AVEROUS, J. and MESTE, M. (1990). Location, skewness and tailweight in $\mathbb{L}_S$ sense: A coherent approach. *Statistics* **21**, 57–74.

AVEROUS, J. and MESTE, M. (1994). Multivariate kurtosis in L_1-sense. *Statistics and Probability Letters* **19**, 281–284.

AVEROUS, J. and MESTE, M. (1997a). Median balls: An extension of the interquantile intervals to multivariate distributions. *Journal of Multivariate Analysis* **63**, 222–241.

AVEROUS, J. and MESTE, M. (1997b). Skewness for multivariate distributions: Two approaches. *Annals of Statistics* **25**, 1984–1997.

BARNETT, V. (1976). The ordering of multivariate data. *Journal of the Royal Statistical Society, Series A* **139**, 318–352. With discussion.

BEBBINGTON, A. C. (1978). A method of bivariate trimming for robust estimation of the correlation coefficient. *Applied Statistics* **27**, 221–226.

BENTLEY, J. L. and FRIEDMAN, J. H. (1975). Fast algorithms for constructing minimal spanning trees in coordinate spaces. Technical report, Stanford Linear Accelator Report (SLAC) PUB-1665.

BILLINGSLEY, P. (1979). *Probability and Measure.* John Wiley & Sons, New York.

BJÖRNER, A., VERGNAS, M. L., STURMFELS, B., WHITE, N. and ZIEGLER, G. M. (1999). *Oriented Matroids.* Cambridge University Press, Cambridge, 2nd ed.

BOKOWSKI, J. (1993). Oriented Matroids. In P. Gruber and J. Wills, eds., *Handbook of Convex Geometry*, 555–602. North-Holland, Amsterdam.

BOLKER, E. (1969). A class of convex bodies. *Transactions of the American Mathematical Society* **145**, 323–346.

BONNESEN, J. and FENCHEL, W. (1934). *Theorie der konvexen Körper.* Springer, Berlin.

BOX, G. E. P. (1949). A general distribution theory for a class of likelihood criteria. *Biometrika* **36**, 317–346.

BRECKLING, J. and CHAMBERS, R. (1988). M-quantiles. *Biometrika* **75**, 761–771.

BROWN, B. and HETTMANSPERGER, T. (1987). Affine invariant rank methods in the bivariate location model. *Journal of the Royal Statistical Society, Series B* **49**, 301–310.

BROWN, B. M., HETTMANSPERGER, T., NYBLOM, J. and OJA, H. (1992a). On certain bivariate sign tests and medians. *Journal of the American Statistical Association* **87**, 127–135.

BROWN, B. M., HETTMANSPERGER, T. P., NYBLOM, J. and OJA, H. (1992b). On certain bivariate sign tests and medians. *Journal of the American Statistical Association* **87**, 127–135.

CARRIZOSA, E. (1996). A characterization of halfspace depth. *Journal of Multivariate Analysis* **58**, 21–26.

CHAUDHURI, P. (1996). On a geometric notion of quantiles for multivariate data. *Journal of the American Statistical Association* **91**, 862–872.

CHEN, Z. (1995). Bounds for the breakdown point of the simplicial median. *Journal of Multivariate Analysis* **55**, 1–13.

CHOQUET, G. (1968). Measures coniques et affines invariants par isométries. Zonoformes, zonoèdres et fonctions de type negatif. *C. R. Acad. Sci. Paris* **266**, 619–621.

CRAMER, K. (2002). Multivariate Ausreißer und Datentiefen. Mimeo, Fakultät für Wirtschafts- und Sozialwissenschaften, Universität zu Köln.

DALL'AGLIO, M. and SCARSINI, M. (2000). Zonoids, linear dependence, and size-based distributions on the simplex. Submitted.

DALL'AGLIO, M. and SCARSINI, M. (2001). When Lorenz met Lyapunov. *Statistics and Probability Letters* **54**, 101–105.

DALTON, H. (1920). The measurement of the inequality of incomes. *Economic Journal* **30**, 348–361.

DHARMADHIKARI, S. and JOAG-DEV, K. (1988). *Unimodality, Convexity, and Applications*. Academic Press, Boston.

DONOHO, D. L. and GASKO, M. (1992). Breakdown properties of location estimates based on halfspace depth and projected outlyingness. *Annals of Statistics* **20**, 1803–1827.

DÜMBGEN, L. (1992). Limit theorems for the simplicial depth. *Statistics and Probability Letters* **14**, 119–128.

DYCKERHOFF, R. (1998). A comparison of multivariate rank tests based on different notions of data depth. In R. Payne and P. Lane, eds., *COMPSTAT 1998. Proceedings in Computational Statistics*, 29–30. IACR-Rothamsted, Harpenden.

DYCKERHOFF, R. (2000). Computing zonoid trimmed regions of bivariate data sets. In J. Bethlehem and P. van der Heijden, eds., *COMPSTAT 2000. Proceedings in Computational Statistics*, 295–300. Physica-Verlag, Heidelberg.

DYCKERHOFF, R. (2002). Datentiefe: Begriff, Berechnung, Tests. Mimeo, Fakultät für Wirtschafts- und Sozialwissenschaften, Universität zu Köln.

DYCKERHOFF, R., KOSHEVOY, G. and MOSLER, K. (1996). Zonoid data depth: Theory and computation. In A. Pratt, ed., *COMPSTAT 1996. Proceedings in Computational Statistics*, 235–240, Heidelberg. Physica-Verlag.

DYCKERHOFF, R. and MOSLER, K. (1997). Orthant orderings of discrete random vectors. *Journal of Statistical Planning and Inference* **62**, 193–205.

EDDY, W. F. (1985). Ordering of multivariate data. In L. Billard, ed., *Computer Science and Statistics: The Interface*, 25–30. North-Holland, Amsterdam.

EFRON, B. (1965). The convex hull of a random set of points. *Biometrika* **52**, 331–343.

ELTON, J. and HILL, T. (1992). Fusions of probability distributions. *Annals of Probability* **20**, 421–454.

FANG, K.-T., KOTZ, S. and NG, K.-W. (1990). *Symmetric Multivariate and Related Distributions*. Chapman and Hall, London.

FLÜCKIGER, Y. and SILBER, J. (1994). The Gini index and the measurement of multidimensional inequality. *Oxford Bulletin of Economics and Statistics* **56**, 225–228.

FRIEDMAN, J. H. and RAFSKY, L. C. (1979). Multivariate generalizations of the Wald-Wolfowitz and Smirnov two-sample test. *Annals of Statistics* **7**, 697–717.

GINI, C. (1912). *Variabilità e Mutabilità: Contributo allo studio delle distribuzioni e delle relazioni statistiche (Variability and Changeability: Contribution to the Study of Distributions and Statistical Relations)*. Cuppini, Bologna.

GIORGI, G. M. (1990). Bibliographic portrait of the Gini concentration ratio. *Metron* **48**, 183–221.

GIORGI, G. M. (1992). *Il rapporto di concentrazione di Gini*. Libreria Editrice Ticci, Siena.

GIOVAGNOLI, A. and WYNN, H. P. (1995). Multivariate dispersion orderings. *Statistics and Probability Letters* **22**, 325–332.

GOODEY, P. and WEIL, W. (1993). Zonoids and generalizations. In P. Gruber and J. Wills, eds., *Handbook of Convex Geometry*, 1297–1326. North-Holland, Amsterdam.

HALL, M. and TIDEMAN, N. (1967). Measures of concentration. *Journal of the American Statistical Association* **62**, 162–168.

HALMOS, P. R. (1948). The range of a vector measure. *Bulletin of the American Mathematical Society* **54**, 416–421.

HANNAH, L. and KAY, J. A. (1977). *Concentration in Modern Industry*. Macmillan Press, London.

HARDY, A. and RASSON, J.-P. (1982). Une nouvelle approche des problèmes de classification automatique. *Statistique et Analyse des Données* **7**, 41–56.

HARDY, G. H., LITTLEWOOD, J. E. and POLYA, G. (1929). Some simple inequalities satisfied by convex functions. *Messenger of Mathematics* **58**, 145–152.

HARDY, G. H., LITTLEWOOD, J. E. and POLYA, G. (1934). *Inequalities*. Cambridge University Press, London, 1st ed.

HE, X. and WANG, G. (1997). Convergence of depth contours for multivariate datasets. *Annals of Statistics* **25**, 495–504.

HELGASON, S. (1999). *The Radon transform*. Birkhäuser, Boston, 2nd ed.

HETTMANSPERGER, T., MÖTTÖNEN, J. and OJA, H. (1998). Affine invariant multivariate rank tests for several samples. *Statistica Sinica* **8**, 785–800.

HETTMANSPERGER, T. P., MÖTTÖNEN, J. and OJA, H. (1997). Affine-invariant multivariate one-sample signed-rank tests. *Journal of the American Statistical Association* **92**, 1591–1600.

HETTMANSPERGER, T. P., NYBLOM, J. and OJA, H. (1992). On multivariate notions of sign and rank. In Y. Dodge, ed., *L_1-Statistical Analysis and Related Methods*, 267–278. North-Holland, Amsterdam.

HETTMANSPERGER, T. P., NYBLOM, J. and OJA, H. (1994). Affine invariant multivariate one-sample sign tests. *Journal of the Royal Statistical Society, Series B* **56**, 221–234.

HETTMANSPERGER, T. P. and OJA, H. (1994). Affine invariant multivariate multisample sign tests. *Journal of the Royal Statistical Society, Series B* **56**, 235–249.

HILDENBRAND, W. (1981). Short-run production functions based on microdata. *Econometrica* **49**, 1095–1126.

HOBERG, R. (2000). Cluster analysis based on data depth. In H. Kiers, J.-P. Rasson, P. Groenen and M. Schader, eds., *Data Analysis, Classification, and Related Methods*, 17–22, Berlin. Springer.

HOBERG, R. (2002). Datentiefe und Klassifikation. Mimeo, Fakultät für Wirtschafts- und Sozialwissenschaften, Universität zu Köln.

HODGES, J. L. (1955). A bivariate sign test. *Annals of Mathematical Statistics* **28**, 523–527.

HOTELLING, H. (1929). Stability in Competition. *Economic Journal* **39**, 41–57.

HUBER, P. J. (1972). Robust statistics: A review. *Annals of Statistics* **43**, 1041–1067.

JEYARATNAM, S. (1991). A robust affine–equivariant median. *Statistics and Probability Letters* **11**, 167–168.

JOE, H. (1997). *Multivariate Models and Dependence Concepts.* Chapman and Hall, London.

JOE, H. and VERDUCCI, J. (1992). Multivariate majorization by positive combinations. In M. Shaked and Y. Tong, eds., *Stochastic Inequalities*, 159–181. Hayward, California.

JOHNSON, T., KWOK, I. and NG, R. (1998). Fast computation of 2-dimensional depth contours. In R. Agrawal and P. Stolorz, eds., *Proceedings of the Fourth International Conference on Knowledge Discovery and Data Mining*, Menlo Park. AAAI Press.

JONES, M. C. (1994). Expectiles and M-quantiles are quantiles. *Statistics and Probability Letters* **20**, 149–153.

KENDALL, D. G. (1974). Foundations of a theory of random sets. In E. Harding and D. Kendall, eds., *Stochastic Geometry*, 322–376. John Wiley & Sons, London.

KOLM, S. C. (1969). The optimal production of social justice. In J. Marjolis and H. Guitton, eds., *Public Economics*, 145–200. MacMillan, New York.

KOLM, S. C. (1977). Multidimensional egalitarianisms. *Quarterly Journal of Economics* **91**, 1–13.

KOLTCHINSKII, V. (1997). M-estimation, convexity and quantiles. *Annals of Statistics* **25**, 435–477.

KOSHEVOY, G. (1997). Multivariate depths and underlying distributions: A uniqueness property. Mimeo. Universität zu Köln.

KOSHEVOY, G. (1999a). Breakdown points. Several variants of definitions. Mimeo.

KOSHEVOY, G. (1999b). Oriented matroids and multivariate ranks and depths. Mimeo.

KOSHEVOY, G. (200xa). Projections of lift zonoids, the Oja depth and the Tukey depth. *Annals of Statistics* . To appear.

KOSHEVOY, G. (200xb). Representations of probability measures. *Bernoulli* . To appear.

KOSHEVOY, G. (200xc). The Tukey depth characterizes the atomic measure. *Journal of Multivariate Analysis* . To appear.

KOSHEVOY, G. and MOSLER, K. (1996). The Lorenz zonoid of a multivariate distribution. *Journal of the American Statistical Association* **91**, 873–882.

KOSHEVOY, G. and MOSLER, K. (1997a). Multivariate Gini indices. *Journal of Multivariate Analysis* **60**, 252–276.

KOSHEVOY, G. and MOSLER, K. (1997b). Zonoid trimming for multivariate distributions. *Annals of Statistics* **25**, 1998–2017.

KOSHEVOY, G. and MOSLER, K. (1998). Lift zonoids, random convex hulls and the variability of random vectors. *Bernoulli* **4**, 377–399.

KOSHEVOY, G. and MOSLER, K. (1999a). Depth of hyperplanes and related statistics. In W. Gaul and M. Schader, eds., *Mathematische Methoden der Wirtschaftswissenschaften. Festschrift für Otto Opitz*. 162–171. Physica-Verlag, Heidelberg.

KOSHEVOY, G. and MOSLER, K. (1999b). Measuring multidimensional concentration: A geometric approach. *Allgemeines Statistisches Archiv* **83**, 173–189.

KOSHEVOY, G. and MOSLER, K. (1999c). Price majorization and the inverse Lorenz function. Discussion Papers in Statistics and Econometrics 3, Universität zu Köln.

KOSHEVOY, G., MÖTTÖNEN, J. and OJA, H. (2002). A scatter matrix estimate based on the zonotope. Mimeo.

LINDENSTRAUSS, J. (1966). A short proof of Liapounoff's convexity theorem. *Journal of Mathematical Mech.* **15**, 971–972.

LIU, R. (1990). On a notion of data depth based on random simplices. *Annals of Statistics* **18**, 405–414.

LIU, R. (1992). Data depth and multivariate rank tests. In Y. Dodge, ed., *L_1-Statistics Analysis and Related Methods*, 279–294. North-Holland, Amsterdam.

LIU, R. (1995). Control charts for multivariate processes. *Journal of the American Statistical Association* **90**, 1380–1388.

LIU, R., PARELIUS, J. M. and SINGH, K. (1999). Multivariate analysis by data depth: Descriptive statistics, graphics and inference. *Annals of Statistics* **27**, 783–858. With discussion.

LIU, R. and SINGH, K. (1993). A quality index based on data depth and multivariate rank tests. *Journal of the American Statistical Association* **88**, 252–260.

LIU, R. and SINGH, K. (1997). Notions of limiting *P* values based on data depth and bootstrap. *Journal of the American Statistical Association* **92**, 266–277.

LOPUHAÄ, H. and ROUSSEEUW, P. J. (1991). Breakdown points of affine equivariant estimators of multivariate location and covariance matrices. *Annals of Statistics* **19**, 229–248.

LORENZ, M. O. (1905). Methods of measuring the concentration of wealth. *Publication of the American Statistical Association* **9**, 209–219.

MAASOUMI, E. (1986). The measurement and decomposition of multidimensional inequality. *Econometrica* **54**, 991–997.

MAHALANOBIS, P. C. (1936). On the generalized distance in statistics. *Proceedings of the National Academy of India* **12**, 49–55.

MARDIA, K., KENT, J. T. and BIBBY, J. (1979). *Multivariate Analysis.* Academic Press, London.

MARFELS, C. (1977). *Erfassung und Darstellung industrieller Konzentration.* Nomos-Verlag, Baden-Baden.

MARSHALL, A. W. and OLKIN, I. (1979). *Inequalities: Theory of Majorization and Its Applications.* Academic Press, New York.

MASSÉ, J.-C. (1999). Asymptotics for the Tukey median. Mimeo.

MASSÉ, J.-C. and THEODORESCU, R. (1994). Halfplane trimming for bivariate distributions. *Journal of Multivariate Analysis* **48**, 188–202.

MATHERON, G. (1975). *Random Sets and Integral Geometry.* John Wiley & Sons, New York.

MEYER, P. A. (1966). *Probability and Potentials.* Blaisdell Publishing Co, Waltham.

MIZERA, I. (1998). On depth and deep points: A calculus. Mimeo.

MOSLER, K. (1994a). Majorization in economic disparity measures. *Linear Algebra and its Applications* **199**, 91–114.

MOSLER, K. (1994b). Multidimensional welfarisms. In W. Eichhorn, ed., *Models and Measurement of Welfare and Inequality*, 808–820. Springer, Berlin.

MOSLER, K. (2001). Indices and orders of multidimensional diversity. *Community Ecology* **2**, 137–143.

MOSLER, K. (2002). Central regions and dependency. Mimeo.

MOSLER, K. and SCARSINI, M. (1991). *Stochastic Orders and Decision Under Risk.* Institute of Mathematical Statistics, Hayward, California.

MÖTTÖNEN, J., HETTMANSPERGER, T., OJA, H. and TIENARI, J. (1998). On the efficiency of affine invariant multivariate rank tests. *Journal of Multivariate Analysis* **66**, 118–132.

MÖTTÖNEN, J. and OJA, H. (1995). Multivariate spatial sign and rank methods. *Journal of Nonparametric Statistics* **5**, 201–213.

MÖTTÖNEN, J., OJA, H. and TIENARI, J. (1997). On the efficiency of multivariate spatial sign and rank tests. *Annals of Statistics* **25**, 542–552.

MOURIER, E. (1956). L-random elements and L^*-random elements in Banach spaces. *Proceedings of the Berkeley Symposium on Mathematical Statistics and Probability* **3**, 231–242.

MOYES, P. (1987). A new concept of Lorenz domination. *Economics Letters* **23**, 203–207.

MULIERE, P. and SCARSINI, M. (1989). Multivariate decisions with unknown price vector. *Economics Letters* **29**, 13–19.

MÜLLER, A. (1997a). Integral probability metrics and their generating classes of functions. *Advances in Applied Probability* **29**, 429–443.

MÜLLER, A. (1997b). Stochastic orders generated by integrals. *Advances in Applied Probability* **29**, 414–428.

MÜLLER, A. and STOYAN, D. (2002). *Comparison Methods for Stochastic Models and Risks.* J. Wiley, New York.

NEWEY, W. K. and POWELL, J. L. (1987). Asymmetric least squares estimation and testing. *Econometrica* **55**, 819–847.

NIINIMAA, A. and OJA, H. (1995). On the influence functions of certain bivariate medians. *Journal of the Royal Statistical Society, Series B* **57**, 565–574.

NIINIMAA, A. and OJA, H. (1999). Multivariate median. In S. Kotz, C. Read and D. Banks, eds., *Encyclopedia of Statistical Sciences, Update*, vol. 3, 497–505. Wiley-Interscience.

NIINIMAA, A., OJA, H. and NYBLOM, J. (1992). Algorithm AS 277. The Oja bivariate median. *Applied Statistics* **41**, 611–617.

NIINIMAA, A., OJA, H. and TABLEMAN, M. (1990). The finite-sample breakdown point of the Oja bivariate median and of the corresponding half-samples version. *Statistics and Probability Letters* **10**, 325–328.

NIINIMAA, A. O. (1992). The calculation of the Oja multivariate median. In Y. Dodge, ed., *L_1-Statistical Analysis and Related Methods Topics*, 379–385. North-Holland, Amsterdam.

NOLAN, D. (1992). Asymptotics for multivariate trimming. *Stochastic Processes and Their Applications* **42**, 157–169.

NYGÅRD, F. and SANDSTRÖM, A. (1981). *Measuring Income Inequality.* Almqvist and Wiksell, Stockholm.

OJA, H. (1983). Descriptive statistics for multivariate distributions. *Statistics and Probability Letters* **1**, 327–332.

OJA, H. (1999). Affine invariant multivariate sign and rank tests and corresponding estimates. *Scandinavian Journal of Statistics* **26**, 319–343.

PIESCH, W. (1975). *Statistische Konzentrationsmaße.* J.C.B. Mohr (Paul Siebeck), Tübingen.

PIGOU, A. C. (1912). *Wealth and Welfare.* MacMillan, New York.

PURI, M. L. and SEN, P. K. (1971). *Nonparametric Methods in Multivariate Analysis.* John Wiley & Sons, New York.

PYATT, G., CHEN, C. and FEY, J. (1980). The distribution of income by factor components. *Quarterly Journal of Economics* **95**, 451–473.

RAO, C. R. (1988). Methodology based on the L_1 norm in statistical inference. *Sankhyā, Series A* **50**, 289–313.

RASSON, J.-P. (1996). Convexity methods in classification. In C. Hayashi, N. Ohsumi, K. Yajima, Y. Tanaka, H.-H. Bock and Y. Baba, eds., *Data Science, Classification, and Related Methods*, 9–108. Springer, Berlin.

REIDEMEISTER, K. (1921). Über die singulären Randpunkte eines konvexen Körpers. *Mathematische Annalen* **83**, 116–118.

RÉNYI, A. and SULANKE, R. (1963). Über die konvexe Hülle von n zufällig gewählten Punkten. *Zeitschrift für Wahrscheinlichkeitstheorie und verwandte Gebiete* **2**, 75–84.

RÉNYI, A. and SULANKE, R. (1964). Über die konvexe Hülle von n zufällig gewählten Punkten II. *Zeitschrift für Wahrscheinlichkeitstheorie und verwandte Gebiete* **3**, 138–147.

RIETVELD, P. (1990). Multidimensional inequality comparisons. *Economics Letters* **32**, 187–192.

ROCKAFELLAR, R. T. (1970). *Convex Analysis*. John Wiley & Sons, New York.

ROHLF, F. J. (1977). A probalistic minimum spanning tree algorithm. Technical report, IBM Research Report C6502.

ROSENBLUTH, G. (1955). Measures of concentration. In *National Bureau Committee for Economic Research*. Princeton University Press, Princeton.

ROUSSEEUW, P. and STRUYF, A. (1998). Computing location depth and regression depth in higher dimensions. *Statistics and Computing* 193–203.

ROUSSEEUW, P. J. and LEROY, A. M. (1987). *Robust Regression and Outlier Detection*. John Wiley & Sons, New York.

ROUSSEEUW, P. J. and RUTS, I. (1996). Algorithm AS 307. Bivariate location depth. *Applied Statistics* **45**, 516–526.

ROUSSEEUW, P. J. and RUTS, I. (1998). Constructing the bivariate Tukey median. *Statistica Sinica* **8**, 823–839.

RUSSELL, W. R. and SEO, T. K. (1978). Ordering uncertain prospects: The multivariate utility functions case. *Review of Economic Studies* **45**, 605–610.

RUTS, I. and ROUSSEEUW, P. J. (1996). Computing depth contours of bivariate point clouds. *Computational Statistics and Data Analysis* **23**, 153–168.

SCARSINI, M. (1985). Stochastic dominance with pair-wise risk aversion. *Journal of Mathematical Economics* **14**, 187–201.

SCARSINI, M. and SHAKED, M. (1990). Some conditions for stochastic equality. *Naval Research Logistics Quarterly* **37**, 617–625.

SCHNEIDER, R. and WEIL, W. (1983). Zonoids and related topics. In P. Gruber and J. M. Wills, eds., *Convexity and Its Applications*, 296–317. Birkhäuser, Basel.

SEN, A. K. (1973). *On Economic Inequality.* Oxford University Press, Oxford.

SHAKED, M. and SHANTHIKUMAR, J. G. (1994). *Stochastic Orders and Their Applications.* Academic Press, Boston.

SHAMOS, M. I. and HOEY, D. (1975). Closest point problems. In *Proceedings of the 16th Annual Symposium of the Foundations of Computer Science*, 151–162. IEEE.

SHEPHARD, G. C. (1974). Combinatorial properties of associated zonotopes. *Canadian Journal of Mathematics* **26**, 302–321.

SINGH, K. (1991). A notion of majority depth. Technical report, Rutgers University, Department of Statistics.

SMALL, C. G. (1990). A survey of multidimensional medians. *International Statistical Review* **58**, 263–277.

STRASSEN, V. (1965). The existence of probability measures with given marginals. *Annals of Mathematical Statistics* **36**, 423–439.

STRUYF, A. and ROUSSEEUW, P. J. (1999). Halfspace depth and regression depth characterize the empirical distribution. *Journal of Multivariate Analysis* **69**, 135–153.

TAGUCHI, T. (1981). On a multiple Gini's coefficient and some concentrative regressions. *Metron* 69–98.

TITTERINGTON, D. M. (1978). Estimation of corrrelation coefficients by ellipsoidal trimming. *Applied Statistics* **27**, 227–234.

TOBIN, J. (1970). On limiting the domain of inequality. *Journal of Law and Economics* **13**, 263–277.

TORGERSEN, E. (1991). *Comparison of Statistical Experiments.* Cambridge University Press, Cambridge, Massachusets.

TSUI, K. (1995). Multidimensional generalizations of the relative and absolute inequality indices: The Atkinson–Kolm–Sen approach. *Journal of Economic Theory* **67**, 251–265.

TUKEY, J. W. (1975). Mathematics and picturing data. In R. James, ed., *Proceedings of the 1974 International Congress of Mathematicians, Vancouver*, vol. 2, 523–531.

VAN DER VAART, A. W. and WELLNER, J. A. (1996). *Weak Convergence and Empirical Processes.* Springer, New York.

VARDI, Y. and ZHANG, C.-H. (2000). The multivariate L_1-median and associated data depth. *Proceedings National Academy of Sciences* 1423–1426.

VITALE, R. A. (1991a). Expectation inequalities from convex geometry. In K. Mosler and M. Scarsini, eds., *Stochastic Orders and Decision Under Risk*, 372–379. Institute of Mathematical Statistics, Hayward, California.

VITALE, R. A. (1991b). Expected absolute random determinants and zonoids. *Advances in Applied Probability* **1**, 293–300.

WALD, A. and WOLFOWITZ, J. (1940). On a test whether two samples are from the same population. *Annals of Mathematical Statistics* **11**, 147–162.

WATSON, G. (1984). *Statistics on Spheres.* Wiley-Interscience, New York.

WEIL, W. (1971). Über die Projektionenkörper konvexer Polytope. *Archiv der Mathematik* **22**, 664–672.

WEIL, W. (1982). An application of the central limit theorem for Banach-space-valued random variables to the theory of random sets. *Zeitschrift für Wahrscheinlichkeitstheorie und verwandte Gebiete* **60**, 203–208.

WEIL, W. and WIEACKER, J. A. (1993). Stochastic geometry. In P. Gruber and J. Wills, eds., *Handbook of Convex Geometry*, vol. B, 1391–1438. Elsevier, Amsterdam.

WILKS, S. S. (1960). Multidimensional statistical scatter. In I. Olkin, ed., *Contributions to Probability and Statistics in Honor of Harold Hotelling*, 486–503. Stanford, California.

YEH, A. B. and SINGH, K. (1997). Balanced confidence regions based on Tukey's depth and the bootstrap. *Journal of the Royal Statistical Society, Series B* **59**, 639–652.

ZUO, Y. (2000). A note on finite sample breakdown points of projection based multivariate location and scatter statistics. *Metrika* **51**. 259–265.

ZUO, Y. and SERFLING, R. (2000a). General notions of statistical depth function. *Annals of Statistics* **28**, 461–482.

ZUO, Y. and SERFLING, R. (2000b). Nonparametric notions of multivariate "scatter measure" and "more scattered" based on statistical depth functions. *Journal of Multivariate Analysis* **75**, 62–78.

ZUO, Y. and SERFLING, R. (2000c). On the performance of some robust nonparametric location measures relative to a general notion of multivariate symmetry. *Journal of Statistical Planning and Inference* **84**, 55–79.

ZUO, Y. and SERFLING, R. (2000d). Structural properties and convergence results for contours of sample statistical depth functions. *Annals of Statistics* **28**, 483–499.

Index

breakdown point, 104, 107, 120, 122

central limit theorem, 9, 69
central regions, 19, 21, 79–104, 208
  dependence order, 199
central regions,
  halfspace, 95, 104
  likelihood, 93
  Mahalanobis, 93, 95
  zonoid, 20–21, 81–90
ceteris paribus property, 197, 225
cluster analysis, 197
combinatorial
  dispersion, 173
  equivalence, 172
  invariance, 123, 124, 127
  invariance in pairs, 172
  structure, 124, 126, 131
computation of
  halfspace central regions, 104
  Mahalanobis depth, 139
  Tukey median, 104
  zonoid depth, 127
  zonoid trimmed regions, 102
concentration
  curve, 249
  function, 248, 252
  hypersurface, 253
  index, 252, 256
  rate, 257
cone, 263
cone,
  simplicial, 263
  solid, 263
contour set, 79
convex function, 262
convex set, 262
convex unimodal, 80

data depth, 21, 105–131, 134
density level set, 79
dependence order, 199
depth of hyperplane, 23, 167
depth order, 135
depth tests, 22–23, 136–141, 150
depth,
  $\mathbb{L}_2$-, 117
  extended majority, 169–172
  halfspace, 119, 131
  Mahalanobis, 106, 118
  majority, 121, 169
  mean hyperplane, 23, 168 174
  Oja, 120, 130
  peeling, 122
  projection, 122
  simplicial, 120
  zonoid, 21–22, 108, 130
dilation, 17, 22, 115
dimension, affine, 261
dispersion index, 197, 247
distribution epigraph, 2, 3
distribution function, 261
distribution mixture, 99
distribution type, 46, 47, 85

distribution,
- binomial, 9
- bivariate exponential, 90
- bivariate normal, 93, 200, 265–273
- egalitarian, 234
- elliptical, 55, 56, 95, 200, 211
- empirical, 4, 89, 100
- exponential, 10
- Marshall-Olkin, 90, 200
- multivariate Cauchy, 150, 175–178
- multivariate normal, 14, 56, 89, 149, 175–178, 220
- scaled, 187
- spherical, 55, 56, 95
- uniform
  - on ball, 90
  - on ellipsoid, 57, 90, 175–178
  - on unit sphere, 14, 57, 90
- univariate normal, 10, 47

diversity, 258
Donsker
- class, 75
- theorem, 4

economic inequality, 230–247
equivariance, affine, 82
extreme point, 33, 263

face, 263
facet, 263
Fréchet bound, 202, 235

Gaussian process on the sphere, 9
general position, 261
Gini index, 18, 183–186, 205
Gini index,
- distance-, 224–227, 247
- volume-, 192–197, 226

Gini mean difference, 18, 38, 183–186, 205
Gini mean difference,
- distance-, 224–227
- volume-, 192–197, 227

Glivenko-Cantelli
- class, 75
- theorem, 3

Hausdorff distance, 8, 27, 260, 262
homogeneity, positive, 48, 262
hull,
- affine, 261
- convex, 261

ILF (see Lorenz function, inverse), 232
increasing risk, 17
industrial concentration, 248–257
interquantile range, 88
invariance,
- affine, 111, 118, 134, 137
- combinatorial, 123

law of large numbers, 9, 68, 102, 238
lift zonoid, 4–14, 40–42, 77
lift zonoid order (see order, lift zonoid), 208
lift zonoid volume, 18, 66, 186, 219
lift zonotope, 8, 43
line segment, 259, 262
linearity, positive, 48
Lorenz curve, 5, 6, 183, 246
Lorenz curve,
- dual, 183
- dual generalized, 8, 44, 183
- generalized, 8, 44, 183

Lorenz function,
- inverse, 232–238, 246
- inverse generalized, 244

Lorenz order, 184
Lorenz order, generalized, 184
Lorenz zonoid, 187, 235
Lyapunov's theorem, 30, 31

measure,
- Dirac, 260
- empirical, 29, 260
- Lebesgue on sphere, 261
- lifted, 40

median, 105
median set, 105
median,
- Oja, 104
- Tukey, 103

metric,
- integral measure, 72
- Kolmogorov-Smirnov, 3
- lift zonoid, 52
- measure, 3

MHD
- -dispersion, 173
- test statistics, 174–178

MHD (see depth, mean hyperplane), 23, 168
Minkowski sum, 4, 260
monotone on rays, 134
monotone, Δ-, 2

order
- of convolutions, 214–216
- of dependence, 198–205
- of marginals, 211

order statistics, 102
order,
- stochastic, 16
- Atkinson-Bourgouignon, 245
- bivariate concordance, 202
- centered, 221–223
- convex, 17, 209, 216–218
- convex-linear, 16, 209
- convex-posilinear, 18, 19
- correlation matrix, 199
- halfspace dependence, 199
- increasing, 220, 223
- increasing convex, 17
- integral, 72
- lift zonoid, 16, 18, 49, 208–219
- lower orthant, 3, 202
- Mahalanobis dependence, 199–205
- price Lorenz, 238–247
- scaled, 221–223
- stochastic, 72, 228
- stop-loss, 17
- usual stochastic, 3
- zonoid, 228
- zonoid dependence, 199–203

outlier, 19, 104, 198, 205

Pigou-Dalton transfer, 246
polyhedron, 261
polytope, 10
price majorization, 18, 243
price submajorization, 243
price supermajorization, 243
projection property, 114
projection to marginals, 53, 83

quantile function, 44, 80

Radon partition, 123
random
- compact set, 7, 264
- convex hull, 188
- determinant, 219
- lift zonotope, 188
- segment, 7, 35, 41

rank tests, 22
relative
- boundary, 261
- interior, 261

representation, statistical, 75, 76, 78
Rosenbluth index, 252, 256

second order stochastic dominance, 17
segment, 262
simplex, 261

starshaped, 135
stochastic dominance, 72
stochastic order (see order, stochastic), 3
support function, 32, 34, 42, 262, 263
supporting point, 263

test,
  Box's M, 141–142, 150–163
  Friedman-Rafsky, 142–144, 150–163
  Hotelling's T^2, 144–145
  Liu, 139–141
  Puri-Sen, 145–147, 150–163
  Wilcoxon distance, 147–149, 151–163
  Wilcoxon rank sum, 136
transform,
  affine, 54
  linear, 53
  shift, 54
transformation,
  doubly stochastic, 251
trimmed region, 95, 107, 135
trimmed region,
  zonoid, 109

uniform integrability, 15, 59–65
unimodal, 80
upper semicontinuity, 134
utility function, 17

vanishing at infinity, 134
vector measure, 30
vertex, 262

zonoid, 8, 27, 263
  clustering, 198
  dependence order, 199–203
  of Dirac measure, 29
  of measure, 28, 42
  order, 228
  trimmed region, 80–85, 88
  volume, 35–37
zonotope, 8, 27, 263

Lecture Notes in Statistics

For information about Volumes 1 to 110, please contact Springer-Verlag

111: Leon Willenborg and Ton de Waal, Statistical Disclosure Control in Practice. xiv, 152 pp., 1996.

112: Doug Fischer, Hans-J. Lenz (Editors), Learning from Data. xii, 450 pp., 1996.

113: Rainer Schwabe, Optimum Designs for Multi-Factor Models. viii, 124 pp., 1996.

114: C.C. Heyde, Yu. V. Prohorov, R. Pyke, and S. T. Rachev (Editors), Athens Conference on Applied Probability and Time Series Analysis, Volume I: Applied Probability In Honor of J.M. Gani. viii, 424 pp., 1996.

115: P.M. Robinson and M. Rosenblatt (Editors), Athens Conference on Applied Probability and Time Series Analysis, Volume II: Time Series Analysis In Memory of E.J. Hannan. viii, 448 pp., 1996.

116: Genshiro Kitagawa and Will Gersch, Smoothness Priors Analysis of Time Series. x, 261 pp., 1996.

117: Paul Glasserman, Karl Sigman, and David D. Yao (Editors), Stochastic Networks. xii, 298 pp., 1996.

118: Radford M. Neal, Bayesian Learning for Neural Networks. xv, 183 pp., 1996.

119: Masanao Aoki and Arthur M. Havenner, Applications of Computer Aided Time Series Modeling. ix, 329 pp., 1997.

120: Maia Berkane, Latent Variable Modeling and Applications to Causality. vi, 288 pp., 1997.

121: Constantine Gatsonis, James S. Hodges, Robert E. Kass, Robert McCulloch, Peter Rossi, and Nozer D. Singpurwalla (Editors), Case Studies in Bayesian Statistics, Volume III. xvi, 487 pp., 1997.

122: Timothy G. Gregoire, David R. Brillinger, Peter J. Diggle, Estelle Russek-Cohen, William G. Warren, and Russell D. Wolfinger (Editors), Modeling Longitudinal and Spatially Correlated Data. x, 402 pp., 1997.

123: D. Y. Lin and T. R. Fleming (Editors), Proceedings of the First Seattle Symposium in Biostatistics: Survival Analysis. xiii, 308 pp., 1997.

124: Christine H. Müller, Robust Planning and Analysis of Experiments. x, 234 pp., 1997.

125: Valerii V. Fedorov and Peter Hackl, Model-Oriented Design of Experiments. viii, 117 pp., 1997.

126: Geert Verbeke and Geert Molenberghs, Linear Mixed Models in Practice: A SAS-Oriented Approach. xiii, 306 pp., 1997.

127: Harald Niederreiter, Peter Hellekalek, Gerhard Larcher, and Peter Zinterhof (Editors), Monte Carlo and Quasi-Monte Carlo Methods 1996. xii, 448 pp., 1997.

128: L. Accardi and C.C. Heyde (Editors), Probability Towards 2000. x, 356 pp., 1998.

129: Wolfgang Härdle, Gerard Kerkyacharian, Dominique Picard, and Alexander Tsybakov, Wavelets, Approximation, and Statistical Applications. xvi, 265 pp., 1998.

130: Bo-Cheng Wei, Exponential Family Nonlinear Models. ix, 240 pp., 1998.

131: Joel L. Horowitz, Semiparametric Methods in Econometrics. ix, 204 pp., 1998.

132: Douglas Nychka, Walter W. Piegorsch, and Lawrence H. Cox (Editors), Case Studies in Environmental Statistics. viii, 200 pp., 1998.

133: Dipak Dey, Peter Müller, and Debajyoti Sinha (Editors), Practical Nonparametric and Semiparametric Bayesian Statistics. xv, 408 pp., 1998.

134: Yu. A. Kutoyants, Statistical Inference For Spatial Poisson Processes. vii, 284 pp., 1998.

135: Christian P. Robert, Discretization and MCMC Convergence Assessment. x, 192 pp., 1998.

136: Gregory C. Reinsel, Raja P. Velu, Multivariate Reduced-Rank Regression. xiii, 272 pp., 1998.

137: V. Seshadri, The Inverse Gaussian Distribution: Statistical Theory and Applications. xii, 360 pp., 1998.

138: Peter Hellekalek and Gerhard Larcher (Editors), Random and Quasi-Random Point Sets. xi, 352 pp., 1998.

139: Roger B. Nelsen, An Introduction to Copulas. xi, 232 pp., 1999.

140: Constantine Gatsonis, Robert E. Kass, Bradley Carlin, Alicia Carriquiry, Andrew Gelman, Isabella Verdinelli, and Mike West (Editors), Case Studies in Bayesian Statistics, Volume IV. xvi, 456 pp., 1999.

141: Peter Müller and Brani Vidakovic (Editors), Bayesian Inference in Wavelet Based Models. xiii, 394 pp., 1999.

142: György Terdik, Bilinear Stochastic Models and Related Problems of Nonlinear Time Series Analysis: A Frequency Domain Approach. xi, 258 pp., 1999.

143: Russell Barton, Graphical Methods for the Design of Experiments. x, 208 pp., 1999.

144: L. Mark Berliner, Douglas Nychka, and Timothy Hoar (Editors), Case Studies in Statistics and the Atmospheric Sciences. x, 208 pp., 2000.

145: James H. Matis and Thomas R. Kiffe, Stochastic Population Models. viii, 220 pp., 2000.

146: Wim Schoutens, Stochastic Processes and Orthogonal Polynomials. xiv, 163 pp., 2000.

147: Jürgen Franke, Wolfgang Härdle, and Gerhard Stahl, Measuring Risk in Complex Stochastic Systems. xvi, 272 pp., 2000.

148: S.E. Ahmed and Nancy Reid, Empirical Bayes and Likelihood Inference. x, 200 pp., 2000.

149: D. Bosq, Linear Processes in Function Spaces: Theory and Applications. xv, 296 pp., 2000.

150: Tadeusz Caliński and Sanpei Kageyama, Block Designs: A Randomization Approach, Volume I: Analysis. ix, 313 pp., 2000.

151: Håkan Andersson and Tom Britton, Stochastic Epidemic Models and Their Statistical Analysis. ix, 152 pp., 2000.

152: David Ríos Insua and Fabrizio Ruggeri, Robust Bayesian Analysis. xiii, 435 pp., 2000.

153: Parimal Mukhopadhyay, Topics in Survey Sampling. x, 303 pp., 2000.

154: Regina Kaiser and Agustín Maravall, Measuring Business Cycles in Economic Time Series. vi, 190 pp., 2000.

155: Leon Willenborg and Ton de Waal, Elements of Statistical Disclosure Control. xvii, 289 pp., 2000.

156: Gordon Willmot and X. Sheldon Lin, Lundberg Approximations for Compound Distributions with Insurance Applications. xi, 272 pp., 2000.

157: Anne Boomsma, Marijtje A.J. van Duijn, and Tom A.B. Snijders (Editors), Essays on Item Response Theory. xv, 448 pp., 2000.

158: Dominique Ladiray and Benoît Quenneville, Seasonal Adjustment with the X-11 Method. xxii, 220 pp., 2001.

159: Marc Moore (Editor), Spatial Statistics: Methodological Aspects and Some Applications. xvi, 282 pp., 2001.

160: Tomasz Rychlik, Projecting Statistical Functionals. viii, 184 pp., 2001.

161: Maarten Jansen, Noise Reduction by Wavelet Thresholding. xxii, 224 pp., 2001.

162: Constantine Gatsonis, Bradley Carlin, Alicia Carriquiry, Andrew Gelman, Robert E. Kass Isabella Verdinelli, and Mike West (Editors), Case Studies in Bayesian Statistics, Volume V. xiv, 448 pp., 2001.

163: Erkki P. Liski, Nripes K. Mandal, Kirti R. Shah, and Bikas K. Sinha, Topics in Optimal Design. xi, 164 pp., 2002.

164: Peter Goos, The Optimal Design of Blocked and Split-Plot Experiments. xiv, 256 pp., 2002.

165: Karl Mosler, Multivariate Dispersion, Central Regions and Depth: The Lift Zonoid Approach. x, 291 pp., 2002.

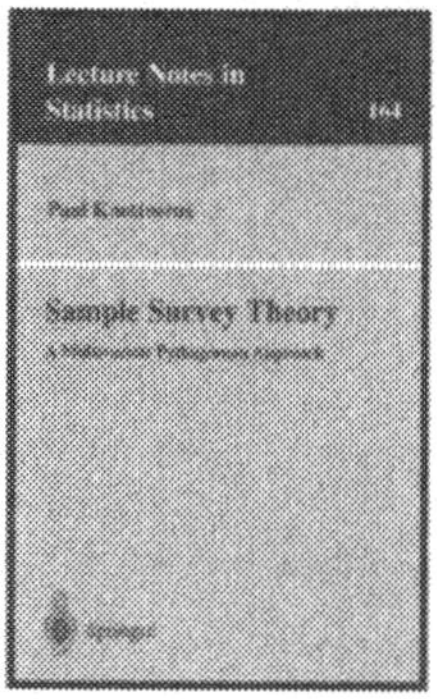
Lecture Notes in Statistics
164
Sample Survey Theory
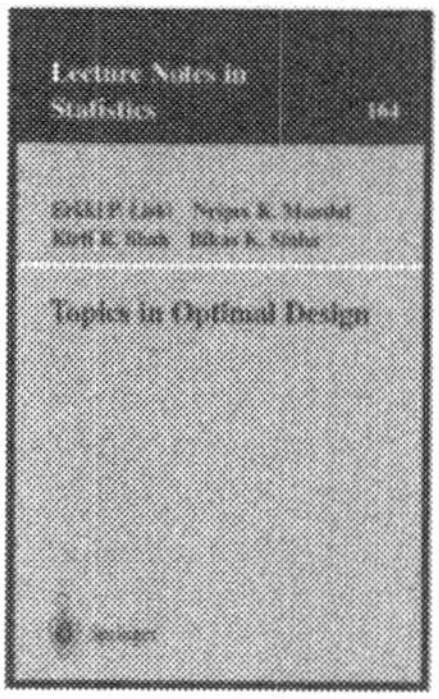
Lecture Notes in Statistics
Topics in Optimal Design

Lecture Notes in Statistics
162
Case Studies in Bayesian Statistics

9 780387 954127